W9-CLS-268

QUALITY MANAGEMENT

IMPLEMENTING THE BEST IDEAS OF THE MASTERS

QUALITY MANAGEMENT

IMPLEMENTING THE BEST IDEAS OF THE MASTERS

Bruce Brocka and M. Suzanne Brocka

BUSINESS ONE IRWIN
Homewood, Illinois 60430

Sponsoring editor: Jean Marie Geracie
Project editor: Paula M. Buschman
Production manager: Ann Cassady
Jacket designer: Image House, Inc.
Designer: Larry J. Cope
Art coordinator: Mark Malloy
Compositor: Carlisle Communications, Ltd.
Typeface: 10.5/12 Times Roman
Printer: Book Press, Inc.

Library of Congress Cataloging-in-Publication Data

Brocka, Bruce
Quality management: implementing the best ideas of the masters / Bruce Brocka and M. Suzanne Brocka.
p. cm.
Includes index.
ISBN 1-55623-540-2
1. Total quality management. I. Brocka, M. Suzanne. II. Title.
HD62. 15.B73 1992
658.4—dc20 92–6945

Printed in the United States of America
3 4 5 6 7 8 9 0 BP 9 8 7 6 5 4 3

This work is dedicated to our children,
Melinda Athena
and
Bennett Paul

PREFACE

Frustration gave birth to this book. We wanted a single-volume source for a broad range of Quality Management topics of sufficient depth to be useful, but not overwhelming and repetitive in details. Something that would bring together the many paths that Quality Management seems to travel. A work that would not intimidate, yet still provide enough information to become familiar with the tools and concepts needed—a desktop adviser, so to speak. Failing to find such a work, we began putting together notes which eventually became this book.

The title of the work recognizes that Total Quality Management, which evolved from Total Quality Control, is evolving into Quality Management. Each of these evolutionary transformations has widened the bounds of Quality Management. It is being transformed from the management of quality into a companywide, top-down, quality of management. Quality principles are applied at all managerial and employee levels, in a variety of departments and disciplines.

We feel that there is a business and moral imperative for the implementation of Quality Management. Global competition is forcing businesses to reexamine how they conduct business. A more educated work force is also compelling employers to look for new ways to tap into a wellspring of talent. The truly good news of the Quality Management approach is that workers feel more fulfilled from a management style and structure that allows a wider range of their capabilities to be exercised, and profitability rises from flatter organizational structures to allow better customer needs satisfaction.

The problem with writing a work like this is that at some point someone has to shoot the authors, or it would never be finished. There is always one more article to look up, or a new and vital book to mine through for its nuggets of information. A work of this sort is never really finished. We are thankful to our editor from Business One Irwin, Jeffrey Krames, for his encouragement.

PURPOSE OF THIS BOOK

The purpose of the book is to familiarize the reader with the best of the broad array of tools, techniques, and philosophies regarding Quality Management. It is a starting point, a springboard for jumping into the world of Total Quality. It is a world that has evolved rapidly in the past 10 years, and has had few guidebooks or handbooks, except in fairly specialized and narrow areas. This book is meant to be an orienteering guide, a desktop adviser, to this new world of management.

What the book is not is an exhaustive handbook. If it were, it would be a multivolume set, and lose its value as an orientation guide. Still, with the exception of the design of experiments and sampling sections, the reader should be able to implement at least an experiment into each of the tools, techniques, and practices presented here.

APPLICATION

This book is for the practicing manager and employee of organizations that face real-business conditions. It is not written for those expert in the field of Total Quality Management, although they will find the text beneficial as well, and may even discover new techniques or look at familiar ones in a new way.

HOW TO USE THIS BOOK

This book does not have to be read in sequential fashion. There are extensive cross references to the other sections of the book, so that needless repetition of concepts was eliminated. However, this does not mean that the book's tenets are to be taken piecemeal. Many of the concepts are strongly interconnected. Without knowledge of Part 3, "Management Dynamics," for example, implementing many of the tools and techniques would be arduous at best. The major sections flow from one into the other. "The Foundations" and "Quality Masters," Parts 1 and 2, present the philosophical underpinnings of Quality Management. This vision must be translated into management principles, discussed in the "Management Dynamics" section. Finally, these principles must be refined into methods, practices, and techniques, which are elaborated in the "Tools and Techniques," Part 4. This tripartite scheme of strategy, and management and operations also reveals the self-reflection and deliberate nature of Quality Management.

It should be noted that occasionally a fictitious company is used, named Widgco. It does not represent any company. The term *widget* has come to represent any sort of product, and therefore *Widgco* is any sort of organization.

REFERENCES

For the most part the references used are recent ones commonly available. New references appear every day, and many may be vital to implementation success. Quality function deployment, for instance, is a new technique that has had few articles on its execution and practice.

The book references are straightforward. Most will be listed in *Books in Print,* and can be ordered from any bookstore. Many can be ordered (with member discounts) from the ASQC Quality Press, whose address is listed in "Resources," Part 5.

ACKNOWLEDGEMENTS

We would like to thank the following for reviewing or contributing to portions of the manuscript: Aquiel Ahmad, Bob Cartwright, Philip Crosby, James Evans, Michael Goodhand, Joseph Juran, Walt Lilius, Gordon McMeekin, Robert Perry, Thomas Ryan, Emily Schlenker, Walter Willborn and Johnathan T. Young.

We would like to recognize the following colleagues who helped shape the authors' precepts of quality in management: Lin Bennett, Wayne Boozer, Barbara Coppens, Mark Newsome, Fred Rockwell, Gary Walton, Jim Wasson, Don Wysoske, the staff at the Army Management Engineering College, and employees of Logistics Support Group.

Bruce Brocka
M. Suzanne Brocka

CONTENTS

PART 5 RESOURCES

PART 6 APPENDICES

ABBREVIATIONS/ACRONYMS

ASQC	American Society for Quality Control
CEO	Chief Executive Officer
DoD	United States Department of Defense
DOE	Design of experiments
EVOP	Evolutionary operation
FMEA	Failure modes and effects analysis
FMECA	Failure modes, effects, criticality analysis
GAO	Government Accounting Office
JIT	Just in time
JUSE	Japanese Union of Scientists and Engineers
MBNQA	Malcolm Baldrige National Quality Award
MBO	Management by objectives
PDCA	Plan-Do-Check-Act cycle (Deming wheel)
PDPC	Process decision program chart
pp.	pages
QC	Quality circles
SPC	Statistical process control
SQC	Statistical quality control
TQC	Total quality control
TQM	Total quality management
TV	Television
VCR	Videocassette recorder

LIST OF FIGURES

PART 1

FOUNDATIONAL ISSUES

1. WHAT IS QUALITY MANAGEMENT?
2. WHY QUALITY MANAGEMENT?
3. PRIMARY ELEMENTS OF QUALITY MANAGEMENT
4. IMPLEMENTATION

CHAPTER 1

WHAT IS QUALITY MANAGEMENT?

What You Would Get From 99.9% Suppliers:
At least 20,000 wrong prescriptions per year
Unsafe drinking water one hour per month
No electricity, water, or heat for 8.6 hours per year
No phone service for 10 minutes each week
Two short or long landings at each major airport per day
500 incorrect surgical operations per week
2,000 lost articles of mail per hour

Original source unknown

DEFINITION

Quality Management or Total Quality Management (TQM) is a way to continuously improve performance at every level of operation, in every functional area of an organization, using all available human and capital resources. Improvement is addressed toward satisfying broad goals such as cost, quality, market share, schedule, and growth. Quality Management combines fundamental management techniques, existing and innovative improvement efforts, and specialized technical skills in a structure focused on continuously improving all processes. It demands commitment and discipline, and an ongoing effort. Operating at a "high level" of quality may not be adequate, as can be observed from the quote above.

Quality Management relies on people and involves everyone. Quality Management is both a philosophy and a set of guiding principles that represent the foundation of a continuously improving organization, all the processes within the organization, and the degree to which present and future needs of the customers are met.

The U.S. Department of Defense uses the following definition of Total Quality Management:

> TQM is both a philosophy and a set of guiding principles that represent the foundation of a continuously improving organization. TQM is the application of quantitative methods and human resources to improve the material and services supplied to an organization, all the processes within the organization, and the degree which the needs of the customer are met, now and in the future. TQM integrates fundamental management techniques, existing improvement efforts, and technical tools under a disciplined approach focused on continuous improvement.

Quality Management could also be defined more simply by one of the following:

> Systematically and continuously improving quality of products, service, and life using all available human and capital resources.
>
> *or*
>
> An organizationwide problem-solving and process-improving methodology.
>
> *or*
>
> A system of means to economically produce goods or services that satisfy customer requirements.

The Quality Management process includes the integration of all employees, suppliers, and customers, within the corporate environment. It embraces two underlying tenets:

- Quality Management is a capability which is inherent in your employees.
- Quality Management is a controllable process, not an accidental one.

These two notions are revolutionary compared to the strict hierarchical, authoritarian organizations that existed in the past. While there may be room for such hierarchical organizations at certain times (e.g., a military unit during war), it is not appropriate for today's and tomorrow's employees in a highly competitive, global economy.

The idea of an integrated, human-oriented systems approach to management is nothing new. W. Edwards Deming, the leading Quality Management master today, was aiding Japanese firms in implementing Quality Management principles and tools in the 1950s. Historically, such authors as Sun-Tzu (a Chinese philosopher of the 2nd century B.C.) have advocated leadership based upon human values. However, U.S. managers are still overcoming 18th-century European and American ideas regarding management, wherein the workers were literally expendable. Figure 1–1, contrasts the old and *new* views of quality. Figure 1–2 shows another way to view the old and new ways via performance.

FIGURE 1–1
Two Views of Quality

Traditional View	*New View*
Productivity and quality are conflicting goals.	Productivity gains are achieved through quality improvements.
Quality defined as conformance to specifications or standards.	Quality is correctly defined requirements satisfying user needs.
Quality measured by degree of nonconformance.	Quality is measured by continuous process/product improvement and user satisfaction.
Quality is achieved through intensive product inspection.	Quality is determined by product design and is achieved by effective control techniques.
Some defects are allowed if product meets minimum quality standards.	Defects are prevented through process control techniques.
Quality is a separate function and focused on evaluating production.	Quality is a part of every function in all phases of the product life cycle.
Workers are blamed for poor quality.	Management is responsible for quality.
Supplier relations are short termed and cost oriented.	Supplier relationships are long term and quality oriented.

Source: U.S. Department of Defense

FIGURE 1–2
Two Views of Quality: Performance Levels

	Old Way	*New Way*
Quality	Parts per hundred If it's not broken, don't fix it Inspection = quality	Parts per million Continuous improvement TQM
Employee involvement	Passive suggestion systems	Proactive quality teams
	Win-lose strategy At most one improvement per employee per year	Win-win strategy Dozen or more improvements per employee per year
Focus	Short-term profits	Long-term survival

After 10 to 20 years of progress as a discipline, the term *Total* in Total Quality Management is becoming duplicative—Quality Management must be total, or companywide, to be optimally effective. Preferring to evade buzzwords as much as possible, this book will instead primarily use the term *Quality Management* rather than TQM, and in this book's context, TQM is equivalent to Quality Management.

HOW QUALITY MANAGEMENT WORKS

Quality Management may be thought of as a whole brained approach to management. Psychologists have found that the right hemisphere of the brain is responsible for affective domain activities: creativity and passion. The left hemisphere controls the cognitive domain: rational thinking. Quality Management combines aspects of both into a coherent, humane way of managing in chaotic times. Quality Management is very difficult to implement correctly, as role models are sparse, and much of the discipline is still new. For every successful implementation, such as at Motorola, there are 20 or more disasters. While it is generally specious to define what something is by what it *is not*, it may be worth stating what mistakes are pulled in the name of TQM (Figure 1–3). The "disputes" which break out among various "disciples" of one Quality Management master versus another, have lent a divisive, even faddish look to Quality Management practice to the casual observer. This is unfortunate, as Dr. XYZ's patented 12-step Total Performance Quality Management process or Dr. ABC's public domain 7-step Quality Management process is probably quite similar, if not nearly identical. The similarity among the quality masters is more striking than their differences.

The greatest resource in business is people—specifically, employees. Managing people demands a lifetime of observation, experimentation, ac-

FIGURE 1–3
What Quality Management Is Not

- Not a management "retreat"
- Not more rewards with little change in current structure
- Not an "I'm OK—You're OK" tonic
- Not a plan for lifetime employment for employees
- Not easy, nor quick to establish
- Not a management fad
- Not Japanese management
- Not just management by walking around
- Not just statistical process control and quality circles

tion, and reflection. If we truly could pick up management skill in a matter of hours, it would not be an endeavor worth devoting a career to. Quality Management is a means of empowering employees, but it also empowers the manager. Quality Management has no room for managers who manage by directive, attendance, standard operating plan, or other means of management by power or fear.

To implement these techniques is a frightening experience. Trusting employees to be capable, intelligent, and compassionate has not been normal management practice over the past 200 years of U.S. business history. The Japanese are, perhaps, the most well-known implementers of Quality Management tools and concepts, yet ironically, much of what we consider to be Japanese management techniques were developed by Americans who were generally unheeded in their own country. Quality Management is not "Japanese Management." It does not depend on a country, language, or even on the type of human resources available, as some texts on Japanese management imply. Quality Management can be successful whenever and wherever the dedication exists to incorporate it into corporate culture.

Quality Management and its empowering of the worker would seem to be at odds with both the U.S. stereotype of the Japanese as kamikaze corporate conformists, and the similar labeling of U.S. workers as incorrigible cowboys yearning for an unexplored horizon. Closer examination reveals something quite different. The Japanese environment is chaotic, cooperative and forgiving; well suited to creativity and psychological fulfillment. Workers in the United States are often preoccupied with a "the way we do things," and "cover yourself" mentality. Creativity and exploration are relegated to mysterious R&D departments. Even the American press ignores or underreports major discoveries in the scientific and engineering world. Comparative studies have shown U.S. children to be the least aware (compared to other industrial nations) in understanding basic scientific facts. Talking to school-age children reveals not a slow and stuporous group, but one that is unaware, unmotivated, and more than a bit shell-shocked. What is needed is a management approach that will unleash this latent, possibly atrophied talent, providing benefit to both the worker and the firm. The goal of Quality Management is continuous improvement, but the threshold of employee motivation and empowerment must be passed through first.

PRINCIPLES AND CONCEPTS

A Quality Management program must:

- Require dedication, commitment, and participation from top leadership.
- Build and sustain a culture committed to continuous improvement.

- Focus on satisfying customer needs and expectations.
- Involve every individual in improving his/her own work processes.
- Create teamwork and constructive working relationships.
- Recognize people as the most important resource.
- Employ the best available management practices, techniques, and tools.

The prism of Figure 1–4 illustrates the notion that Quality Management commences with the "white light" of a strategic vision which is then transformed by management dynamics, the prism, into component colors, the tools and methods needed to implement this vision.

1. Process orientation, rather than solely result-oriented orientation. By being process oriented, we can affect results in an early stage. Process orientation demands a reexamination of why things are done the way they are. Improving the quality of the process improves the quality of the result. Figure 1–5 illustrates this view.

2. Cascade the implementation and involve everyone. Quality Management will first be implemented by top leadership and flow through the management structure similar to a waterfall. This cascading deployment ensures that leaders understand, demonstrate, and can teach Quality Management principles and practices before expecting them from, and evaluating them in, their staff. The cascade effect flows to the suppliers as well.

FIGURE 1–4
Quality Management Prism

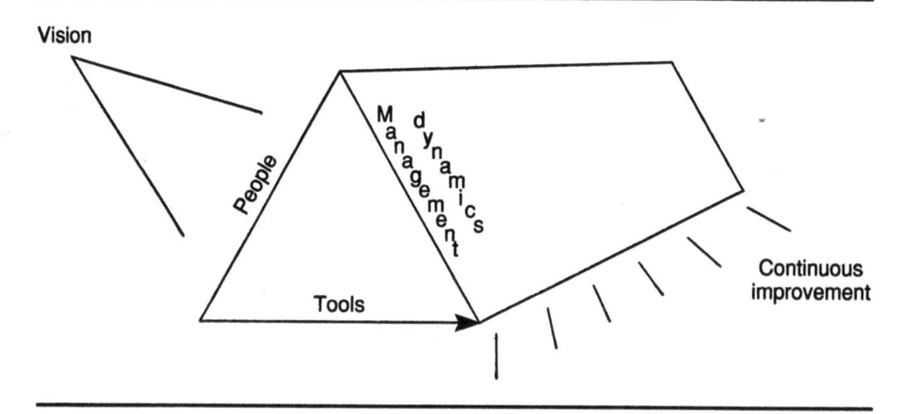

FIGURE 1–5
Process Oriented versus Results Oriented

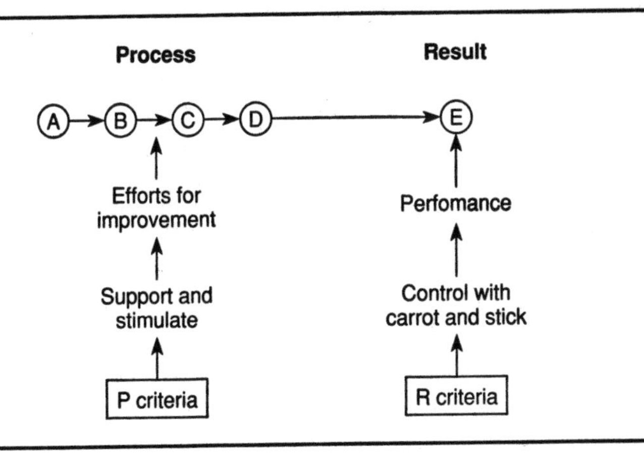

Source: Masaaki Imai, *Kaizen: The Key to Japan's Competitive Success* (New York: Random House, 1986), p. 18.

3. Commitment from top leadership. This leadership ensures strong, pervasive commitment to continuous improvement. Cost reduction, schedule compliance, customer satisfaction, and pride in workmanship all flow from an overt dedication to continuous improvement. Acting on recommendations to make positive changes demonstrates this commitment.

4. Effective, unfettered vertical and horizontal communication. Using this type of communication is essential to continuous improvement efforts. Quality Management practices aim at removing communication blocks, facilitating bidirectional communication between leaders and subordinates, and ensuring that the firm's goals and objectives are clearly delineated and disseminated throughout. A broad array of tools and techniques are available for enhancing vertical and horizontal communication. Virtually every technique in Part 3, "Management Dynamics" and Part 4, "Tools and Techniques" is communication oriented or enhancing.

5. Continuous improvement of all processes and products, internal and external. The primary Quality Management objective is an unending improvement of every aspect of one's work. That objective is implemented through a structured, disciplined approach that improves each process. With Quality Management, emphasis is placed on preventing defects through problem identification and problem-solving tools.

6. Constancy of purpose and a shared vision. A common purpose or set of principles must guide the organization. This can be as simple as making leakproof buckets, or as complex as the principles pervading a monastery. Whatever this purpose is, all personnel must know it and work to fulfill it. Consistency is vital; dissonant goals will result in frustration. To be courteous and informative to all callers and to process 45 calls per employee hour are goals completely at odds. When conflicting goals arise, select the higher quality alternative. In this case, eliminate the 45 calls per hour requirement, or hire the additional staff necessary to disseminate the work load reasonably.

7. The customer is king. The customer must rule, whether this is an internal customer (a customer is the next person in the process—internal customers are peers and supervisors) or external customer (the traditional customer). Every worker has a customer of some sort. Customers or users must be identified, and their needs, wants, expectations, and desires clearly delineated and served. Customers and their needs are the only reason a business exists.

8. Investment in people. Any firm's largest and most valuable investment is in its people. They are the most essential component in continuous process improvement. Training, team building, and work life enhancements are important elements in creating an environment in which employees can grow, gain experience and capability, and contribute to the firm on an ever-increasing scale. Rewards may be established, but only if it is clear to the employee what performance is expected to attain the reward, and if they are given the tools necessary to achieve this goal.

9. Quality management begins and ends with training. Training is constantly required for all staff. It may be to increase affective domain skills such as writing, team building concepts, or verbal skills; or it may be cognitively oriented such as statistical quality control. Training *given* today must be *used* today. Internal trainers can be invaluable, bringing a camaraderie and unique slant to the information. Unlike education, training should have an observable outcome. However, education is valuable in providing a background necessary for divergent and innovative thinking.

10. Celebrate success and accentuate the positive. Negative reinforcements have been found to be ineffective motivators. Establish a system of fair rewards that everyone can achieve, and everyone knows what is required to achieve the reward.

11. Two heads Are better than one. Without teamwork, Quality Management is finished before it can start. The modern team works together as a single entity, and not as a "committee" where one or a few members do or direct the work. This may require dramatic rethinking by management and employees as well. Teamwork is essential for continuous improvement. Team activities build communication and cooperation, stimulate creative thought, and provide an infrastructure supporting Quality Management practices.

12. Goal setting is communicated and determined by everyone. Employees must participate in establishing their goals. Others must be aware of goals that may impact them.

Figure 1–6 categorizes and summarizes the primary Quality Management principles. This list is not all-inclusive, but represents a distillation of many authors. These principles should be redefined and made specific for each company as it develops its own Quality Management philosophy.

The idea behind continuous improvement, or *kaizen* as the Japanese call it, is that small improvements, done continuously, will amount to major changes in an over time, and not necessarily a long period of time. Japanese manufacturing firms employing this technique during the late 1970s and early 1980s *leapfrogged* U.S. manufacturers into a position of worldwide technological dominance. Management techniques can evolve too, and Figure 1–7, from Rossier and Sink provides an evolutionary view of Quality Management. Understanding where your organization is in the evolutionary avenue will lead to an understanding of the Promethean in undertaking Quality Management.

BIBLIOGRAPHY

Ackoff, R., and F. Emery. *On Purposeful Systems*. Chicago: Aldine/Atherton, 1972.

Badiru, Adedeji B. "A Systems Approach to Total Quality Management." *Industrial Engineering* 22, no. 3 (March 1990), pp. 33–36.

Deming, W. Edwards. *Out of the Crisis*. Cambridge, Mass.: MIT Press, 1986.

Ernst & Young Quality Improvement Consulting Group. *Total Quality: An Executive's Guide for the 1990s*. Homewood, Ill.: Richard D. Irwin, 1990.

Feigenbaum, Armand V. *Total Quality Control*. New York: McGraw-Hill, 1983.

Nemoto, Masao, and David Lu, trans. and ed. *Total Quality Control for Management: Strategies and Techniques from Toyota and Toyoda Gosei*. Englewood Cliffs, N.J.: Prentice Hall, 1987.

Rossier, Paul E., and D. Scott Sink. "What's Ahead for Productivity and Quality Improvement." *Industrial Engineering* 22, no. 3 (March 1990), pp. 25–31.

FIGURE 1–6
Quality Management Principles Summary

Management Vision and Commitment

Process must begin with topmost management.
Problems often must be solved by changing or renewing the process or system.
Quality is a way of life.

Barrier Elimination

Rampant involvement, including suppliers and customers, is essential.
Yield authority to the lowest possible level to resolve problems.
Change must be the norm, not the exception.

Communication

Communication and information dissemination are vital.
Inform the end user of the information as quickly as possible.
It is more important to be clear than correct.

Continuous Evaluation and Measurement

Identify customer requirements.
Constantly use feedback.
Self-assess and reflect continuously.

Continuous Improvement

Quantify and measure.
Measure the cost of quality.
Continuously monitor vital measurements of a product.
Reduce variation.

Customer/Vendor Involvement

Customer must be king.
Vendors are part of the solution, not the problem.
Customer requirements, desires, hopes, and fears must be continuously monitored.
Customers may be internal.

Empowerment

Management style must be actively participative.
Employees must be actively involved.
Authority and autonomy must be commensurate with duties.

Training

Emphasize that long-term success is survival.
Quality is conformance to customer requirements.
Enhance skills to measure quality and identify problems.
Training must be at all levels.

FIGURE 1-[illegible]

Major Present, Emerging, and Future Improvement Strategies and Techniques

Planning	*Measurement*	*Total Quality Management*	*Management of Participation*	*Reward System*
Present: Business Facilities Capital Product	Statistical quality control. Financial ratios. Management by objectives. Performance by appraisal. Discounted cash flow. Project management. Cost accounting. Individual oriented. Tool driven. Analyst developed and maintained.	Quality assurance. Inspection. Quality circles. Statistical quality control. Streamlining. Value engineering. Buzzwords, no operational definitions. Could cost.	Quality circles. Individual suggestion systems. Education, training, and development as expense. Brainstorming. "Participative management." Task forces. Committees. Employee = workers only as used in employee involvement. Education, training, development is up front, rigid not flexible.	Incentive systems. Piece rate. Merit pay. Across-the-board-pay raises. Rigid benefits. Rewards not linked to performance. Individual focus. Profit sharing. Labor focus.
Emerging: Human resource Performance improvement Quality	Input output analysis. Statistical process control. Total factor productivity measurement model. Objectives matrix. Cost driver analysis (i.e., IBM common staffing study). Cost schedule control system. Competitive benchmarking. Management systems analysis. Decision support system.	Quality management. Assurance. Less reliance on inspection. Statistical process control. Design to production transition. Customer focus. Subcontractor/vendor control. Operational definitions TQM as buzzword.	Performance action teams. Group suggestion systems. Small-group activities. Education, training, and development as investment. Nominal group technique. "Management of participation." Employee = everyone including top management. Education, training, development is on as-needed basis, and flexible.	Gainsharing systems. Award fees. Industrial modernization incentives program. Performance based. Flexible benefits. Rewards linked to performance. Group focus. Performance improvement sharing. All factor (i.e., labor, capital, material, energy) focus
Future: Total integrated and highly participative	Total integrated measurement system. Management support systems. Improvement-oriented measurement. Statistical performance control. Total system oriented (individual, group, organization) and integrated. Management team/user driven. User developed and maintained.	Design of experiments. Quality function deployment. Real TQM = total life-cycle quality management. Total performance management.	Semiautonomous work groups, autonomous work groups. Self-managed work groups. Top-down and bottom-up strategic performance improvement planning.	Employee stock ownership. Integrated individual group and organization performance appraisal and reward systems. Total compensation system management.

Source: Paul E. and D. Scott Sink, "What's Ahead for Productivity and Quality Improvement." *Industrial Engineering* 22, no. 3 (March 1990), pp. 25–31.

Schultz, Louis E. *The Role of Top Management in Effecting Change to Improve Quality and Productivity.* Minneapolis, Minn.: Process Management Institute, 1985, pp. 1–8.

Schultz, Louis E. *Overview of Quality Management Philosophies.* Minneapolis, Minn.: Process Management Institute, 1986.

Shores, Dick. "TQC: Science, Not Witchcraft." *Quality Progress* 22, no. 4 (April 1989), pp. 42–45.

Strickland, Jack, and Peter Angiola. *Total Quality Management in the Department of Defense.* Washington, D.C.: U.S. Government Printing Office, 1989.

Strickland, Jack. "Key Ingredients to Total Quality Management." *Defense,* March/April 1989, pp. 17–21.

Sullivan, Edward, and Douglas D. Danforth. "A Common Commitment to Total Quality." *Quality Progress,* April 1986.

Tribus, M. *Deming's Redefinition of Management.* Cambridge, Mass.: MIT Press, 1985.

Tribus, M. *Reducing Deming's 14 Points to Practice.* Cambridge, Mass.: MIT Press, 1984.

Vansina, Leopold S. "Total Quality Control: An Overall Organizational Improvement Strategy." *National Productivity Review* 9, no. 1 (Winter 1989/90), pp. 59–73.

Walsh, Loren M.; Ralph Wurster; and Raymond J. Kimber, eds. *Quality Management Handbook.* New York: Marcel Dekker, 1986.

Willoughby, W. J. *Best Practices: How to Avoid Surprises in the World's Most Complicated Technical Environment.* Department of the Navy, Washington, D.C.: U.S. Government Printing Office, March 1986.

CHAPTER 2

WHY QUALITY MANAGEMENT?

For every complex question there is a simple answer, and it is wrong.

H. L. Mencken

INTRODUCTION

"My business is doing fine—Why bother with Quality Management if it's so much work?" A reasonable question, but it ignores the following:

Your product could be wiped out overnight by:

- Technology: There are no Fortune 500 candlestick makers—but everybody needed this product at one time.
- A new law: Environmental laws are squeezing more and more chemicals off the shelf, and changing the way business conducts day-to-day operations.
- A competitor's newfound superiority: The word processor WordStar® was once dominant; its market share eroded quickly when WordPerfect came on the scene and WordStar® was slow to respond.
- A change in lifestyle: Better find a way to microwave your food product, remove most of its fat, salt, cholesterol, and calories, *and* have it be as tasty as if it were prepared conventionally.

Even if your product is not threatened with extinction wouldn't you like to:

- Go from 92 percent (about one out of a dozen) defect free to 99.97 percent (one out of 3,333) defect free? Xerox did—in just six years. Their goal is no defects per million. It may be that you would have better odds in a lottery than finding a defective Xerox part.

- Cut order processing times from 55 days to 15? Motorola did.
- Reduce setup times from hours to one minute? Toyota did.
- Create, define, and dominate a market as did Fred Smith of Federal Express?
- Materialize from Bentonville, Arkansas, to challenge and eventually overtake all discount retail stores (including 100-year-old giant Sears)? Sam Walton locates most of his company's Wal-Mart stores in the sparsely populated areas of the "fly over land" of the Midwest and South.

Or do you already match the firm of the 90s that the chronicler of excellence Tom Peters describes in *Thriving on Chaos:*

> Flatter, populated by autonomous units, oriented towards product differentiation. Quality conscious, service conscious, more responsive, much faster at innovation, and a user of highly flexible people.

A number of works explore at great length the perils facing business. *Fortune, Forbes, BusinessWeek,* and *The Wall Street Journal* do so on a regular basis. A condensation of the major issues is presented in this chapter. Readers interested in either depressing or inflaming (or both) themselves should read any or all of the works referenced at the end of this chapter, or in viewing the PBS television documentary, "Quality or Else," aired in October 1991.

LESSONS FROM HISTORY

In their mesmerizing work, *American Business: A Two-Minute Warning,* Grayson and O'Dell offer 10 lessons from the history of leading economic nations (leader) and newly industrialized nations (challengers) that can guide our motivation in adopting Quality Management:

1. Complacency is the cancer of leadership.
2. Leaders overlook growth rates of challengers.
3. The growth rates are small and incremental, and not realized until too late.
4. Size is a poor estimator of success.
5. Challengers have the "eye of the tiger" (desire), the leaders have lost it.
6. Challengers stress education and improvement; leaders chop training when the budget gets tight.

7. Challengers copy strategies; leaders find it beneath them.
8. Challengers are customer oriented, leaders become producer oriented.
9. Protectionism hurts leaders and helps challengers.
10. The leader's ability to change and respond wanes with time.

LESSONS FROM TODAY

A survey of 700 British CEOs by Lascelles and Barrie found the following motivators for quality improvement programs:

- Demanding customers—73 percent
- Need to reduce costs—63 percent
- CEO initiative—59 percent
- Competitors—34 percent

They also discovered the disconcerting fact that personnel in fewer than half of what the companies surveyed had received training in Quality Management techniques. Not surprisingly, it would seem that customers are driving the need for continuous improvement. There are also societal factors that motivate us to find new ways to conduct work as the needs and expectations of new generations emerge.

The U.S. Government Accounting Office (GAO), the investigative service of Congress, analyzed 20 of the highest scoring companies applying for the Malcolm Baldrige National Quality Award (Chapter 18, section "Benchmarking") and found the following benefits from instituting quality management:

- Quality improved and costs decreased. Reliability and on-time delivery increased; errors, lead time, and complaints were reduced.
- Customer satisfaction increased, and the overall perception of quality increased.
- Profitability and market share increased. Quality Management practices lead to improved profitability.
- Employee relations improved somewhat. Absenteeism and turnover was reduced, and employees experienced increased job satisfaction.

The GAO found the following distinctive features of the Quality Management efforts:

- Attention focused on meeting customer requirements.
- Senior management demonstrates quality values by incorporating them into daily operations.
- Systematic processes promoting continuous improvement were woven through the organization.
- Training and empowering processes were instrumental in quality improvement efforts.

The GAO also found that the performance improvement efforts required about 2.5 years.

MODERN FORCES

Just as classrooms have changed from places where reading, writing, and arithmetic were taught to school systems that provide for the health, education, welfare, and well-being of their charges, so has the work force needed to evolve from a place to earn a wage for food and shelter into a complex socioeconomic system. Workers now seek fulfillment from their work, not just a paycheck. The transformation of the work force is made even more difficult by the forces of global competition, technological change, environmental change, social forces, and changing work ethics. These external forces are changing forever the way in which we conduct business.

The effect of these forces is not magically cured by Quality Management, but Quality Management will allow all available company resources to develop alternatives to mitigate the impacts. Some of these forces may demand Quality Management as part of their resolution.

Global Competition

In the past, lack of communication and education often made possible a company's success. Markets of broad scope could be targeted with little thought to unknown or small competitors. Today, competition is truly global. One never knows who or where the next competitor will be. Quality Management aids in anticipating competition by continuously striving for quality improvements; there is no room for complacent leadership. For example, in the same quarter, three major computer manufacturers (IBM, Apple, and Compaq) announced difficulties in coping with what they saw as a flat marketplace, while at least three upstart companies (Dell, Zeos, and Gateway) announced another recordbreaking quarter of sales.

Technological Change

The very foundations of how and why we do what we do will change in ways unknown to us now. Yet these forces can wipe out an entire mature industry in a few years (e.g., compact discs versus vinyl record albums). The older adult generation of today may in fact have more in common with the writers of ancient papyri than with the nearly illiterate future world of audiovisual images. If this seems unlikely to you, consider the following:

- Videotaped courtroom proceedings as legal record.
- Fast-food cash registers with keys representing the food (no numbers).
- Videotaped high school yearbooks.
- Universal traffic symbols.
- Videos outsell books four to one.

There does not seem to be a technology plateau in sight; new generations of computer hardware that used to appear every three to five years, to every two years, are now appearing about every 10 to 12 months.

Quality Management reduces turbulence caused by new technology by embracing it, rather than ignoring it. New technology can be stimulating and is constantly opening new business opportunities. What would the state of American business be today if it were not for its seminal and dominant (or codominant) role in computer hardware and software technology?

Social Forces

We spend many of our waking hours at our workplace, and several more hours preparing, commuting, and thinking about our work. For most adults between the ages of 21 and 65, work is the primary feature of their life. We may get married, divorced, have custody of children for shorter or longer times, but during that span, we are almost always employed. With such a focus, it is not surprising that people look to work for fulfillment and enrichment; some may even find a calling in their work. Quality Management can provide the galvanizing impulse that an organization needs to provide management that dissipates frustrations, and capitalizes upon the pent-up energy of its work force.

The social fabric has been torn—the evidence assails us in the daily newspapers:

- Children without fathers are rapidly becoming a social norm, even a majority. Women who head single parent households are generally living in poverty as are their children.

- Twenty-three percent of the children in the United States under the age of six live in poverty.
- The United States spends more on social security for elderly who have incomes of over $50,000 per year than we do for the entire food stamp program.
- Manipulative "information" from broadcast and print media cheats viewers and readers of realistic views of how life should be lived.

Thus it is little wonder, then, that the typical worker trudges into work embittered, his mind wandering to anything except the tasks before him. Quality Management cannot fix any of these problems, but it can help to provide a workplace where humane and honest treatment can be found, and perhaps provide a measure of fulfillment.

Work Ethic

The value of work is praised in countries that are economically successful regardless of the era. Germans pride themselves on their discipline. So do the Japanese. Work ethic used to be equated with loyalty to a company in the United States. This has changed. The new breed of professional feels loyal to their discipline (mechanical engineering, accounting, writing, and so on), not to any company. It may be easy to misinterpret this new work ethic. It manifests itself in workers seeking new projects and opportunities, and having authority over these projects. The old conformance work ethic operated under a code of unquestioning obeisance, stifling creativity and optimum performance.

Work value cannot be directed from the CEO or a minister. It stems from the powerful force of peer values. Peer values are not limited to human-to-human interaction. Television, for example, can exert a considerable influence, since we average some 20 to 40 hours per week watching it. Quality Management can provide a work atmosphere that nurtures a work ethic and allows workers to achieve their potential as workers as well as humans. The president of Johnsonville Sausage, Ralph Stayer, says that it is immoral (a rare word these days) not to allow workers to reach their full potential, at least while on the job. Rigid hierarchies that allow no questioning and teamwork are demeaning, and criminally waste the human resources of society. However, the message is often the opposite of this ideal: CEO compensation ratios to that of a laborer can be as high as 9,000:1.[1] When a society values individuals solely on income, it is easy to see that the breakdown of societal values has either happened or is looming.

---

[1]Steven Ross, CEO of Time Warner, received $78 million in total compensation in 1990. "Corporate America's Most Powerful People," *Forbes*, May 27, 1991.

The work ethic prescribes perseverance. In an era of 44-minute life-and-death decisions (the actual length of most hour-long television shows), apparently instant riches with no evident physical or mental effort, it is understandable why perseverance is an anachronism.

The work ethic problem affords Quality Management a great challenge, as it cannot be controlled outside of the workplace, where many attitudes are formed. A company can, nevertheless, provide examples and reward individuals who display such a work ethic. A corporate culture cannot entirely supplant an external culture, however, it can reshape attitudes and reinforce positive traits from the external culture.

If the above forces have not been sufficient to stimulate a desire to engage in a Quality Management program, then consider that your competitor is probably instituting just such an approach, viewing it as a necessary means for survival. Quality Management is not the simple answer to the problems facing management, but it can be the right one.

BIBLIOGRAPHY

Bennis, Warren. "The Coming Death of Bureaucracy." *Think* 32 (November/December 1966), pp. 30–35.

Bhote, Keri. "America's Quality Health Diagnosis: Strong Heart, Weak Head." *Management Review,* May 1989.

Brown, J. H. U., and J. Comola. *Educating for Excellence: Improving Quality and Productivity in the 90's*. New York: Auburn House, 1991.

Dobyns, Lloyd, and Clare Crawford-Mason. *Quality or Else*. Boston: Houghton Mifflin, 1991.

Drucker, Peter F. "The Coming of the New Organization." *Harvard Business Review,* January/February 1988, pp. 45–53.

Drucker, Peter F. *The New Realities*. New York: Harper & Row, 1989.

Grayson, C. J., and C. O'Dell. *American Business—A Two-Minute Warning: Ten Tough Issues Managers Must Face*. New York: Free Press, 1988.

Harrington, H. James. *Excellence—The IBM Way.* Milwaukee, Wis.: ASQC Quality Press, 1988.

Lascelles, David, and Barrie Dale. "Quality Management: The Chief Executive's Perception and Role." *Journal of European Management* 8, no. 1 (March 1990), pp. 67–75.

Townsend, Patrick L., and Joan A. Gebhardt. *Commit to Quality.* New York: John Wiley & Sons, 1990.

U.S. Government Accounting Office. NSIAD 91-190, as reported in *On Q,* September 1991.

CHAPTER 3

PRIMARY ELEMENTS OF QUALITY MANAGEMENT

My mistake was buying stock in the company.
Now I worry about the lousy work I'm turning out.

Marvin Townsend

The "Pillars of TQM" or the primary elements of Quality Management philosophy vary from author to author, and their number may vary, but their marrow is the following:

1. Organizational vision.
2. Barrier removal.
3. Communication.
4. Continuous evaluation.
5. Continuous improvement.
6. Customer/vendor relationships.
7. Empowering the worker.
8. Training.

This set was selected because it applies to organizations initially making the move to a Quality Management program. As organizations advance, training, barrier removal, and communication can be subsumed into empowering the worker. Continuous evaluation and continuous improvement could perhaps be combined into continuous analysis.

Thus distilled, this leaves us with four essential ingredients from which the other concepts flow. These concepts are presented in Figure 3–1. These components make up the strategic portion of the quality pyramid developed by Ronald Snee in Figure 3–2. The managerial aspects are detailed in Part 3, and the operational aspects may be found in Part 4. Chapter 4, "Implementation,"

FIGURE 3–1
Four Key Components

Top-down strategic vision demonstrated daily via leadership
Continuous analysis and product/service improvement
Empower and liberate employees
Listen and react to customers and vendors

FIGURE 3–2
Quality Management Pyramid

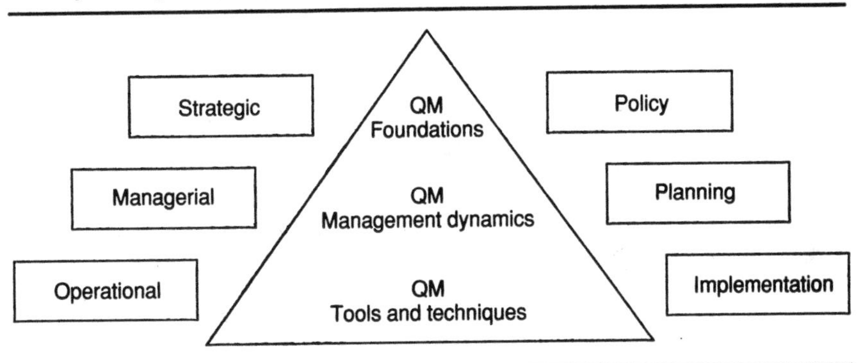

relates to the strategic, managerial, and operational elements. Each of the eight foundational elements is discussed below.

1. ORGANIZATIONAL VISION

Organizational vision provides the framework that guides a firm's beliefs and values. This can be as simple as "making the best widgets at the lowest cost to the consumer," or as structured as the organizational culture of IBM. The gist of the corporate vision should be a simple, one sentence guide or motto that every employee knows, and more important, believes. If well crafted, the vision statement (which is typically one paragraph to no more than two pages) can serve through a torrent of change in product and service technology. For example, if you were a coach builder in the year 1910, and decided that you made the finest horse coaches available, the subsequent decade or two would have bankrupted you. If you had decided that you provided fine coach work, irrespective of the driving engine, you may have adapted quickly to the new challenges. Technology is rapidly making many things obsolete. Broad vision state-

FIGURE 3–3
The General Motors Quality Network Process Model

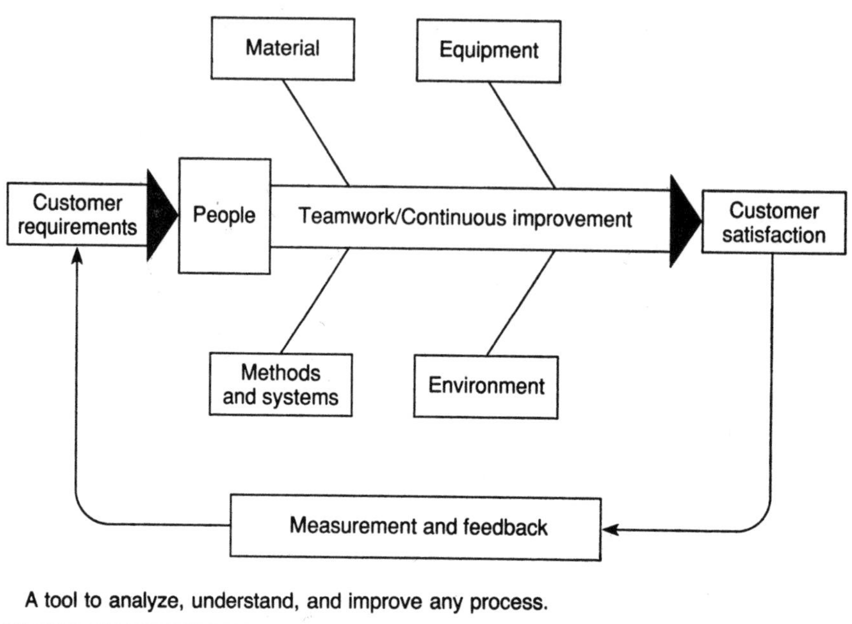

A tool to analyze, understand, and improve any process.

ments are now the norm, although this sometimes leads to such stilted phrases as "music delivery system" for audiotapes, CDs, and videos of musicians.

General Motors provides all employees a card with its strategic vision, including a cause-effect diagram indicating the teamwork necessary. This card is reproduced in Figure 3–3. John Hartley, CEO of Harris Corporation, formulated this statement:

> Harris must be perceived as a company of the highest quality in every aspect of its business activity.
>
> This would seem rather broad, although Harris makes an incredibly diverse array of products.

This strategic vision needs to consider both the external customer and the internal customer, or the employees, but lacks a defining or differentiating phrase. We all like to be the best but this needs to be combined with a goal such as "fastest to market" or "lowest cost," or some other distinguishing phrase. As younger, more consumer-oriented workers gain dominance in the work force, a firm's identity, ethics, and beliefs will become increasingly important to the work force and shareholders alike. A sense of purpose must guide our actions if our lives are to have any sense of meaning and fulfillment. In the past, the utility of the work we did was subsumed by an external goal. We did the work to bring ourselves or our children out of poverty. The work could be degrading because our survival was at risk. This is no longer the case in most businesses, as in many areas of the world. We are ascending Maslow's hierarchy of needs[1] from the survival stage and are nearer the actualization stage. As work occupies most people's waking hours, it is only logical for them to seek fulfillment while on the job. This does not (necessarily) translate to the need to implement all manner of costly *wellness* programs, but presents a need to let the work force know that what they do is important and vital to the community in some way.

Simply stating a vision is not enough. It needs to be demonstrated by the actions of the executives, managers, supervisors, foremen, and individuals. It is done continuously in all of their actions and initiatives. Employees know the difference between an "open-door" policy and a "slightly ajar" actuality.

Deliberation must be exercised in developing these goals and strategies. They must reflect the values and culture of the work force. While top-management commitment is essential, it is also essential to know when to lead and when to get out of the way (the old saw, "If they throw you out of town, make it look like a parade," has relevancy here). In a sense, Quality Management is management from the bottom up, as latent talents are coaxed out. Quality must be infused into all processes the organization engages in, whether it is managerial, administrative, functional, or some other process. An atmosphere must be created in which each individual feels responsible to the customer for whatever product is produced or service rendered, and a responsibility exists to the customer, whether that customer is internal or external to the organization. This may well require a gut-wrenching change in the way day-to-day business is conducted.

Inspiring an Organizational Vision

While a vision or mission statement is a product of the strategic vision process, unless it is transformed into action, the statement is useless. There are four keys to successful vision implementation:

1. Total involvement. Every level of the organization, including senior management, must be involved in quality improvement activities.

2. Communication. It is essential that everyone in the organization understand the specifications of his or her customers, and be conscious of how well he or she is meeting the customers' needs.

---

[1]Maslow developed a hierarchy of needs ranging from food and shelter to family needs to "self-actualization" or spiritual enlightenment. This hierarchy is illustrated in Figure 3–10.

FIGURE 3–4
Strategies in Successful Vision Implementation

Demonstrate commitment.	Inform suppliers.
Maintain a constancy of purpose.	Take a long-term view.
Create more leaders.	Establish meaningful goals.
Examine your mission.	Discuss TQM with peers.
Behavior and actions must be consistent with goals.	Build awareness.

3. Barrier removal. Structures, policies, and procedures must be implemented to encourage quality. All those that restrict progress toward Quality Management must be removed. Quality Management must be part of the strategic plan, the budget process, and the employee reward system. Barrier removal is the first step in empowering employees. Constantly ask why it should be done this way, and constantly ask why am I deciding this, versus the people directly involved.

4. Continuously improve and evaluate. Keep looking for a better way, even if your customers are satisfied with the current product or service. Quality improvement can be the ultimate integrator of the organization, helping to achieve critical Quality Management objectives: improved product quality, lower costs, stronger customer loyalty, increased employee morale and lower unwanted turnover.

2. BARRIER REMOVAL

It is inevitable that change will be resisted. In fact, a great deal of effort in Quality Management is expended in overcoming such resistance, usually by allowing change to come from individuals directly involved, rather than as a directive from management. The whole idea of continuous improvement leads to continuous change. Ideas for adapting to this change are discussed in Chapter 13, section on Change Management. This section will focus on overcoming barriers associated with the implementation of the continuous improvement process.

The following strategies are recommended throughout the barrier removal process:

- Drive out fear.
- Encourage and reward creative thinking—even if the ideas are not implemented.
- Share the credit for success.
- Revise and renew performance measurement systems.

FIGURE 3–5
Some Common Barriers

We know what they really want (without asking them).
Quality is not a major factor in decisions—low initial cost mentality prevails.
Creative accounting can increase corporate performance.
Can't manufacture competitively at the low end.
The job of senior management is strategy, not operations.
Success is good, failure is bad.
If it ain't broke, don't fix it.
The key disciplines from which to draw senior management are finance and marketing.
Increase in quality means increase in cost.
Thinking that time, quality, cost are at worst mutually exclusive, at best we can only choose two out of three.

- Justify cost over the life cycle, not just initial cost.
- Establish ownership of tasks and projects.

The following are the steps to barrier removal:

1. Identify barriers. Anything that stands in the way of implementing and realizing continuous improvement should be considered a barrier. This means examining internal procedures, customer relations and concerns, and personnel issues. Anything that is perceived to be a barrier deserves further consideration. At this initial stage, no judgment as to priority or validity should be made. Generation of the list can be accomplished by several of the techniques described in Part 4 "Tools and Techniques." Perhaps brainstorming (Chapter 19) would be the most effective at this stage. David A. Nadler of Delta Consulting group provided the examples of corporate culture barriers listed in Figure 3–5 during a Xerox Quality Education Seminar.

2. Place into categories. Related barriers and their systemic causes may now be analyzed. Validity judgment should still be held in abeyance at this stage. Categorization may be facilitated by using cause-effect diagrams or other organizing tools (Chapter 16). Be alert for barriers that mask or cause one another. It is not unusual for a myriad of problems to be caused by a few difficulties. Quality function deployment is another useful technique that may be used in conjunction with, or instead of cause-effect diagrams.

3. Establish priority. This should be done by using a tool such as Pareto analysis (Chapter 20), cause-effect diagrams, or by the Delphi technique (Chapter 19). Care must be taken to establish an objective process, whatever that process might be. It should not be influenced by management, or by a hidden agenda. At this stage barriers are judged on their validity in accordance with the severity of the problem. It can be difficult to compare relative barriers at this stage, if a common denominator, such as dollars or number of defective units is not used. In companywide searches for barriers, it may be necessary to find more than one denominator and deal with the problem accordingly.

4. Problem solve. This means more than symptom removal! In medicine, symptom alleviation allows the patient to think that he is cured, even though the patient, if untreated, will recover in the same amount of time, or may not recover at all. Sick organizations do not recover for the long term if symptoms are masked. At best, symptom masking makes one quarter's report better. Akio Morita, chairman of Sony corporation, has chastised American managers for planning ahead for only 10 minutes rather than 10 years.

It is vital to address the root cause of the problem. By using cause-effect diagrams and quality function deployment it may become apparent that the elimination of one barrier may solve many problems. Do not be surprised if this "master cause" looks intractable, such as "poor communication among management and workers." These "soft" problems are the ones that plague us for years, and may take years to solve. While this sounds expensive to cure, be assured that the competition will ultimately solve any such long-term illnesses for you. Analysis of the problem should include estimates of resources required for its resolution. A cause-effect diagram or force field analysis will be useful in identifying the nature of the solutions, and potential hindrances in successful problem resolution.

5. Goals and strategies for resolution. Resolution of problems may entail goals over a period of months or years. Goals should be realistic and attainable with the given resources. Strategies ensure that goals can be accomplished. Bear in mind that numerical goals as such may not be what is required. A 15 percent improvement with no strategy is meaningless, if not insulting. Numerical goals may also limit the amount of growth, particularly in organizations used to working up to an "average," as occurs in many piecework situations. Allowing people to work to their optimum, without harming other workers, will provide measurable improvements without numerical quotas. Attaining a short-term goal may be possible by altering the natural rhythm of a process, but may not be workable over an extended period of time.

Who Should Do This?

The general supposition is that volunteers in a cross-functional team should identify proposed solutions for barrier removal. No one should be tasked to do it. However, in the early stages of Quality Management implementation, a valuable individual may ignore such a call for volunteers, having been beaten down by the existing system. This person may need to be asked. If he refuses, he should not be co-opted. As Quality Management is implemented properly, this type of volunteer problem goes away. People see that they can indeed make a difference. A balance should be achieved among departments, that is, shop floor versus nonshop floor workers, and so on. Do not consider an area where representation is not present. Do not suppose that union workers necessarily have a built-in adversarial role, or that less-educated workers will be unable to articulate their concerns and analyze problems. After all, a union electrical worker has led Poland into a new era of social democracy that embraces capitalism.

3. COMMUNICATION

Communication is the glue that binds all of the techniques, practices, philosophies, and tools. Ineffective communication will doom the most clever of Quality Management initiatives. Communication may be: written, verbal, or nonverbal. Understanding and refining skills for each of the main types of communication is an ongoing process for everyone. Recurring training in each of the areas is a must to develop and retain communication skills.

All forms of communication involve four elements: the sender, the receiver, the message, and the medium. The medium is the method of delivery, and can influence the message. One pop media guru of the 1960s declared that "the medium was the message," referring in part to the hypnotic entrancement of television. Because of the filtering effects that can happen to a message (Figure 3–6), it is important to understand how communication works, and how personality factors influence our understanding. An excellent method on understanding cross-generational differences is Massey's idea of *what you are now is what you were when.* His thesis is that people are value programmed by 10 years of age, and that understanding the receiver's value system relative to the sender's value system is vital. For example, someone who turned 10 during the Depression would view a situation differently than someone who grew up during the more comfortable 50s. Understanding how past experiences color present situations can overcome the differences in value systems, one of the most difficult to overcome barriers listed in Figure 3–6.

Written Communication

Principally the domain of office workers, written skills take decades to hone. Office memos and reports are often the results of hundreds of hours (studies indicate anywhere from 21 percent to 70 percent of office workers' time is spent manipulating written information) of work, and their final form should be worthy of spending some time to get the words right. Shun bureaucratic

FIGURE 3–6
Barriers to Communicating Effectively

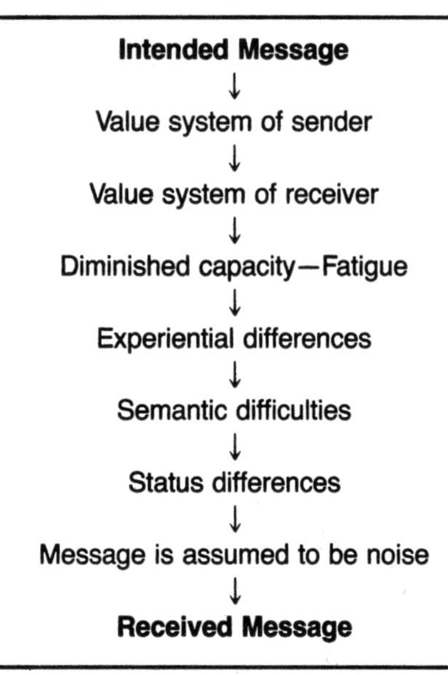

language and write in the active voice as much as possible when preparing memos and reports. The use of white space and graphical elements such as figures and charts enhances the readability of any written piece.The ability to write is directly correlated with reading. The more we read, the better our writing becomes. Given the vast amount of time spent on reading and creating memos, letters, proposals, and the like, the byword on written communication should be more is better, and the less permanent (memos sent electronically, faxes, hand notes on the bottom of letters, rather than a typed, recorded reply) the better.

Verbal

Verbal communication takes place in a variety of settings, and the form of the communication will vary. One sort of vocabulary may be used in addressing shareholders; a different idiom may be used altogether when chatting with the loading dock crew. The skills principally lacking in verbal communication are:

- Public speaking.
- Small group interaction.

Public speaking scares people more than death, if one believes the *Book of Lists*. This fear is not ameliorated by speaking to a group of known peers.

In fact, it can be worse. Three things can help overcome the fear of public speaking:

Training provides the framework for developing a public or small group speech. Organization and practice are essential ingredients in preparing a presentation.

Videotaping of presentations may be embarrassing, so allow the speech giver to review the videotape in private, with the reviewer's comments in a written form. Videotaping makes abundantly clear every "and-umm, uhhh, and y'know." Alternative means of alleviating stress may need to be developed.

Practice. Join a group such as Toastmasters that requires regular public speaking in a friendly environment. As confidence grows, so does ability.

Small group interaction is not always identified as a separate type of speech, but when implementing the myriad Quality Management tools that require teams, it is vital to understand how small groups interact. Small groups are discussed in Chapter 14, section on "Team Building" and Chapter 19, "Group Techniques."

Nonverbal

Humans infer a great deal of information from nonverbal clues. This nonverbal information includes *body language* as well as such things as *dress for success*. Anthropologists have discovered that human emotions are registered on the face in the same way, irrespective of cultural origin. Nonverbal clues lead to "gut feels" about the how to interact with another person. Despite the similarities of nonverbal communication, there are cultural differences, and is probably most important to understand these, rather than "reading" an individual's body language. It is easy to fall into the trap of overanalyzing nonverbal clues and infusing them with meaning, when, for example, someone may simply be hard of hearing or near/far-sighted rather than being inattentive (or too attentive).

Conflict Resolution

Communication can be the cause and cure of conflicts that arise. A conflict resolution process needs to identify the problem by identifying the who, what, why, when, and how, of each side, and treat both complaints as legitimate. Determine the common goal and causes. What are the underlying interests of each party? A root cause should be searched for, as there may be a systemic problem. An approach to solving the dispute must then be described, being as fair as possible to both sides. A facilitator may be required to aid in the conflict resolution. Most organizations have formal resolution mechanisms available, but have few or no informal mechanisms. An ombudsman or designated facilitator may assist in resolving conflicts prior to their reaching a point where a formal process is required.

4. CONTINUOUS EVALUATION

Feedback is essential to continuous improvement. How else would we know if our goals are coming to fruition or if the variation has been reduced? How else can we implement corrective action in a timely fashion? It is too late to find after the scheduled completion date that the project has run aground. These feedback mechanisms may be simple oral or written reports, information systems, or complex automated statistical analyses integrated with expert systems. The key is to receive the information in time to allow initiating corrective action.

Not only is it important to have timely information, but to deliver that information to someone who can initiate action. In a manufacturing envi-

FIGURE 3–7
Improvement versus Innovation

	Continuous Improvement	*Innovation*
Effect	Long term and long lasting but undramatic.	Short term, but dramatic.
Pace	Small steps.	Big steps.
Time frame	Continuous and incremental.	Intermittent and nonincremental.
Change	Gradual and constant.	Abrupt and volatile.
Involvement	Everybody.	Select few "champions."
Approach	Collectivism, group efforts, systems approach.	Rugged individualism, individual ideas and efforts.
Mode	Maintenance and improvement.	Scrap and rebuild.
Spark	Conventional know-how and state of the art.	Technological breakthroughs, new inventions, new theories.
Practical requirements	Requires little investment but great effort to maintain it.	Requires large investment but little effort to maintain it.
Effort orientation	People.	Technology.
Evaluation criteria	Process and efforts for better results.	Results for profits.
Advantage	Works well in slow-growth economy.	Better suited to fast-growth economy.

Source: Masaaki Imai, *Kaizen* (New York: Random House, 1986), p. 24.

ronment, this means access to quality control information by shop floor workers, not quality inspectors. How can the inspectors correct the problem? Certainly they can assist in the design and analysis of control charts, but they have no direct responsibility for product manufacture. It is essential to supply the shop floor worker real-time information to correct or prevent defects. Summaries and trends should also be analyzed by the shop floor worker, as well as by quality control, manufacturing engineering, and management.

Be sure to understand and separate assignable causes from chance causes. Assignable causes have distinct reasons for their existence. Chance causes are those causes over which we have no control. The employee can hardly be blamed for chance causes. Be sure of what you are trying to measure. It is unrealistic to tell the clerk to service 20 employees an hour, and provide quality service to all, no matter what it takes. What if "no matter what it takes" requires more than three minutes? The worker must not be demeaned while implementing, revamping, or evaluating a performance measurement tool.

Models of feedback can be based upon the Shewhart or Deming wheel (Chapter 17).

5. CONTINUOUS IMPROVEMENT

Engineers often employ a process called *derating* to achieve superior reliability when designing electronic devices. The idea is to select a component that can more than handle the rated voltage and operating conditions, but not greatly add to the cost of manufacture. The idea is similar to one in body-building: it is easier, and more effective, to lift 50 pounds 10 times, than to move 500 pounds all at once! Continuous improvement is similar; small improvements done continuously arrive at the same point as a major innovation.[2]

Unlike innovation, which can require great resources, and no small amount of serendipity, continuous improvement is easier to manage and utilizes everyone's talents. Japanese companies have used this idea for some time, and call this approach *kaizen*. This idea fits hand in hand with team building approaches such as quality circles and brainstorming, and can be inexpensively managed. *Kaizen* and innovation are compared in Figure 3–8.

[2]There are two types of innovation. One type fills in the missing pieces of a "knowledge matrix." The other type of innovation overturns the matrix entirely, and asks very different questions. Einstein's theory of relativity was an innovation of the second kind. Refinements to this theory are of the first kind. The sort of innovation treated here is of the first kind. Highly creative and divergent thinking is required for the second kind, which should be cultivated, but is perhaps not as important to daily business operations as the first sort of innovation.

Traditionally, in a manufacturing environment, a quality control department inspects products against a set of requirement specifications. In a service or white-collar setting this function is often performed by an audit organization. After such inspections, defective items (or reports) may then have to be either scrapped or reworked. This sequential processing leads to quality issues being addressed after the fact. The damage has already been done; a defective product or service is completed. Time to correct the problem must now be invested—time that should have been spent doing it right the first time.

In software development it has been shown that the cost to correct an error increases exponentially with the life-cycle phase. This means that to correct a requirements error after deployment costs 100 times (or more) than to correct the error when it occurred. Quality is, therefore, achieved at increased cost and decreased productivity, the antithesis of Quality Management.

To reduce cost and increase productivity, the focus must be projected on the process that produces the product. Improving the process reduces or eliminates variation and increases the uniformity of the product. This results in lower costs through the reduction of scrap, rework, and complexity. This method applies to administrative processes also—a report format may be reused, one spreadsheet may serve many different departments. Process improvement involves everyone in the organization and becomes a part of everyone's job, rather than the responsibility of just a few members of the organization. Through the inspection and analysis of the process, everyone shares a common learning experience and the accumulated knowledge and understanding of the process become the basis for improving it. Figure 3–8 contains 10 precepts of quality improvement by Motiska and Schilliff.

Strategies

Some strategies to bear in mind when implementing continuous improvement are:

Start with an example project. Small is beautiful when initiating radical new ideas into the workplace. Prove that it can work in a department of a division before exporting it companywide.

Analyze variation of all processes. This means administrative as well as on the production line. Are quarterly reports constantly being reinvented? Develop and use templates as much as possible.

Recognize the process, not just results. How can results be changed if we ignore the process? The process is the key to improving the results.

FIGURE 3–8
Ten Precepts of Quality Improvement

1. Quality leadership must begin with top management.
2. The most important aspect of the quality process is identifying the activities within the organization that affect quality.
3. Written procedures are one of the necessary communications media by which the management functions of directing and controlling are exercised.
4. One of the most critical activities in quality improvement is preparing a clear, concise description of the product or service to be acquired or produced.
5. The cost, time, and effort devoted to evaluating and selecting suppliers must be commensurate with the importance of the goods or services to be procured.
6. Quality audits must determine the adequacy of, and compliance with, established policies, procedures, instructions, specifications, codes, standards, and contractual requirements. Quality audits must also assess the effectiveness of their implementation.
7. The simple objective of most quality audits is to gather enough reliable data through inspection, observation, and inquiry to make a reasonable assessment of the quality of the activity being audited.
8. The foundation of quality control is having timely and accurate information so that systems that are not capable of producing consistent quality can be identified and improved.
9. An effective quality cost program can help the management team allocate strategic resources for improving quality and reducing costs.
10. Productivity, profit, and quality are the ultimate measures of the success of the production system. However, it is impossible to increase productivity, profit, and quality in the long run without exemplary programs for human resources.

Source: Paul J. Motiska and Karl Shilliff, "10 Precepts of Quality," *Quality Progress* 23, no. 2 (February 1990), pp. 27–28.

Simplify, simplify, simplify. Thoreau's injunction is truer today than it was in the mid-19th century. Constantly ask what is the value added for each work step, each form, and each line on the form.

Expect to constantly reinvest in new technology. Things don't stay done anymore. This may be why women managers are beginning to excel in the work force. Traditionally, "women's tasks" are never finished—laundry, cooking, child rearing are all ongoing. There is no real notion of being finished with a task. "Men's tasks" traditionally stay done and have a definite end point—fixing the car, bagging the meat, painting the house. Like traditional women's work, business today is messy—there is no end point, no time to relax.

Failures and problems are opportunities. Perfection is boring—there is no opportunity to learn new things. We learn from our mistakes as individuals, and we can learn from organizational mistakes as well. Be stubborn—make the same mistake several times before moving on. The typical entrepreneur starts and fails at two to three enterprises before finding the right match. Restaurants also fail at a given location at about the same rate before an appropriate match to the location is made.

Reorganize in order to bring about improvement. If the strategies advocated are too difficult to bring about under the current organizational structure, a change in organization may be needed. Self managing teams may need to be established, and layers of middle managers reduced.

6. CUSTOMER/VENDOR RELATIONSHIPS

The "hearing the voice of the customer" has become a key phrase in the past five years. That companies could do anything other than listen closely to customers' needs may puzzle casual observers of business. It would seem to be an obvious point. To many American companies, however, it is not. After World War II, the United States was the only major country that did not have a devastated economic infrastructure. Therefore, we could produce items of any quality and sell them. The needs of the customer seemed irrelevant, and industries were internally driven, and not customer oriented, or customer driven.

As other players have entered the field, management styles did not adapt. This reached crisis proportions in the 1970s and 1980s. Large companies take a very long time to change. Deming compares it to turning around a very large ship, and estimates it will take America about 30 years to change. Listening to the customer requires listening throughout a product life cycle, from requirements definition to maintenance after the sale. The customer also means anyone to whom you give your work, whether or not it is a "public" customer. This greatly broadens the scope of the "customer" and assessing "customer satisfaction." Listening to the customer entails surveys, research, and the implementation of tools such as quality function deployment.

Consumer research can be made very difficult by the inability of the customer or user to articulate and separate wants, needs, and desires. Even performing such research does not guarantee stellar success. Who can predict such successes in the children's toy market such as Cabbage Patch Kids dolls, Barbie dolls, and Teenage Mutant Ninja Turtles®? Yet there is always a niche to be found and money to be made even if your product does not get splashed on the front page of the newspaper's *Lifestyles* section. Little Tikes® toy company makes high-quality products that are useful although not exactly glamorous—but they are likely to stay in business for a long time, if they actively pursue their niche. Uncontrolled growth does not equate to unlimited profits.

Some strategies for improving customer and vendor relations are:

- Link organizational vision to customer satisfaction.
- Reward suppliers.
- Move to single sourcing.
- Minimize the overall number of vendors.
- Identify internal and external customers.
- Identify end users and distributors.
- Establish routine dialogue with customers.
- Involve the customer in planning and development.

Become a Customer of Your Own Product or Service

If practical, the employee who provides a service or product should also be a customer, if only for a short time. This means that administrative employees should understand how their reports are going to be used, and they should understand how the product works. A small arms unit of the Canadian Army had all of its personnel, including civilian administrative staff, take a course in firing and maintaining the weapon. This increased the ability of the staff to respond to questions and the secretaries could better proofread correspondence, since they now had an idea of the meaning of the content. All employees should be part of an enthusiastic (and un-co-opted) sales force.

Vendors Are Partners

Procurement systems spend some 60 percent of the sales dollar, and therefore is 60 percent of the quality problem or solution. By viewing purchasing as a strategic function, procurement becomes an essential link in the Quality Management chain. Most Quality Management coaches advise reducing the number of suppliers, and establishing long-term partnerships with those that remain. The result of not assuming a partnership role and focusing on quality is evident in the *BusinessWeek* article quoting Joseph M. Juran (in 1982):

> The automakers turn the screws to the point where it's almost impossible to make money selling to the auto companies. So the vendors have to make it on spare parts in the aftermarket. That gives them a vested interest in failures, a miserable arrangement.

This arrangement has since changed, and the quality of American cars has increased dramatically since that time.

Viewing vendors as partners, rather than as adversaries leads to the ability to implement successfully such cost-saving measures as just-in-time, whereby materials arrive as needed for the production line—eliminating inventory almost entirely. Vendors must be qualified and have policies compatible with your statistical process control and production program. This means moving away from the "low bidder" concept to one that builds long-term relationships. This means developing supplier certification processes. These may vary from commodity to commodity, and may require modifying as process variance is reduced. This also means that some suppliers need to be educated. Some companies find that it is to their advantage to educate their own suppliers.

Figure 3–9 outlines some criteria for vendor selection. Supplier certification is a field of growing interest and has yet to see any major industry-wide standards emerge. If you have a supplier certification program, be sure to include training, or arrange to have low-cost training provided. Otherwise, you may be slighting a superb vendor who is new, or who is completely unfamiliar with Quality Management tools and techniques.

Integrate the Customer into Development

Taking the requirements definition further involves reducing change proposals late in the development life cycle, and in reducing the time from concept

FIGURE 3–9
Vendor Selection Criteria

1. Quality
2. Delivery
3. Performance history
4. Warranties and claim policies
5. Production facilities
6. Price
7. Technical capability
8. Financial position
9. Communication system
10. Reputation
11. Desire for business
12. Management and organization
13. Operating controls
14. Repair service
15. Attitude
16. Impression
17. Packaging
18. Labor relations
19. Geographical location

to showroom floor. American industry still requires many years to reach the showroom floor. Many Japanese companies talk about development times of months, not years. To reduce development time requires linking the voice of the customer with barrier removal, training, empowering the worker, and continuous improvement. In other words, improving requirements definitions may dramatically improve the process, but implementing the other foundational considerations are important to become and remain competitive. A key tool in achieving this reduced development time, and integrating the customer into the development process, is quality function deployment, discussed in Chapter 17.

Concurrent engineering is an emerging multidisciplinary approach to reducing development time dramatically. It seeks to remove disciplinary boundaries and make a "without walls" engineering approach to new product development.

7. EMPOWERING THE WORKER

Empowering the employee means enabling a worker to achieve his or her highest potential. For most American companies, this is new, and may be the most powerful and useful concept in Quality Management. Allowing and facilitating workers to achieve their highest potential may seem obvious or impossible, but it is in fact neither. Empowerment requires turning the organization chart upside down, recognizing that management is in place to aid the worker in overcoming problems they encounter, not to place new roadblocks in the way.

Tapping into optimum individual performance is a holistic endeavor, which most American businesses have been slow to attempt to do. Yet, how can a worker plagued by concerns over a child's day care and older parent's care be devoting all of their energies to their job? The successful company will address and help to adequately resolve these issues.

Empowerment Strategies

Empowering strategies will include:

Ownership. A key strategy in empowering employees is to allow them ownership of a tasking, project, or division. Ownership implies trust and it requires a delegation of authority commensurate with the responsibility of the task. Ownership can also be granted to a team. Ownership also demands that the final resolution of the tasking be in the hands of the owner.

Nitpicking, rearranging, and otherwise finding fault with the tasking upon completion will undermine any attempt at empowerment via ownership.

A simple concept, but hard to do. Just as it is hard for a parent not to correct a child's first attempt at making a bed, putting things away, or cleaning the table. Any correction may ruin weeks of encouragement. Besides, to the child, the bed looks just fine. They've made it to the best of their abilities. If your employees' abilities are none too impressive, it's time to train them.

Value all contributions. Whether or not we appreciate them, it is important to enhance self-esteem of the contributor to accept their contribution and evaluate it. Try it—even if you think it is a goofy idea.

Listen to the least voice. Sometimes the newest and the least have invaluable contributions. It was an upholsterer working on a psychologist's patient's chair who inquired why only the front edge of the chair was worn, which was unusual in the upholsterer's experience. The psychologist deeply pondered the upholsterer's suggestion, and developed the theory of type A personalities.

Everyone has a value. If they didn't why would they be employed? Treat everyone with respect. All work has dignity to it.

Teams must own the problem. Teams are a waste of time if management vetoes or substantially changes their recommendation. Teams must be allowed autonomy. If management is unable to trust the recommendations that come from the team, then management by fear rules, and will spiral to lower and lower productivity.

Give quality awards to customers who have improved their business. Prompt payment to a vendor is their due, not their reward. Reward vendors with exceptional service or greatly improved service by giving them more business, or acknowledging an award in the media.

Delegate authority to the lowest possible organizational level. Constantly ask: Why should I do it? If you've hired competent people, let them do their job. No one knows more about the job than the person directly involved with it. Giving advice on what it was like when you were a neophyte 20 years ago will fall on deaf ears. They will learn how to act within the new environment.

Employee Involvement—A Pleonasm?

A pleonasm is a redundancy in a phrase, such as "large professional football player." If our employees are not involved, who is?[3] Can management alone run the company effectively? Can the union staff? Of course not. Yet millions of dollars are spent every year "motivating" employees, and "involving" employees, or getting them to "participate." The only barriers to worker participation, with rare exception, are those that management has established.

The underlying principle involved in fostering worker participation or empowering is trust. Management must trust its employees. Employees must trust management. Trust is easily erased. One study suggested that a boss was perceived as being negative if she was not complimentary four times more than she was negative. In other words, one "aw shoot" wipes out four "attagirls." Even indirect negative messages come across loudly. Giving an award to one worker may tell the other workers that they didn't perform well. This is why rewards or incentives often fail in a professional setting.

Ironically, promotions can be another way to tell other employees that you don't trust them or that you find them below par. This is due to the large talent pool available. The baby boom has left a large number of competent people plateaued in their career. If the traditional supervisory roles are seen as the leadership track, and these are becoming fewer as organizations become flatter, then crises accompanying promotions are likely to become more common. Those companies that foster growth on the job as a measure of success will retain high-caliber people. Those that maintain strict line orientations as a measure of success will keep experiencing high turnover rates.

Money Is Irrelevant—Almost

Money is becoming less important with the advent of two-income families. Work satisfaction is very important. Few American companies have been able to perceive this and capitalize on it. Money is simply no longer a prime motivator. Maslow's hierarchy (Figure 3–10) provides a look at why this is so. Most professionals have ascended to the Belonging or Esteem stages. Money is a suitable motivator at a lower stage, Safety.

---

[3]An employment ad in a newspaper had as one of its qualifications: "must be able to answer the telephone without being told to do so."

FIGURE 3–10
Maslow's Hierarchy of Needs

Stage	*Process*	*Needs*
5th stage	Self-actualization	To achieve one's best.
4th stage	Esteem	To be held in high regard.
3rd stage	Belonging	To be accepted by family and friends.
2nd stage	Safety	To have economic and physical security.
1st stage	Physiological	To eat, sleep, and have shelter.

This is not to say that workers are beginning to refuse raises, but that many other motivators exist other than money, as long as there is enough money to maintain a lifestyle suitable to the Safety stage. Motivation is discussed at length in Chapter 14.

8. TRAINING

Each year, about $210 billion is spent on corporate education, and $230 billion is spent on education from kindergarten through the doctorate level. This training can have a high payoff: Harris Corporation used training to reduce work-in-process inventory by 60 percent, and reduced cost of quality 15 percent. This demonstrates what the president of Harvard, Derek Bok, meant when he said "If you think education is expensive, try ignorance." If spent properly, training can return the investment many times.

The outcome of training is modified behavior. It may be enhanced interpersonal skills or a specific manual skill, but there is a direct, identifiable modification. Training need not consist solely of traditional classroom instruction. Dedicating time to learn how to use a software package could be considered training, especially if the trainee could stop external interruptions, as in a classroom environment. Employees can train other employees very effectively. Erwin Schroedinger, an eminent physicist, once said that the best way for him to learn anything was to have to teach it.

Training someone forces the instructor to consider the task from a different viewpoint. However, to conduct all training by fellow employees or even in-house can cause a stagnation effect. Conferences and seminars are especially good to refresh workers and overcome blindsightedness. Many seminars and courses are conducted at community colleges or local colleges, and are inexpensive. Don't overlook trainers in your own company. Anyone who is unafraid of talking to groups has the potential to train or facilitate training. Training should not be meted out as a punishment or reward. All employees need and deserve training. Training needs and results should be evaluated with the employee to gain insight. Of course, there will be, at times, resistance to certain specific training. It is then useful to send one or a few employees to the training. If the training is successful, the changed behavior will be apparent to fellow workers, and will help convince the skeptical employees of the utility of this training.

Education

Unlike training, education has no such immediately identifiable outcome. The utility of education may not be discoverable for a long period of time, if ever. However, education is vital in promoting a divergent look at the way things are done. Training focuses primarily on the event at hand, thus filling in empty or fuzzy spots on an information matrix (or puzzle). Education may lead us to determine that we are working the wrong puzzle. For example, at a candlemaking factory around the turn of the century, the convergent thinking worker would have considered quality improvements such as making the candle drip less, last longer, and burn more smoothly. The divergent thinker might have suggested manufacturing light bulbs as an improvement.

Divergent thinking is necessary for a business to survive in the long term. Some businesses, as well as government agencies, have instituted the idea of a senior level staff member being a new technology specialist and a technology insertion facilitator. While it would be beneficial if line staff would initiate such actions, their duties often preclude the "stare-at-the-wall" time necessary to consider new technology implementation.

A SAMPLE CURRICULUM

Figure 3–11 shows a sample curriculum. A companywide curriculum should be developed that addresses needs of individual departments. To train everyone, it may well pay to provide a Train the Trainer course to prepare peers to teach some of the courses. Courses should be just long enough to be effective. Anything over three or four days is unlikely to immediately be absorbed into daily work habits. Immediate reinforcement of the training is necessary for it to be effective. After a course in brainstorming, students should have a need to conduct or participate in such a session within two weeks of the training.

THE OMNIPRESENT QUESTION

Quality Management is quite complex. Unfortunately it is difficult to recall all the scores of principles and techniques without referring to a text. Yet it

FIGURE 3–11
Sample TQM Curriculum

First 1–2 Years:
- TQM at XYZ Company.
- Continuous improvement.
- Human behavior in organizations.
- Team development.
- Problem solving.
- Statistical measures.
- Organizational communications.
- Specific tools training:
  - Brainstorming.
  - Quality function deployment.
  - Cause-effect diagrams.
  - Pareto analysis.
  - Etc.

Years 2–5:
- Advanced team building.
- Advanced problem solving.
- Specific tools training (continued).

After 5 Years—Ongoing Training:
- Team skills workshops.
- Problem solving seminar.
- Communications workshop.
- Quality improvement seminar.
- Human behavior workshop.

is still pleasant to have a phrase that will serve us like a Swiss Army knife—easy to remember, yet useful for almost any of the situations we encounter. For Quality Management advocates, this phrase is: *What Is the value added?*

This can be the omnipresent question with which we prune old bureaucracies, work rules, and product designs. It does not demand a complex answer, requiring consultants and reports and spreadsheets. The answer does not have to be numeric or measurable in a traditional sense. This question gives us insight into continuous improvement while maintaining a return on investment strategy.

BIBLIOGRAPHY

American Productivity Center. *Allen-Bradley: First Line Supervisors Play Pivotal Role in Employee Communication Program Aimed at Boosting Productivity.* Case study no. 49. Houston, Tex.: American Productivity Center, 1985.

Aubrey, Charles A. II, and Patricia Felkins. *Teamwork: Involving People in Quality and Productivity Improvement*. Milwaukee, Wis.: ASQC Press, 1988.

Baumgarten, S., and J. S. Hensel. "Add Value to Your Service," ed. C. Surprenant. Chicago: *American Marketing Association*, 1987, pp. 105–10.

Booher, Diane. "Quality or Quantity Communication." *Quality Progress* 21, no. 6 (June 1988), pp. 65–68.

Bossert, James L., ed. *Procurement Quality Control*. 4th ed., Milwaukee, Wis.: ASQC Quality Press, 1988.

Ertel, Danny. "How to Design a Conflict Management Procedure That Fits Your Dispute." *Sloan Management Review* 32, no. 4 (Summer 1991), pp. 29–42.

Imai, Masaaki. *Kaizen: The Key to Japan's Competitive Success*. New York: Random House, 1986.

Kaplan, Robert S. "Measuring Manufacturing Performance: A New Challenge for Managerial Accounting Research." *The Accounting Review*, February 1985.

Kiechel, Walter. "Visionary Leadership and Beyond." *Fortune*, July 21, 1986, pp. 127–28.

Kurokawa, Kaneyuki. "Quality and Innovation." *IEEE Circuits and Devices*, July 1988, pp. 3–8.

Lammermeyr, Horst U. *Human Relations—The Key to Quality*. Milwaukee, Wis.: ASQC Quality Press, 1990.

Liswood, Laura A. *Serving Them Right: Innovative and Powerful Customer Retention Strategies*. Milwaukee, Wis.: ASQC Quality Press, 1990.

Luthans, Fred. *Organizational Behavior*. New York: McGraw-Hill, 1973.

Maass, Richard A.; John O. Brown; and James L. Bossert. *Supplier Certification—A Continuous Improvement Strategy*, Milwaukee, Wis.: ASQC Quality Press, 1990.

Main, Jeremy. "Detroit Is Trying Harder for Quality." *BusinessWeek*, November 1, 1982.

Maslow, Abraham H. *Motivation and Personality*. New York: Harper & Row, 1954.

Massey, C. "What You Are Now Is What You Were When." Videotape.

McGregor, Douglas. *Leadership and Motivation*. Cambridge, Mass.: MIT Press, 1983.

Motiska, Paul J., and Karl A. Schilliff. "10 Precepts of Quality." *Quality Progress*, 23, no. 2 (February 1990), pp. 27–28.

Newman, R. G. "Insuring Quality: Purchasing's Role." *Journal of Purchasing and Materials Management*, 1988, pp. 14–20.

Newman, R. G. "The Buyer-Supplier Relationship under Just-in-Time." *Production and Inventory Management Journal*, 1988, pp. 45–50.

Oakland, John S., and Ric Grayson. "Quality Assurance Education and Training in the U.K." *Quality and Reliability Engineering International* 3 (1987), pp. 169–75.

Rout, L. "Hyatt Hotel's Gripe Sessions Help Chief Maintain Communication with Workers." *The Wall Street Journal* 27 (July 16, 1981).

Sandholm, L. "Management Training—A Prerequisite of TQC," *EOQC* 23, no. 4 (December 1989), pp. 5–10.

Scholtes, Peter R., and Heero Hacquebord. "Beginning the Quality Transformation, Part I; and Six Strategies for Beginning the Quality Transformation, Part II." *Quality Progress,* July/August 1988.

Schultz, Louis E. "Creating a Vision for Strategy and Quality: A Way to Help Management Assume Leadership." *Concepts in Quality Proceedings,* November 1988.

Schultz, Louis E. *Pathway to Continuous Improvement.* Bloomington, Minn.: Process Management Institute.

Sink, S. Scott and Thomas C. Tuttle. *Planning and Measurement in Your Organization of the Future.* Atlanta, Ga.: IIE Press, 1991.

Sloan, David, and Scott Weiss. *Supplier Improvement Process Handbook,* ASQC, 1987.

Snee, Ronald D. "Statistical Thinking and Its Contribution to Total Quality." *The American Statistician,* May 1990.

Squires, Frank H. "Who Is Responsible for Quality?," *Quality,* December 1987, p. 73.

Staveley, J. C., and B. G. Dale. "Some Factors to Consider in Developing a Quality-Related Feedback System." *Quality and Reliability Engineering International* 3, no. 4 (1987), pp. 265–71.

Tichy, Noel M., and Mary Anne Devanna. *The Transformational Leader.* New York: John Wiley & Sons, 1986.

Vroom, V. *Work and Motivation.* New York: John Wiley & Sons, 1964.

Zaremba, Alan. *Management in a New Key: Communication in the Modern Organization.* Atlanta, Ga.: IIE Press, 1991.

Zeithaml, Valerie A.; L. L. Berry; and A. Parasuraman. "Communication and Control Processes in the Delivery of Service Quality." *Journal of Marketing* 52 (April 1988), pp. 35–48.

CHAPTER 4

IMPLEMENTATION

Even if you're on the right track, you'll get run over if you just sit there.

Will Rogers

INTRODUCTION

The philosophies, practices, tools, methods, and techniques must be integrated into a coherent implementation plan. There are perhaps as many implementation schemes as there are Quality Management practitioners. The reason for this lies in the individual needs of the organization. Some desire to achieve companywide quality control within a short time frame, others wish to experiment with various techniques and refine them before integrating them throughout the company.

Doing something is not the same thing as doing it well. In the case of Quality Management, not doing it well may be worse than doing nothing, as it sullies the reputation of a fine approach to management. It is difficult to implement Quality Management because it is complex, and because it is very difficult to change attitudes formed over years. Quality Management evolved from the management of quality control, and to many, TQM concepts must include a very heavy dosage of statistical tools. This is a narrow vision of Quality Management, which is attempting to improve the quality of management as well as the product or service. Tools are needed, but they are sterile if not coupled with dynamic management involvement. The tools can be learned over the course of a few days or weeks. Changing a culture of status quo will take months, if not years. The larger the organization, the longer it will take to change.

TQM often espouses a top-down commitment. Top-level managers become enlightened, and then pass the lamp of wisdom onward. Nice theory, but few of us are CEOs or business owners or otherwise at the top of the pyramid. Fortunately, it is not true that Quality Management *must* come from the top—it is the ideal, but not a requirement. Each of us can institute Quality Management on whatever sphere of influence we have. As our success increases, so will interest in exporting Quality Management to other areas of the organization.

This bottom-up approach may not be as awkward as it seems. After all, how can the work force change unless it is either already willing, or encounters a significant or enabling event brought about by top management? If the work force is willing, but was unaware of how to proceed, Quality Management can quickly revolutionize the way work is done. If the work force is not willing, coaching the work force has a slim chance of working. A significant event, such as laying off 25 percent of the work force, may be necessary.

However Quality Management evolves, it is instructive to have some sort of implementation model. There are simply too many tools and concepts to implement to proceed with all of it at once. There are many implementation models. The one presented here is one that is based upon the author's observation of Quality Management principles in action. No magic is involved in picking a model. Adapt one for your unique set of circumstances.

PROCESS

1. Develop a compelling vision. Leaders must have a compelling reason that will sustain them and the company for years to come. Thomas Berry describes Quality Management as a journey, not a destination. Quality Management is not another program. TQM is not an office or a department. Quality Management is a way of organizational life. It is a revolutionary way to invert the organizational hierarchy, put customers first, eliminate managerial deadwood, and overcome whatever stands in the way of fulfilling customer needs.

2. Start small. Implementing a companywide Quality Management plan all at once is suicidal. A single division, department, or branch must first serve as a test site. In this stage vision leadership is articulated and implemented. Try to transform the test site completely before transporting the plan companywide. This means at least 12 to 18 months, and encountering and surmounting at least one crisis.

Don't pick a certain set of members within the test site. Pick everyone in an easily identifiable group, or pick another group. Quality Management will have to work on all employees, not just the best or the worst. Use the PDCA (Plan-Do-Check-Act) cycle (Chapter 17) at all times. Constantly communicate and provide feedback.

Pursue the right way, not the quick way. Visit a Baldrige Award winner. Take courses on the Malcolm Baldrige National Quality Award and benchmarking (see Chapter 18). Read every relevant case study available.

3. Become obsessed. Plan strategically. Become obsessed with implementing the vision. Sweat out the details. These are what the customer sees, not wondrous corporate platitudes. Make abundantly clear that the customer is welcome. This ranges from making the bathrooms spotless to friendly and helpful staff to producing an item or service of world-class quality that the customer will be proud to own and find a pleasure to use. Obsession is imbued from top management. Employees must know who the top managers are—personally. They cannot remain faceless.

4. Celebrate success. Show how well the test site did by making a video, conduct tours of the test site, and allow test site employees to host discussion groups. Rewards are nice, but take it easy. Rewards are often monetary in nature, and prove to be poor motivators (especially among knowledge workers), and actually demotivate others. A job well done is its own reward—work can be a fulfilling and rewarding experience when employees are involved and empowered.

5. Export results to the rest of the organization. Taking the Quality Management process organizationwide is a big step, and will require several years to implement. And that's just the beginning. Employees must understand that Quality Management never stops, and that this is not a program or push. It is way of life to be applied with religious zeal.

A SPIRAL MODEL

The spiral model depicted in Figure 4–1 relates the concepts and principles of Quality Management. While most models appear in linear form, the spiral model serves as a reminder that Quality Management implementation needs to be cascaded through the company, and done iteratively. From the center of the spiral emanates the vision of the organization. The first layer consists of the foundational principles, the second the management dynamics required by midlevel managers and supervisors, and the third layer, or the implementation layer, contains some suggested tool kits. There are three shell layers:

- Vision leaders—top management.
- Vision articulators—middle and supervisory management.
- Vision implementors—supervisors and individuals.

There are four slices that correspond to the four key principles of Quality Management:

- Vision.
- Empowerment.

FIGURE 4–1
Spiral Quality Management Model

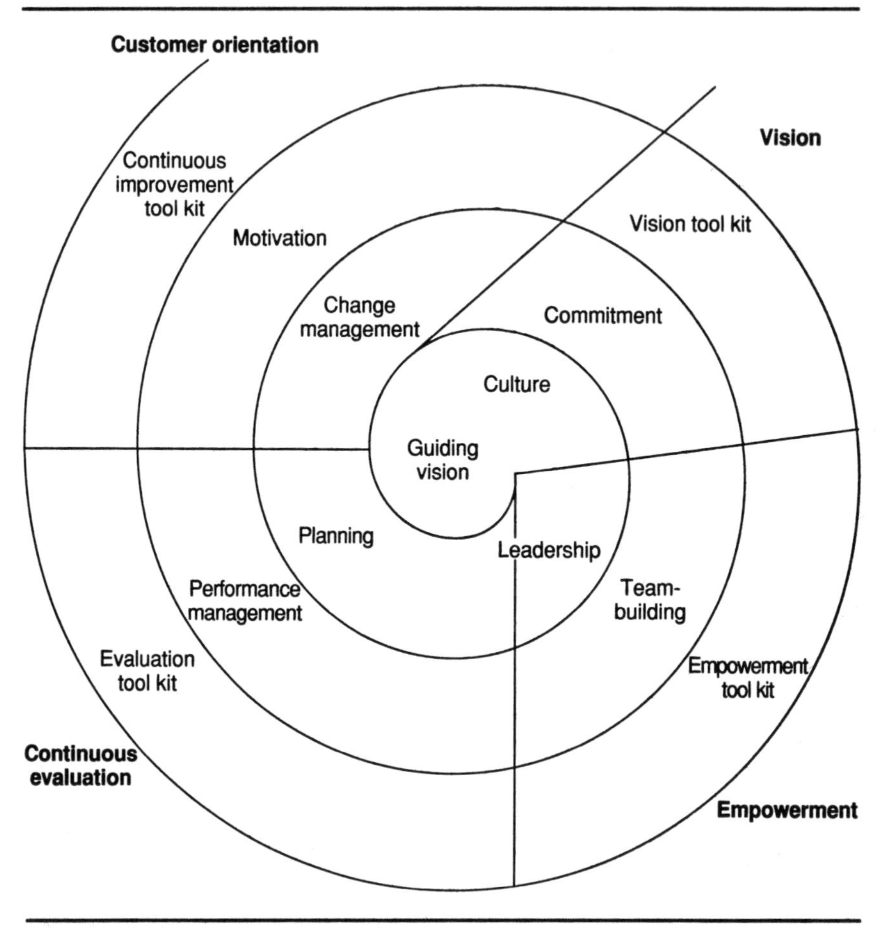

- Continuous evaluation.
- Customer orientation.

Some suggested tools are listed in Figure 4–2. The model is generic enough to be adapted by any organization, and be made specific quite readily.

IMPLEMENTATION ISSUES AND TRAPS

Management must not react to problems with a "How the hell did that happen?," but with a helping attitude. After all, shouldn't the job of man-

FIGURE 4–2
Spiral Model Outline

Vision
Leader task:
- Vision and culture definition

Articulator task:
- Employee involvement

Implementor tool kit:
- Benchmarking
- Force field analysis
- Goal setting
- Systematic diagram

Empowerment
Leader focus:
- Leadership

Articulator task:
- Team building

Implementor tool kit:
- Auditing
- BrainStorming
- Cause-effect diagrams
- Creativity
- Data collection
- Nominal group technique
- Pareto analysis
- Process decision program chart
- Quality circles
- Service quality
- Time management
- Work flow analysis

Continuous Evaluation
Leader focus:
- Strategic planning

Articulator task:
- Performance management

Implementor tool kit:
- Auditing
- Benchmarking
- BrainStorming
- Cause-effect diagrams
- Control charts
- Data collection
- Delphi technique
- Design of experiments
- Evolutionary operation
- Failure modes, effects, and criticality analysis (FMECA)
- Flowcharts
- Nominal group technique
- Quality costs
- Sampling
- Statistical measures
- Five Ws, one Y

Customer Orientation
Leader focus:
- Change management

Articulator task:
- Motivation

Implementor tool kit:
- Benchmarking
- Data collection
- Delphi technique
- Foolproofing
- Quality function deployment
- Service quality
- Nominal group techniques
- Quality function deployment
- Sampling

agement be to help with exceptions to the routine? If things always worked smoothly, with no unusual circumstances or problems, why bother with managers?

Douglas Patterson points out a number of traps to avoid when implementing Quality Management:

Delegating Quality Management authority. Quality Management must be the responsibility of top management—first. Then it must become everyone's responsibility. As Deming, Juran, and others point out, the management culture must change in order for Quality Management to be properly instituted.

Quality Management is a new name for existing programs. Statistical process control, analysis of variance, quality circles, cross-functional teams are all part of Quality Management, but do not constitute it in its entirety. Quality Management requires a vision and principles, as well as tools.

Do it right the first time. Too vigorous and literal an implementation of this concept will stifle creativity. Failures are more than OK; they are opportunities for learning.

Quality Management is statistical quality control and quality circles. Tools are only a part of the picture. Humanistic management principles, discussed in Part 3, Management Dynamics, and the principles discussed in this section are necessary for achieving the goals of Quality Management.

Barry Sheehy has developed some guidelines on surviving the inevitable: the first crisis of the Quality Management or quality improvement program:

1. Acknowledge the crisis.
2. Consider it a hidden opportunity.
3. Make sure everyone understands that there is no going back.
4. Give voice to your fears/concerns—but don't back down in the face of naysayers.
5. Recall accomplishments that have occurred so far.
6. Get counsel from workers and suppliers, not your fears.
7. Ask for advice. No voice is too small.
8. Revise your plan and inform everyone of the changes.

Sheehy also recommends the following preventative or mitigating factors:

1. Build backsliding/setbacks into your plan.
2. Underpromise and overdeliver.

3. Plan a renewal at about month 12.
4. Review goals for their attainability.
5. Record all accomplishments.

While a crisis is unavoidable, preparing for it can considerably lessen the damage to the credibility of the program.

BIBLIOGRAPHY

All of the following are highly recommended. Thomas Berry's work is particularly engaging.

Berry, Thomas H. *Managing the Total Quality Transformation*. New York: McGraw-Hill, 1990.

Docstader, S. L. "Managing TQM Implementations: A Matrix Approach." Unpublished manuscript. San Diego: Navy Personnel Research & Development Center, 1987.

Hunt, Daniel V. *Quality in America: How to Implement a Competitive Quality Program*. Homewood, Ill.: Richard D. Irwin, 1991.

Mansir, Brian E., and Nicholas R. Schacht. *Total Quality Management: A Guide to Implementation,* Bethesda, Md., 1989.

Patterson, Douglas O. "Saying Is One Thing, Doing Is Another!" *Journal of the Institute of Environmental Sciences,* January/February 1991, pp. 17–20.

Scholtes, Peter R., and Heero Harquebord. "Six Strategies for Beginning the Quality Transformation, Part II." *Quality Progress,* August 1988.

Sheehy, Barry. "Hitting the Wall: How to Survive Your Quality Program's First Crisis." *National Productivity Review,* 9, no. 3 (Summer 1990), pp. 329–35.

U.S. Department of Defense. *TQM Implementation Guide, Volumes I and II*. U.S. GPO, February 1990.

Weaver, Charles N. *TQM: A Step-by-Step Guide to Implementation*. Milwaukee, Wis.: ASQC Quality Press, 1991.

PART 2

QUALITY MASTERS

INTRODUCTION

Charismatic pacesetters such as Deming, Juran, or Crosby are often identified (or even equated) with the Quality Management movement. Their magnetism has resulted in passionate devotees and "disciples" of the various masters, each proclaiming their pundit to have revealed the one true path to total quality enlightenment. The squabbling that occasionally occurs between each guru's camp sometimes resembles religious "heresy" disputes, where much arguing proceeds over differences without distinction. Because their writings are motivational, it is important to read and reflect upon more than one of the masters prior to embarking on a total Quality Management program. While all of their platforms agree to within 95 percent, that last 5 percent difference may be appealing.

Because most of the masters have written several volumes on their thoughts on TQM, it is useful to have a condensed version of their major tenets readily available. Of course, in the space available it is impossible to convey all of the profound notions contained in these volumes, or to do justice to the subtlety of many of their points. However, it should aid in understanding the differences and similarities of their respective approaches—just as you wouldn't mistake a field guide description for a tree, so this section should not be mistaken as an exhaustive catalog of a master's canon. Presented in this section are thumbnail sketches of the quality masters and their primary teachings, punctuated with some of their words of wisdom. The reader is strongly encouraged to read the major works of the quality masters described. An essential bibliography is provided in Part 5, Resources. While some may seem bulky (*Thriving on Chaos* by Tom Peters runs more than 700 pages in paperback; Feigenbaum's magnum opus *Total Quality Control* runs over 800), they can be read very quickly, and are worth reading several times.

Why this particular set of seven leaders? These seven are the most popular and are commonly identified with the Quality Management effort today. There are many other masters in the quality field, such as Shigeo Shingo, Shewhart, Dodge, A. J. Duncan, Mizuno, Ohno, Godfrey, Martin Smith, Scherkenbach, and others who advance and encourage the state of the art of quality management techniques. Others may be gleaned from the quality-related service directory listed annually in the August issue of *Quality Progress*. Many of these experts offer training and consulting, and may be far more accessible, while as illuminating, as the masters profiled here. Space limits force consideration of a small number of leaders, and this set has succeeded in appealing to an unusually large audience, often far outside the traditional bailiwick of quality control specialists. Yet these masters are closely identified with quality management per se, rather than any and all aspects of management; otherwise Peter Drucker would figure prominently here. His influence on management has been extraordinary, and will reach well into the next millennium. No Quality Management library should be without his works, and he is frequently referred to in this book.

A QUICK FIELD GUIDE

Crosby is closely associated with the zero defects concept, but in later years has shifted more toward the mainstream of Quality Management thinking. Deming is a godlike figure of quality, and his "14 Points" pop up everywhere. Feigenbaum's fairly early work on total quality control is well worth reading; he has fallen out of the limelight somewhat as he does not seem to seek publicity. Ishikawa was the aristocrat of Japanese quality, and is associated with his "Seven Tools." Juran is an indefatigable promoter of Quality Management, and is famous for his indispensable *Quality Control Handbook*. Peters is an annalist of business excellence, taking an empirical and anecdotal approach. Taguchi focused narrowly on design of experiments, but his influence in Japan has been dramatic, and his work may present the "next phase" beyond statistical quality control.

Deming and Peters are the most revolutionary, demanding a managerial transformation. Crosby and Feigenbaum see quality fitting in more as a quality promotion department that acts as a facilitator/consultant. Ishikawa works within an existing organizational framework, but his ideas benefit from Japan's historically better access to top management. Lately, Juran has also called for a revolution.

Most quality masters have the context of the medium to large organizations in mind. However, almost all masters' tenets can be applied to small businesses as well. In fact, especially when one reads Peters's works, it would seem that the intent is to achieve integrated small business units within one corporation. Many of the human aspects of the tenets are readily achievable in small businesses. Small businesses have the ability to transform themselves quickly. Their disadvantage can be a lack of resources, both human and capital.

Which master is for your organization depends upon the corporate culture (Chapter 13) and the level of top-managerial commitment. Assiduously following any one of them would lead to a total quality transformation. There is no *best* master, although Deming is highly favored by many, and provides a comprehensive philosophy, which is essential to Quality Management implementation.

IMPLEMENTING A MASTER'S STRATEGY

It is becoming apparent that while it would be nice to perform a transition into Quality Management, what is really needed is the phoenixlike transformation of Peters or Deming. Like a sports coach, Quality Management masters offer a package, and must be dealt with whole, not in pieces. While you may be able to synthesize your own game plan from the coaches here, be wary of "cafeteria management"—taking only those aspects that appeal to you from each master. If there is something unpalatable in their dicta (such as giving up management parking spots), this may expose a management weakness (such as perpetuating an Us-versus-Them management to employee climate) that needs to be corrected, and may actually be crucial to successfully executing the plan in your organization. Partial implementation is often a mistake. View it as if you had built a race car without an engine. While it may look nice, it'll never cross the finish line.

A more common mistake made in implementing a Quality Management approach is to espouse it loudly from the top, somehow expecting everyone to become wildly motivated by meetings that may look the same as always, and memos that, while more enthusiastic, still lack specifics. Employees perceive that this is another lip service program, and that by saying the right buzzwords, management will soon move on to another program, and employees will continue with the status quo. Extraordinary energy is called for to decimate the status quo. Examples and demonstrations will have to proceed daily at all levels. Genuine shifts in an organization's culture take time. Deming compares the United States to a great ship that takes a long time to turn around. He estimates it will take the United States some 30 years to turn its ship around. While tools and other techniques can be implemented quickly and easily, an organization's beliefs are articulated by its culture and behavior. Hence it is vital for a manager to understand the management dynamics involved (Part 3), the tools and techniques available (Part 4) and have a firm grasp of the philosophical base (this part and Part 1).

Start small. Establish Quality Management in a division or department first. Work out the bugs in instituting the techniques, practices, and tools prior to companywide deployment. Companywide deployment will bring about its own unique problems, aside from learning the tools. It also provides a model for others to follow, and shows that Quality Management can be done in your organization, and separates operational problems (proper implementation of the techniques) from integrative and strategic problems encountered in scaling the methods to the entire work force. It should allow time to highlight the scaling up problems, instead of a broad range of troubles in understanding the tools, mastering the new management practices, changing organizational policies, and confronting inertia and resistance.

Try a single approach, then modify. Use common sense. TQM is sometimes mistaken as some sort of Panglossian "I'm OK/You're OK" self-esteem booster. It is not. Implementing TQM may mean eliminating deadwood. Perhaps entire layers of management need to be removed or rearranged. Quality Management means taking a tough, proactive stand for continuous improvement.

QUO VADIS?

Can a synthesis be achieved? Or are the approaches too disparate to unite? Does one assume people are inherently bad, while another assumes rather that they are good? Is Quality Management just a trend, such as *One Minute Management* in the 1980s or communication in the 1950s? A synthesis was presented in Part 1, Foundational Issues. It combines elements from all the primary leaders. The approaches are not really that disparate and would, in fact, seem to be converging. The bottom line of Quality Management is common sense. If one dictum diligently applied failed to work, perhaps it can't work in your industry, or perhaps the company wasn't ready for it. Don't abandon the entire Quality Management framework. The snares are many, and various roadblocks and pitfalls have been identified throughout this book. Change cannot come without mistakes along the way.

Quality Management is definitely not a fad, having been around in some form for 50 years, or, if we may challenge the reader's belief, some 2,500 years (see Chapter 12, Historical Masters). This is because Quality Management is not driven by present economic forces, but founded on such questions as: What is human nature? How must we manage? What is the simplest, yet most powerful tool I can find?

Situational economics has created trendy management guides such as *The Peter Principle* and *Leadership Secrets of Attila the Hun*. Quality Management also transcends single personal experiences, such as *Swim with the Sharks*. These works do not provide us with a permanent foundation on which to build management skills, but they make for enjoyable reading, and indeed, can even help us understand the times in which they were created and provide valuable personal and managerial insight. For example the Peter principle presents an interesting thesis for the U.S. manufacturing capability erosion of the 1950s and 1960s. Today however, in the face of massive layoffs of skilled managers and a highly talented pool of professionals in the unemployment line, his thesis seems cruel and untrue. Peter Drucker provides a reflective and cogent discussion of today's condition in his work *The New Realities*. Lloyd Dobyns and Clare Crawford-Mason also provide an intriguing look at the present global market in their work *Quality or Else*.

BIBLIOGRAPHY

Blanchard, Kenneth, and Spencer Johnson. *The One Minute Manager*. ed. Pat Golbitz. New York: William Morrow, 1982.

DeYoung, H. Garrett. "Preachings of Quality Gurus: Do It Right the First Time." *Electronic Business*, October 16, 1989, pp. 88–94.

Dobyns, Lloyd, and Clare Crawford-Mason. *Quality or Else*. Boston: Houghton Mifflin, 1991.

Drucker, Peter. *The New Realities*. New York: Harper & Row, 1989.

Lodge, Charles. "Six Gurus Show the Way to Improved Product Quality." *Plastics World*, August 1989, pp. 29–40.

Main, Jeremy. "Under the Spell of Quality Gurus." *Fortune*, August 18, 1986, pp. 30–34.

McKay, Harvey. *Swim with the Sharks*. New York: William Morrow, 1988.

Peter, Laurence J., and Raymond Hull. *The Peter Principle: Why Things Always Go Wrong*. New York: William Morrow, 1969.

Roberts, Wess. *Leadership Secrets of Attila the Hun*. New York: Warner Books, 1989.

Wood, Robert Chapman. "The Prophets of Quality." *Quality Review* 2, no. 4 (Winter 1988), pp. 18–25.

CHAPTER 5

PHILIP B. CROSBY*

BRIEF BIOGRAPHY

Philip B. Crosby was born in 1926 in Wheeling, West Virginia. Crosby has a degree in podiatry (his father's profession) but decided he didn't like it. In 1952 he became a reliability engineer for Crosley Corporation in Richmond, Indiana. He later worked for the Martin Corporation from 1957 to 1965. Crosby was in charge of quality on the Pershing missile project. From 1965 to 1979 he was the director (vice president status) of quality for ITT. In 1979, he founded Philip Crosby Associates (PCA) in Winter Park, Florida. In 1991 he retired from PCA and began Career IV, Inc. to help grow executives.

MAJOR TENETS

Philip B. Crosby is most closely associated with the idea of zero defects which he created in 1961. To Crosby, quality is conformance to requirements, which is measured by the cost of nonconformance. Poor or high quality has no meaning, only nonconformance and conformance. Using this approach means that one arrives at a performance goal of zero defects.

Crosby equates quality management with prevention. Therefore, inspection, testing, checking, and other nonpreventive techniques have no place. Statistical levels of compliance program people for failure. Crosby maintains that there is absolutely no reason for having errors or defects in any product or service.

Companies should adopt a quality "vaccine" to prevent nonconformance. The three ingredients of this vaccine are: determination, education,

*The authors would like to thank Dr. Crosby for his review and comments on this chapter.

and implementation. Quality improvement is a process, not a program; it should be permanent and lasting.

Supplier quality audits are nearly useless, unless the vendor is totally incompetent. It is impossible to know if the supplier's quality system will provide the required quality merely by auditing their plan.

Zero defects is not a slogan. It is a management performance standard. Further demotivating employees by constant exhortation is not the answer. Crosby believes that in the 1960s various Japanese companies properly applied zero defects, using it as an engineering tool, with responsibility of proper implementation left to management. By contrast, zero defects was used as a motivational tool in the United States, with responsibility left to the worker, where it failed. This strategy requires management commitment and technical direction. Crosby's 14 steps to quality improvement and his four absolutes are provided below (from *Quality Is Free* and *The Eternally Successful Organization*).

CROSBY'S 14 STEPS TO QUALITY IMPROVEMENT

1. Make it clear that management is committed to quality.
2. Form quality improvement teams with representatives from each department.
3. Determine how to measure where current and potential quality problems lie.
4. Evaluate the cost of quality and explain its use as a management tool.
5. Raise the quality awareness and personal concern of all employees.
6. Take formal actions to correct problems identified through previous steps.
7. Establish a committee for the zero defects program.
8. Train all employees to actively carry out their part of the quality improvement program.
9. Hold a "zero defects day" to let all employees realize that there has been a change.
10. Encourage individuals to establish improvement goals for themselves and their groups.
11. Encourage employees to communicate to management the obstacles they face in attaining their improvement goals.
12. Recognize and appreciate those who participate.

13. Establish quality councils to communicate on a regular basis.
14. Do it all over again to emphasize that the quality improvement program never ends.

ABSOLUTES OF QUALITY MANAGEMENT

- Quality means conformance to requirements. If you intend to do it right the first time, everyone must know what *it* is.
- Quality comes from prevention. Vaccination is the way to prevent organizational disease. Prevention comes from training, discipline, example, leadership, and more.
- Quality performance standard is zero defects (or defect-free). Errors should not be tolerated. Errors are not tolerated in financial management; why should they be in manufacturing?
- Quality measurement is the price of nonconformance.

BIBLIOGRAPHY

Crosby, Philip B. *Quality Is Free*. New York: McGraw-Hill, 1979.

Crosby, Philip B. *Quality without Tears*. New York: McGraw-Hill, 1984.

Crosby, Philip B. "Quality—Management's Choice." *Quality,* Anniversary Issue, 1987.

Crosby, Philip B. *The Eternally Successful Organization*. New York: McGraw-Hill, 1988.

CHAPTER 6

W. EDWARDS DEMING

BRIEF BIOGRAPHY

W. Edwards Deming was born on October 14, 1900 in Sioux City, Iowa. Shortly thereafter, his family moved to Powell, Wyoming. Deming graduated with a B.S. in physics from the University of Wyoming in 1921, and graduated from Yale with a Ph.D. in mathematical physics in 1928. He worked for the U.S. Census Bureau during and after World War II. In 1950, Deming went to Japan to help conduct a population census, and lectured to top business leaders on statistical quality control. Deming told the Japanese they could become world-class quality leaders if they followed his advice. During the 1950s, Deming again traveled to Japan at the behest of the Japanese Union of Scientists and Engineers (JUSE). Because of his refusal of payment for his lectures (Japan at the time was quite impoverished), JUSE used the funds to establish the Deming Prize, which is the most honored quality award in Japan today. In the 1980 NBC White Paper, "If Japan Can, Why Can't We," he was called the "founder of the third wave of the Industrial Revolution." Today Deming is generally regarded as the top leader in quality management, and is still cited as the founder of the third wave of the Industrial Revolution (the first wave occurred in the early 19th century with simple automation; the second wave occurred with assembly concepts in the late 19th century, and the third wave is occurring with the information/computer revolution).

BASIC TENETS

Quality does not mean luxury. Quality is a predictable degree of uniformity and dependability, at low cost, suited to the market. In other words, quality is whatever the customer needs and wants. And since the customer's needs and desires are always changing, the solution to defining quality in terms of the customer is to redefine requirements constantly.

Productivity improves as variability decreases. Since all things vary, quality control is needed. Statistical control does not imply absence of defect goals and services, but rather, it allows prediction of the limits of variations. There are two types of variation: chance and assignable. It is futile to attempt to eradicate defects caused by chance. However, it can be very difficult to distinguish between the two, or to determine assignable causes. It is not enough to meet specifications; one has to reduce variation as well.

Deming is extremely critical of U.S. management and is an advocate of worker participation in decision making. He claims that management is responsible for 94 percent of quality problems, and points out that it is management's task to help people work smarter, not harder. Deming insists that one of the first steps is for management to remove the barriers that rob the workers of their right to do a good job. Motivational programs which offer lip service have no place here; workers distinguish between sloganeering and commitment.

Inspection of incoming or outgoing goods is too late, ineffective, and costly. Inspection neither improves quality, nor guarantees it. Additionally, inspection usually allows a certain number of defects. The best recognition one can give a quality vendor is to award the vendor more business. Deming advocates sole sourcing, believing that multiple sourcing for protection is a costly practice. The advantages of sole sourcing include better vendor commitment, eliminating small differences between products from two suppliers, and simplifying accounting and paperwork. Counter to the argument that single source can mean paying a higher price, Deming believes that the policy of always trying to drive down the price of purchased items, without regard to quality and service, can drive good vendors and good service out of business.

Deming's celebrated 14 Points, seven deadly diseases, and a number of obstacles are summarized below. They are elaborated at length in Deming's work *Out of the Crisis,* and in several of the works listed in the bibliography, most notably the works by Scherkenbach, Tribus, and Walton.

DEMING'S 14 POINTS

1. Create constancy of purpose for improvement of product and service. An organizational vision must guide the corporate culture and provide a focus to the organization. This vision equips the organization with a long-term perspective. Measure management commitment, and benchmark how the organization is doing relative to other related firms.

2. Adopt the new philosophy. Western management must awaken to the challenge, and assume a new leadership role. The quality revolution is equal in economic import as the Industrial Revolution. It is concurrent with the globalization of the economy.

3. Cease dependence upon inspection to achieve quality. Introduce modern quality tools such as statistical process control, evolutionary operation, design of experiments, and quality function deployment. Inspection only measures a problem, and does not allow any correction of the problem. It is often said that one cannot "inspect in quality."

4. Minimize total cost by working with a single supplier—end the practice of awarding business on the price tag alone. Don't blindly award business to the low bidder. Instead, minimize total cost. Move toward a single supplier for any one item, establishing a long-term relationship of loyalty and trust. Vendor certification programs and total life-cycle cost analysis play a role here.

5. Improve constantly and forever every process. Simply fixing problems is no longer enough. Constantly improve quality and productivity, thus constantly decrease costs. Prevent defects and improve the process. Don't fight fires—this is not quality improvement, it's management by crisis. Improvement requires feedback mechanisms from customers and vendors.

6. Institute training on the job. Training applies to all levels of the organization, from the lowest to the highest. Do not overlook the possibility that the best trainers may be your own employees.

7. Adopt and institute leadership. Leadership emanates from knowledge, expertise, and interpersonal skills, not level of authority. Everyone can and should be a leader. Leadership qualities are no longer mysterious and innate—they can be learned (Chapter 13). Leaders remove barriers that prevent people and machines from achieving the optimum.

8. Drive out fear. Fear stems fom insecure leadership which must rely on work rules, authority, punishment, and a corporate culture based upon internal competition—grading on a curve has no place within a business. It may also come from physical and emotional abuse by peers and superiors. Fear snuffs out creativity, which is the engine for quality improvement. This fear can be defeated by identifying and overcoming gaps in communication, culture, and training. Systemic factors may also promote management by fear, such as performance evaluations, bonus programs, and work quotas.

9. Break down barriers between staff areas. Everyone must work as a team, working toward the good of the team. Teamwork is imperative in modern management. New organizational structures may be needed (Chapter 13). Turning the organizational chart upside down is a frightening experience, but may well be required to achieve the proper balance and perspective.

10. Eliminate slogans, exhortations, and targets for the work force. Programs or campaigns which command a task but leave the worker powerless to achieve the objective constitute management by fear.

11. Eliminate numerical quotas for the work force and numerical goals for management. Eliminate management by objective, or more precisely, management by numbers. Substitute leadership. Numerical quotas disregard statistical notions which impact all workers. Not all workers can be above average; nor can they all be below. Traditional industrial engineering practice is "management by numbers" and this is precisely what Deming is referring to. Work measurement worked well at a certain stage in industrial development, but society and work have evolved beyond that. Today work quotas can impose a quality and production ceiling, rather than a target. Natural variation is ignored in these systems, and the numbers game takes precedence over all other business concerns.

12. Remove barriers that rob people of pride of workmanship. Eliminate the annual rating system. Remove barriers that rob the hourly worker of their pride of workmanship. The responsibility of supervisors must be changed from volume and bottom line to quality. Remove barriers that rob people in management and in engineering of their right to pride of workmanship. This means abolishment of the annual or merit rating and of management by objective.

13. Institute a vigorous program of education and self-improvement for everyone. Training provides an immediate change in behavior. Results of education may not manifest themselves immediately, but it can have far-reaching long-term effects. Self-improvement is an ongoing task of education and self-development. This may mean offering courses in time management, stress reduction, allowing employees work time to do physical activity if they have a sedentary job, allowing employees who have active jobs to partake in mentally challenging tasks or education.

14. Put everybody in the company to work to accomplish the transformation. Top-management commitment is required to put everybody in the company to work to accomplish the transformation. The transformation is everybody's job.

DEMING'S SEVEN DEADLY DISEASES

1. Lack of constancy of purpose. Lack of vision results in a lack of focus and a lack of discipline, which can lead to a deterioration of the job environment and the organization itself.

2. Emphasis on short-term profits; short-term thinking. This is actually saying the same thing as above, but is so common among American businesses that it deserves a separate entry. Amplifying the quarterly reoort into the be-all and end-all of the business is organizational suicide. Unfortunately there are many institutional mechanisms working against this aspect of the transformation.

3. Annual performance reviews. The effects of performance appraisals are devastating. Management by objective, on a go, no-go basis, is the same thing. Management by fear would be better than these extremely demotivational tools.

4. Mobility of management; job hopping. Little value is placed in Western society on staying in the job for years, and performing at one's peak.

5. Use of visible figures only for management. If information is relevant to their work, they need to be informed.

6. Excessive medical costs. Books on "stress" abound. The reason for this is the intense dissatisfaction in working in the contemporary corporate workplace; a corollary to this is the entrepreneurial boom. Simply put, people who enjoy their work stay healthy (see Chapter 15.1 on Stress Management). Health plans which cover preventive measures must be selected over those that merely react to problems.

7. Excessive costs of liability. This is fueled by lawyers that work on contingency fees, in a society that highly values a profession that provides little or no added value.

DEMING'S OBSTACLES

1. Neglect of long-range planning and transformation.
2. The idea that problems are solved with automation, gadgets, and other "things."
3. Partaking of a smorgasbord approach to implementing quality improvements without basic principles will prove disastrous.
4. The attitude that "Our problems are different" leads to ignoring basic principles.
5. The obsolescence in schools (grade school through graduate school) must be overcome.
6. Reliance on quality control departments to "take care of all our problems of quality." Quality must become part of everyone's job.
7. Blaming the work force for problems. There must be improvement of the system as well as a product. Defect free workmanship means nothing if the wrong product is being made.
8. Quality by inspection. Quality cannot be inspected in. Meeting specifications is not quality, either—they don't tell the whole story.
9. False starts can result from mass teaching with little guidance in implementation. Other false starts happen when the idea to be implemented will require years of cultural change. Deming points to the example of Quality Control circles being poorly implemented in the United States due to a lack of understanding and action on the part of management.
10. The unstaffed computer. Computers can take tedium out of calculations, but not the need for interpretation. Expert systems have not advanced to the point of a "Lights Out" factory of robots, and it is quite likely this will not happen for a long time.
11. Inadequate testing. Prototypes are a lot cheaper than a massive production failure. Computer-aided manufacturing allows "soft"

prototypes that are easy to change, and do a fair job at mimicking reality. A "hard" prototype can be built after experimenting with many different soft ones.

12. "Anyone that comes to try to help us must understand all about our business" is an arrogant attitude that leads to failure. Answers can be found within the organization and from outside consultants and other sources.

BIBLIOGRAPHY

Aguayo, Rafael. *Dr. Deming: The American Who Taught the Japanese about Quality.* New York: Fireside Press, 1991.

Baillie, Allan S. "The Deming Approach: Being Better than the Best." *Advanced Management Journal* 51 (Autumn 1986), pp. 15–19.

Butterfield, Ronald W. "Deming's 14 Points Applied to Service." *Training: The Magazine of Human Resources Develop.* 28 (March 1991), pp. 50–56.

Deming, W. Edwards. *Japanese Methods for Productivity and Quality.* Washington, D.C.: George Washington University, 1981.

Deming, W. Edwards. *Quality, Productivity and Competitive Position.* Cambridge, Mass.: Center for Advanced Engineering Study, MIT Press, 1982.

Deming, W. Edwards. "Quality: Management's Commitment to Quality." *Business,* 35 (January 1985), pp. 50–55.

Deming, W. Edwards. *Out of the Crisis.* Cambridge, Mass.: Center for Advanced Engineering Study, MIT Press, 1986.

Deming, W. Edwards. "New Principles of Leadership." *Modern Materials Handling* 42 (October 1987), pp. 37–41.

Deming, W. Edwards. "Out of the Crisis," *Journal of Organizational Behavior Management* 10 (Spring 1989), pp. 205–13.

Duncan, W. Jack, and Joseph G. Van Matre. "The Gospel according to Deming: Is It Really New?" *Business Horizons* 33 (July/August 1990), pp. 3–17.

Gabor, Andrea. *The Man Who Discovered Quality: The Management Genius of W. Edwards Deming.* Milwaukee, Wis.: ASQC Quality Press, 1991.

Gitlow, Howard, and Shelly Gitlow. *The Deming Guide to Quality and Competitive Position.* Englewood Cliffs, N.J.: Prentice-Hall, 1986.

Rosander, A. C. *Deming's 14 Points Applied to Services.* Milwaukee, Wis.: ASQC Quality Press, 1991.

Scherkenbach, William. *The Deming Route to Quality and Productivity: Road Maps and Roadblocks.* Washington, D.C.: CeePress, George Washington University, 1986.

Scholtes, Peter. *An Elaboration of Deming's Teachings on Performance Appraisal.* Madison, Wisconsin: Joiner Associates, Inc., 1987.

Tribus, M. *Deming's Redefinition of Management*. Cambridge, Mass.: Center for Advanced Engineering Study, MIT Press, 1985.

Tribus, M. *Reducing Deming's 14 Points to Practice*. Cambridge, Mass.: Center for Advanced Engineering Study, MIT Press, 1984.

Walton, Mary. *The Deming Management Method*. New York: Putnam, 1986.

Walton, Mary. *Deming Management at Work*. Milwaukee, Wis.: ASQC Quality Press, 1990.

CHAPTER 7

ARMAND V. FEIGENBAUM

BRIEF BIOGRAPHY

Armand Vallin Feigenbaum was born in 1922. In 1944 he was the top quality expert for General Electric in Schenectady, New York. He received a Ph.D. from the Massachusetts Institute of Technology in 1951. While there he authored his magnum opus, *Total Quality Control* (now in its third edition). In 1958 he was made executive of manufacturing operations for General Electric worldwide. In 1968 Feigenbaum founded General Systems in Pittsfield, Massachusetts, where he serves as president.

BASIC TENETS

Feigenbaum championed the phrase *total quality control* in the United States. Total quality control approaches quality as a strategic business tool that requires awareness by everyone in the company, just as cost and schedule are in most companies today. Quality reaches far beyond defect management on the shop floor; it is a philosophy and commitment to excellence.

Quality is a way of corporate life, a way of managing. Total quality control has an organizationwide impact that involves implementation of customer-oriented quality activities. This is a prime responsibility of general management, as well as the mainline operations of marketing, engineering, production, industrial relations, finance, and service, and also of the quality control function itself at the most economical levels. Feigenbaum's definition of total quality control is: Total quality means being excellence-driven, rather than defect-driven.

An overview of Feigenbaum's approach is given in the Three Steps to Quality and The Four Deadly Sins. These and other ideas are explored further in the Nineteen Steps to Quality Improvement, derived from several of Feigenbaum's works.

THREE STEPS TO QUALITY

1. Quality leadership. There must be continuous management emphasis and leadership in quality. Quality must be thoroughly planned in specific terms. This approach is excellence-driven rather than the traditional failure-driven approach. Attaining quality excellence means keeping a constant focus on maintaining quality. This sort of continuous approach is very demanding on management. The establishment of a quality circle program or a corrective action team is not sufficient for its ongoing success.

2. Modern quality technology. The traditional quality department cannot resolve 80 to 90 percent of quality problems. In a modern setting, all members of the organization must be responsible for quality of their product or service. This means integrating office staff into the process, as well as engineers and shopfloor workers. Error-free performance should be the goal. New techniques must be evaluated and implemented as appropriate. What may be an acceptable level of quality to a customer today may not be tomorrow.

3. Organizational commitment. Continuous motivation is required, and more. Training that is specifically related to the task at hand is of paramount importance. Consideration of quality as a strategic element of business planning needs to occur in the United States.

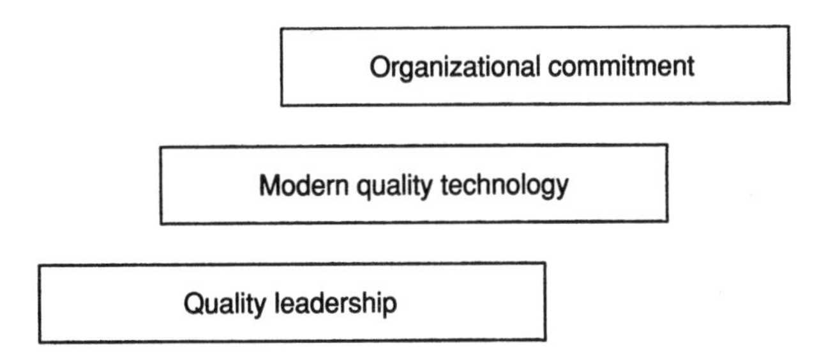

FOUR DEADLY SINS

1. Hothouse quality. Quality gets top-level attention in a "fireworks display" manner. These programs disappear from view when production demands become heavy, or something else captures top-level attention.

2. Wishful thinking. The federal government cannot wave a wand and make imports go away, nor should it engage in protectionist activity. This is complacency that will be costly later.

3. Producing overseas. A competitive advantage cannot be gained by having someone else fight our "quality war." The radio, television, auto, and consumer electronics industries have proven this.

4. Confining quality to the factory. Quality achievement is for everyone in every sector of the company.

NINETEEN STEPS TO QUALITY IMPROVEMENT

1. Total quality control defined. TQC may be defined as: An effective system for integrating the quality development, quality maintenance, and quality improvement efforts of the various groups in an organization so as to enable marketing, engineering, production, and service at the most economical levels which allow for full-customer satisfaction.

2. Quality versus quality. "Big Q" or Quality refers to luxurious quality whereas "little q" refers to high quality, not necessarily luxury. Regardless of an organization's niche, little q must be closely maintained and improved.

3. Control. In the phrase "quality control," the word *control* represents a management tool with four steps:

1. Setting quality standards.
2. Appraising conformance to these standards.
3. Acting when the standards are exceeded.
4. Planning for improvements in the standards.

4. Integration. Quality control requires the integration of often uncoordinated activities into a framework. This framework should place the responsibility for customer-driven quality efforts across all activities of the enterprise.

5. Quality increases profits. Total quality control programs are highly cost effective because of their results in improved levels of customer satisfaction, reduced operating losses and field service costs, and improved utilization of resources. Without quality, customers will not return. Without return customers, no business will long survive.

6. Quality is expected, not desired. Quality begets quality. As one supplier becomes quality oriented, other suppliers must meet or exceed this new standard.

7. Humans impact quality. The greatest quality improvements are likely to come from humans improving the process, not adding machines.

8. TQC applies to all products and services. No person or department is exempted from supplying quality services and products to its customer.

9. Quality is a total life-cycle consideration. Quality control enters into all phases of the industrial production process, starting with the customer's specification, through design engineering and assembly to shipment of the product and installation, including field service for a customer who remains satisfied with the product.

10. Controlling the process. These controls fall into four natural classifications: new design control, incoming material control, product control, and special process studies.

11. A total quality system may be defined as. The agreed companywide and plantwide operating work structure, documented in effective, integrated technical and managerial procedures, for guiding the coordinated actions of the people, the machines, and the information of the company and plant in the best and most practical ways to assure customer quality satisfaction and economical costs of quality. The quality system provides integrated and continuous control to all key activities, making it truly organizationwide in scope.

12. Benefits. Benefits often resulting from total quality programs are improvement in product quality and design, reduction in operating costs and losses, improvement in employee morale, and reduction of production-line bottlenecks.

13. Cost of quality. Quality costs are a means for measuring and optimizing total quality control activities. Operating quality costs are divided into four different classifications: prevention costs, appraisal costs, internal failure costs, and external failure costs. These costs are discussed at length in Chapter 18.

14. Organize for quality control. It is necessary to demonstrate that *quality is everybody's job*. Every organizational component has a quality-related responsibility; for example, marketing for determining customers' quality preferences, engineering for specifying product quality specifications, and shop supervision for building quality into the product. Make this responsibility explicit and visible.

15. Quality facilitators, not quality cops. The quality control organization acts as a touchstone for communicating new results in the company, providing new techniques, acting as a facilitator, and in general resembles an internal consultant, rather than a police force of quality inspectors.

16. Continuous commitment. Management must recognize at the outset of its total quality control program that this program is not a temporary quality improvement or quality cost reduction project.

17. Use statistical tools. Statistics are used in an overall quality control program whenever and wherever they may be useful, but statistics are only one part of the total quality control pattern. They are not the pattern itself. The development of advanced electronic and mechanical test equipment has provided order of magnitude improvements to this task.

18. Automation is not a panacea. Automation is complex and can become an implementation nightmare. Be sure the best human-oriented activities are implemented before being convinced that automation is the answer.

19. Control quality at the source. The creator of the product or the deliverer of the service must be able to control the quality of their product or service. Delegate authority, if necessary. Norton Stores has the simple company policy of "Use your own best judgment," and allows its employees the authority and freedom that this policy requires.

BIBLIOGRAPHY

Feigenbaum, Armand V. *Total Quality Control*. 3rd ed. New York: McGraw-Hill, 1983.

Feigenbaum, Armand V. "Total Quality Leadership," *Quality* 25, no. 4 (April 1986), pp. 18–22.

Feigenbaum, Armand V. "America on the Threshold of Quality." *Quality* 29, no. 1 (January 1990), pp. 16–18.

CHAPTER 8

KAORU ISHIKAWA

BRIEF BIOGRAPHY

Kaoru Ishikawa was born in 1915, and earned a degree in applied chemistry from the University of Tokyo in 1939. After the war he became involved in JUSE's fledgling efforts to promote quality. Later he became president of the Musashi Institute of Technology. Until his death in 1989, Dr. Ishikawa was the foremost figure in Japan regarding quality control. He was the first to use the term *total quality control,* and developed the "Seven Tools" that he thought any worker could use. He felt that this distinguished him from other approaches, which he thought placed quality in the hands of specialists. He received many awards during his life, including the Deming Prize and the Second Order of the Sacred Treasure, a very high honor from the Japanese government.

BASIC TENETS

The Seven Tools of Ishikawa are:

1. Pareto charts.
2. Cause-effect diagrams (fishbone or Ishikawa diagrams).
3. Histograms.
4. Check sheets.
5. Scatter diagrams.
6. Flowcharts.
7. Control charts.

These are all discussed in their own chapters and sections in Part 4, Tools and Techniques (histograms and scatter diagrams are both found in the section "Data Presentation," Chapter 16). While Ishikawa realized that not all problems could be solved by these tools, he felt that 95 percent could be, and that any factory worker could effectively use them. While some of the tools had been well known for some time, Ishikawa organized them specifically to improve quality control. Ishikawa originated the cause-effect diagram, descriptively called the *fishbone diagram,* and sometimes called the *Ishikawa diagram* to distinguish it from a different form of cause-effect diagram used in computer programming.

Perhaps the most far reaching of the tools was the idea of a quality control (QC) circle. Its success surprised even him, especially when exported beyond the shores of Japan. He assumed that any country which did not have a Buddhist/Confucian tradition would be inhospitable to QC circles. Today there are over 250,000 QC circles registered with Japan's QC Circle Headquarters, and more than 3,500 case study reports have been filed. This essential aspect of Quality Management was responsible for much of the increase in quality of Japanese products during the past three decades. Ishikawa sees that QC circles are more important to service industries than to manufacturing, since they work much closer to the customer.

ISHIKAWA'S QUALITY PHILOSOPHY

As industry progresses and the level of civilization rises, quality control becomes increasingly important. Some basic tenets of Ishikawa's quality philosophy are summarized here:

1. Quality begins with education and ends with education.
2. The first step in quality is to know the requirements of customers.
3. The ideal state of quality control is when inspection is no longer necessary.
4. Remove the root cause, and not the symptoms.
5. Quality control is the responsibility of all workers and all divisions.
6. Do not confuse the means with the objectives.
7. Put quality first and set your sights on long-term profits.
8. Marketing is the entrance and exit of quality.
9. Top management must not show anger when facts are presented by subordinates.
10. Ninety-five percent of the problems in a company can be solved by the seven tools of quality control.
11. Data without dispersion information (Chapter 20) is false data—for example, stating an average without supplying the standard deviation.

BIBLIOGRAPHY

Ishikawa, Kaoru. *Guide to Quality Control*. 2nd rev. ed. White Plains, N.Y.: UNIPUB–Kraus International, 1976.

Ishikawa, Kaoru. *Quality Control Circles at Work*. Tokyo: JUSE, 1984.

Ishikawa, Kaoru. *What Is Total Quality Control? The Japanese Way*. Englewood Cliffs, N.J.: Prentice Hall, 1985.

CHAPTER 9

JOSEPH M. JURAN*

BRIEF BIOGRAPHY

Joseph M. Juran was born in 1904 in Romania, and came to the United States in 1912. A holder of degrees in engineering and law, he advanced to the positions of quality manager at Western Electric Company, government administrator and professor of engineering at New York University before embarking on a consulting career in 1950. Juran is regarded as one of the architects of the quality revolution in Japan, where he lectured and consulted frequently, starting in 1954. However, he feels that the people mainly responsible for the Japanese quality revolution have been the Japanese operating managers and quality specialists. In 1979, he founded the Juran Institute, which conducts quality training seminars and publishes quality-related works.

BASIC TENETS

Juran defines quality as consisting of two different, though related concepts:

> One form of quality is income oriented, and consists of those features of the product which meet customer needs and thereby produce income. In this sense higher quality usually costs more.
>
> A second form of quality is cost oriented and consists of freedom from failures and deficiencies. In this sense higher quality usually costs less.

Juran points out that managing for quality involves three basic managerial processes: quality planning, quality control, and quality improvement. (These processes parallel those long used to manage for finance.) His "Trilogy," Figure 9–1, shows how these processes are interrelated.

---

*The authors would like to thank Dr. Juran for his review and comments on this chapter.

FIGURE 9–1
Juran's Trilogy

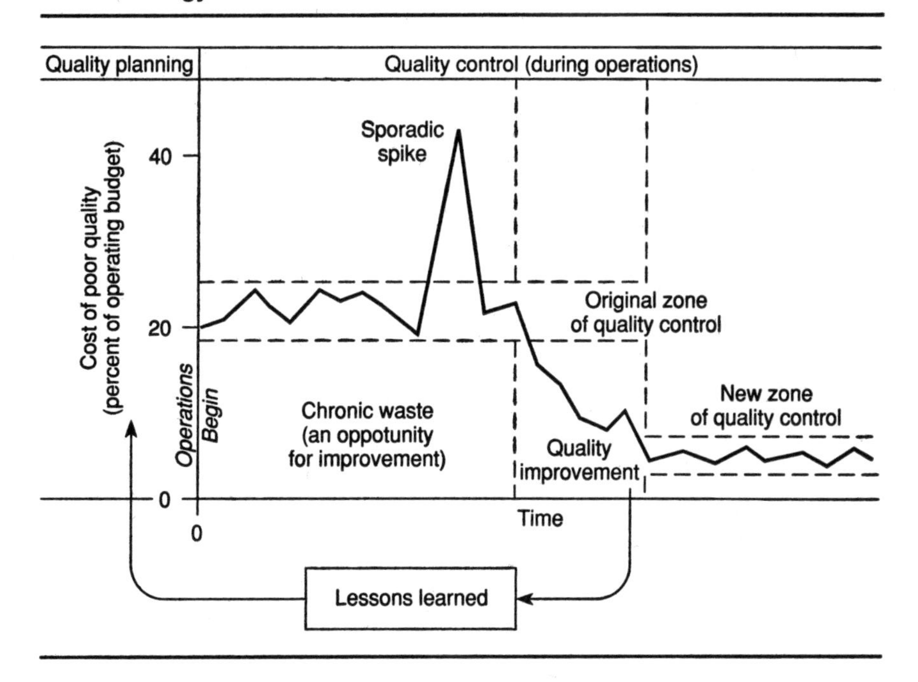

Juran identifies the ingredients of the Japanese quality revolution as follows:

1. The upper managers took charge of managing for quality.
2. They trained the entire hierarchy in the processes of managing for quality.
3. They undertook to improve quality at a revolutionary rate.
4. They provided for work force participation.
5. They added quality goals to the business plan.

Juran feels that the United States and other Western countries should adopt similar strategies in order to attain and retain world-class quality status.

JURAN'S APPROACH TO QUALITY IMPROVEMENT

In Juran's priority list, quality improvement comes first. He has a structured approach for this, which he first set out in his book *Managerial Breakthrough* in 1964. This approach includes a list of nondelegable responsibilities for the upper managers:

1. Create awareness of the need and opportunity for improvement.
2. Mandate quality improvement; make it a part of every job description.
3. Create the infrastructure: establish a quality council; select projects for improvement; appoint teams; provide facilitators.
4. Provide training in how to improve quality.
5. Review progress regularly.
6. Give recognition to the winning teams.
7. Propagandize the results.
8. Revise the reward system to enforce the rate of improvement.
9. Maintain momentum by enlarging the business plan to include goals for quality improvement.

In Juran's view a major, long neglected opportunity for improvement lies in the business processes.

JURAN'S APPROACH TO QUALITY PLANNING

Juran also has identified a universal process for planning to meet quality goals:

1. Identify the customers. Anyone who will be impacted is a customer, whether external or internal.
2. Determine the customers' needs.
3. Create product features which can meet the customers' needs.
4. Create processes which are capable of producing the product features under operating conditions.
5. Transfer the processes to the operating forces.

Juran feels that quality planning should provide for participation by those who will be impacted by the plan. In addition, those who plan should be trained in the use of modern methods and tools of quality planning.

JURAN'S APPROACH TO QUALITY CONTROL

Here Juran follows the familiar feedback loop:

1. Evaluate actual performance.

2. Compare actual with the goal.
3. Take action on the difference.

Juran favors delegating control to the lowest levels in the company through putting workers into a state of self-control. He also favors training workers in data collection and analysis to enable them to make decisions based on facts.

JURAN AND TOTAL QUALITY MANAGEMENT (TQM)

Juran is a strong proponent of TQM. He defines TQM as a collection of certain quality related activities:

1. Quality becomes a part of each upper management agenda.
2. Quality goals enter the business plan.
3. Stretch goals are derived from benchmarking: focus is on the customer and on meeting competition; there are goals for annual quality improvement.
4. Goals are deployed to the action levels.
5. Training is done at all levels.
6. Measurement is established throughout.
7. Upper managers regularly review progress against goals.
8. Recognition is given for superior performance.
9. The reward system is revised.

JURAN'S VIEWS ON WORKER PARTICIPATION

Juran takes a dim view of campaigns to exhort workers to solve the company's quality problems. He discovered many decades ago that over 85 percent of quality problems had their origin in the managerial processes.

Juran feels that the Taylor (Part 3 introduction) system of separating planning from execution has largely become obsolete due to the dramatic rise in worker education. This same rise now makes it possible to delegate to workers many functions previously carried out by planners and supervisors. He feels that the Taylor system should be replaced, and favors experimenting with various options such as: self-control, self-inspection, self-supervision, and self-directing teams of workers.

He believes that self-directing teams of workers will most likely become the dominant form of successor to the Taylor system.

JURAN ON OTHER MAJOR ISSUES

In Juran's view certain major practices of the past should undergo extensive change:

1. The product development cycle should be shortened through participative planning, concurrent engineering, and training the planners in the methods and tools of managing for quality.
2. Supplier relations should be revised. The number of suppliers should be reduced. A teamwork relation should be established with the survivors, based on mutual trust. The traditional adversary approach should be abolished. The duration of contracts should be increased.
3. Training should become results oriented rather than tool oriented. The main purpose of training should be to change behavior rather than to educate. For example training in quality improvement should be preceded by assignment to an improvement project. The training mission should then be to help the team complete the project.

BIBLIOGRAPHY

Juran, J. M. *Managerial Breakthrough*. New York: McGraw-Hill, 1964.

Juran, J. M. and Frank M. Gryna, Jr. *Quality Planning and Analysis*. New York: McGraw-Hill, 1980.

Juran, J. M. *Juran on Quality Improvement Workbook*. New York: Juran Enterprises, 1981.

Juran, J. M. *Quality Control Handbook*. New York: McGraw-Hill, 1988.

Juran, J. M. *Juran on Planning for Quality*. New York: Free Press, 1988.

CHAPTER 10

TOM PETERS

BRIEF BIOGRAPHY

Thomas J. Peters was born in Baltimore, Maryland. Peters has a B.C.E. and M.C.E. in civil engineering, and an M.B.A. and Ph.D. in business from Stanford University. He was a principal of the consulting firm McKinsey & Company, and later established his own consulting firm, Palo Alto Consulting Center. He presently writes a syndicated newspaper column and is a regular commentator on the "McNeil/Lehrer News Hour" on PBS.

BASIC TENETS

Tom Peters is the consummate chronicler of excellence in business. His first work, *In Search of Excellence,* was a major best seller. Peters takes an empirical approach to Quality Management. He is interested in what has worked for whom, and why it was successful. This makes for highly absorbing and inspirational reading. Some have criticized this approach as being primarily anecdotal in nature, and lacking a strong framework. Peters has attempted an answer to these charges in his third work, *Thriving on Chaos: Handbook for Management Revolution.* In this work he provides 45 specific prescriptions in transforming an organization. These points are summarized below, as are the nine primary observations from *In Search of Excellence.*

NINE ASPECTS OF EXCELLENT COMPANIES

1. Managing ambiguity and paradox. Chaos is the rule of businesses, not the exception. The business climate is always uncertain and always ambiguous. The rational, numerical approach does not always work because we live in irrational times.

2. A bias for action. Do it, try it, fix it. The point is to try something, without fear of failure. Sochiro Honda, founder of Honda, said that only 1 out of 100 of his ideas worked. Fortunately for him, he kept trying after his 99th failure.

3. Close to the customer. Excellent companies have an almost uncanny feel for what their customer wants. This is because they are a customer of their own product, or they closely listen to their customer.

4. Autonomy and entrepreneurship. Ownership of a department, tasking, or problem is essential in motivating employees. It is the most cited reason for entering into self-employment. Excellent companies allow and encourage autonomy and within company entrepreneurship.

5. Productivity through people. Not surprisingly, people act in accordance with their treatment. Treat them as being untrustworthy, and they will be. Treat them as business partners, and they will be. Excellent companies have taken the leap of faith required to trust your employees to do the right thing right.

6. Hands-on, value-driven. Practice management by walking around. Constantly ask what the value added is of every process and procedure.

7. Stick to the knitting. Stay close to the basic industry of your organization. The skills or culture involved in a different industry be may a shock that is fatal to the organization.

8. Simple form, lean staff. Flat organizations unencumbered by a heavy headquarters characterized the excellent companies.

9. Loose-tight properties. Tight control is maintained while at the same time allowing staff far more flexibility than is the norm.

PETERS'S PRESCRIPTIONS FOR MANAGEMENT REVOLUTION

1. Create total customer responsiveness. Customer responsiveness requires listening to the customer at every available opportunity. Be extraordinarily responsive. Create a niche and differentiate your product from your competitors.

2. Pursue fast-paced innovation. Never let up on innovating new projects. Don't be concerned over failure, and don't worry about being original. If failure occurs, make it happen quickly.

3. Empower people. Trust your people. Train them. Use self-managing teams, involving everyone in everything. Eliminate management by fear and edict.

4. Love change. Create a vision and demonstrate this vision by example. Delegate authority to the lowest practical level.

5. Rebuild systems for a chaotic world. Revise and reexamine what is measured. Decentralize information, providing it in real time to those who need it to perform better. Set conservative goals and demand integrity.

BIBLIOGRAPHY

Peters, Thomas J., and Robert H. Waterman, Jr. *In Search of Excellence*. New York: Harper & Row, 1982.

Peters, Thomas J., and Nancy Austin, *A Passion for Excellence*. New York: Random House, 1985.

Peters, Thomas J. "It's Time to Get Back to Basics." *Quality*, May 1986, pp. 14–20.

Peters, Thomas J. *Thriving on Chaos: Handbook for Management Revolution*. New York: Alfred A. Knopf, 1987.

CHAPTER 11

GENICHI TAGUCHI

BRIEF BIOGRAPHY

Genichi Taguchi is a four-time winner of the Deming Prize in Japan. He received the first award ever given, in 1960, for the development of practical statistical theory. Dr. Taguchi worked for Nippon Telegraph and Telephone. While controversial, many companies have used his ideas to advantage in designing experiments and reducing process and product variation. His son, Shin Taguchi, carries on his father's work at the American Supplier Institute in Dearborn, Michigan.

BASIC TENETS

Taguchi's philosophy involves the entire manufacturing function from design through manufacture. His methods focus on the customer by using the loss function (Figure 11–1). Taguchi describes quality in terms of the loss generated by that product to society. This loss to society can be from the time a product is shipped until the end of its useful life. The loss is measured in dollars and therefore allows engineers to communicate to nonengineers the magnitude of this loss in a recognizable, common term. This is sometimes referred to as a *bilingual* mode, meaning that one can talk to upper-level managers in terms of dollars, and to engineers and those who work on the product or service in terms of things, hours, pounds, and so on. With the loss function, the engineer is able to communicate in the language of things and the language of money.

> The key to loss reduction is not meeting specifications, but reducing variance from the nominal or target value.

The Taguchi method has been described as *the most powerful tool* for achieving quality improvements, according to Jim Pratt, director of ITT's statistical programs. Pratt estimates that ITT saved some $60 million in an 18-month period.

FIGURE 11–1
The Loss Function

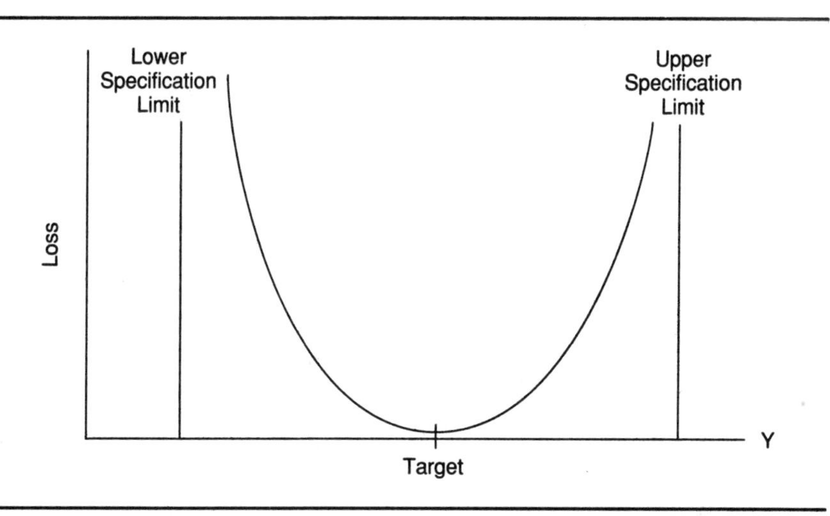

Many practitioners of Taguchi methods in this country feel that the on-line quality control practices described below will eventually supplant statistical quality control as it has to a large extent in Japan.

An overview of Taguchi's quality philosophy is provided below.

TAGUCHI'S QUALITY PHILOSOPHY

1. An important dimension of the quality of a manufactured product is the total loss generated by that product to society.
2. In a competitive economy, continuous quality improvement and cost reduction are necessary to stay in business.
3. A continuous quality improvement program includes incessant reduction in the variation of product performance characteristics about their target values.
4. The customer's loss due to a product performance variation is often approximately proportional to the square of the deviation of the performance characteristics from its target value. Thus, a quality measure quickly degrades with large deviation from the target.
5. The final quality and cost of a manufactured product are determined to a large extent by the engineering designs of the product and its manufacturing process.

6. A performance variation can be reduced by exploiting the nonlinear effects of the product (or process) parameters on the performance characteristic.
7. Statistically planned experiments can be used to identify the settings of product (and process) parameters that reduce performance variation.

ON-LINE AND OFF-LINE QUALITY CONTROL

Taguchi methods of off-line and on-line quality control provide a unique approach to reducing product variation. On-line methods include the various techniques of maintaining target values and the variation about the target in the manufacturing environment. These techniques would include such methods as statistical control charts. It is the off-line quality control techniques that make the Taguchi methods distinctive. Off-line quality control involves the design or quality engineering function and consists of three components:

System design. System design is the selection and design of a product that will satisfy the requirements of the customer. The design should be functional and robust[1] to changes in environmental conditions during service. The product should have minimal variation and should provide the most value for the price. Additionally the product should provide minimal functional variation due to use factors such as wear. The techniques employed here are various methods of establishing customer requirements and translating those requirements into engineering terms. Quality Function Deployment methods and the loss function are often employed in system design.

Parameter design. Parameter design is the identification of key process variables that affect product variation and the establishment of parameter levels that will impart the least amount of variation into the product's function. This is accomplished through the use of statistical experimental designs. Taguchi methods depart somewhat from classical experimental design in that the Taguchi approach uses only a small fraction of all possible experimental combinations and extracts the "right" conditions in a very efficient manner.

Tolerance design. Tolerance design is the determination of which factors contribute the most to end product variation and the establishment of the appropriate tolerances for those factors required to bring the final product

---

[1]A product is said to be robust when it has a lower degree of sensitivity to variation within the manufacturing process, although quality controls try to minimize variation.

into specification. Tolerance design is used only if the product variability is not confined to some "tolerable" level. The advantage of these methods is efficiency; rather than tightening tolerances across the board, only those that will have the most impact are tightened.

These three functions may be thought of as the definition of quality, the design engineering of quality, and the engineering of the production process. The traditional approach has been to design a product more or less independently of the manufacturing processes, and then to go about the reduction of variability in those processes to enhance the product quality. Taguchi methods attempt to design products that are robust to variation in the manufacturing process.

This requires analyzing two variables that can affect the performance of the product or process: design parameters and noise. The design parameters can be selected by the engineer. Such a parameter forms a design specification.

The noise consists of all those variables that cause the design parameter to deviate from the target value. Whether these noise causes are assignable is of small importance. Noise that can be identified should be included as an element in the experiment. Outer noise is caused by such things as variation in operating environments and human errors. Inner noise is caused by such things as deterioration. Between product noise is caused by manufacturing imperfections. Outer and inner noise is controllable by off-line methods such as parameter design. Between product noise can be controlled by on-line and off-line techniques.

The point of the experiment is to identify the parameter settings at which the results of noise are minimal. Through development of a design parameter matrix and a noise matrix, either physical experiments or computer simulations may be conducted to determine optimal settings of design parameters. Orthogonal arrays (see Chapter 20, Design of Experiments section) are suggested for constructing the matrices. These arrays are eventually used to determine signal-to-noise ratios as performance statistics.

Successes have been achieved using these techniques (see Schmidt's work and almost any issue of *Technometrics*), their application is somewhat controversial, and demands a specialist in statistics. Some of the technique is still being formulated in detail. Despite the apparent technical complexity, the design of experiments methodology provides a rich framework upon which to base quality and producibility determinations.

According to Raghu Kackar (an advocate and articulator of the Taguchi method), the four primary reasons for using statistically planned industrial experiments are:

1. To identify settings of design parameters at which the effect of the noise source on performance characteristics is minimized.
2. To identify settings of design parameters that reduce cost without impairing quality.

3. To identify parameters that have a large influence on the mean value of the performance characteristic but have no effect on its variation.
4. To identify parameters which have no detectable influence on performance characteristics and on which tolerances can be relaxed.

BIBLIOGRAPHY

Kackar, Raghu N. "Off-Line Quality Control, Parameter Design, and the Taguchi Method." *Journal of Quality Technology* 17, no. 4 (October 1985), pp. 176–209.

Kackar, Raghu N. "Taguchi's Quality Philosophy: Analysis and Commentary." *Quality Progress* 19, no. 12 (December 1986), pp. 21–29.

Kackar, Raghu N. "Quality Planning for Service Industries." *Quality Progress* 21, no. 8. (August 1988), pp. 39–42.

Schmidt, Michael S., and Larry C. Meile. "Taguchi Designs and Linear Programming Speed New Product Formulation." *Interfaces* 19, no. 5 (September/October 1989), pp. 49–56.

Taguchi, Genichi. *Introduction to Quality Engineering*. Dearborn, Mich.: American Supplier Institute Inc., 1986.

Taguchi, Genichi. *System of Experimental Design*. Dearborn, Mich.: American Supplier Institute, 1987.

Taguchi, Genichi and Don Clausing. "Robust Quality." *Harvard Business Review,* January/February 1990, pp. 65–75.

Taguchi, Genichi, and Yu-In Wu. *Introduction to Off-Line Quality Control Systems*. Central Quality Control Association, 1980.

CHAPTER 12

HISTORICAL MASTERS

INTRODUCTION

The phrase *historical masters* may bring to mind such leading quality figures of the early part of the 20th century such as Shewhart, creator of control charts. The group of scientists who developed quality control at AT&T in the first quarter century are often cited as *the originators* of quality. While this may be true of 20th-century United States, it certainly is not true worldwide. China had quality standards some 2,000 years or more ago. If we broaden our search for "original" Quality Management masters, and engage in some divergent thinking (Chapter 15, Creativity and Innovation section), where should we start?

Surely the first quality engineer was some unknown chariot maker in Sumeria who observed that the smoother the wheel, the faster the chariot, or some swordsmith farther north in the land of the Hittites who scrutinized his work in order to free it of all blemishes, thereby increasing its cutting power. Or was it a Neanderthal, who plied his craft with flint knapping tools, being attentive and conscientious in shaping a tool for a precise use? Modern archaeologists have demonstrated that these skills are difficult to acquire. Yet we know that the skills were practiced by Neanderthals and were passed on through generations. So perhaps quality control predates *Homo sapiens,* and originated with *Homo sapiens neanderthalensis*.

While this *gedanken* (thought experiment) is entertaining, we have no written record to guide us. Written history, however, provides a grand procession of characters and legends to emulate and avoid. Some would seem to presage Total Quality Management principles. While others can undoubtedly be found[1] from nearly every era and area, the spotlight here will be on three of the authors' favorites: Sun-Tzu, Aesop, and Socrates.

---

[1]Clemens and Mayer's work, *The Classic Touch: Leadership Lessons from Homer to Hemingway,* dispenses management lessons from historical figures, and provides a diverting outlook.

SUN-TZU

Sun-Tzu lived sometime during the warring states period (480–221 B.C.) in China. During this time, the entire world, as the Chinese knew it, was in constant war. Sun-Tzu observed the political and leadership struggles and attempted to codify how best to approach these disciplines. An analogy has been made between Sun-Tzu's process and those of the immune system[2]: it wards off disease and attackers without damaging the underlying support system. In this respect it also embodies principles of TQM: survival, growth, and continuous improvement in a chaotic world.

The work is only some 5,600 words long, and can be read easily in an evening. There are a number of excellent translations in English, most notably the one by Samuel B. Griffith. A novel translation, by R. L. Wing, casts the terminology into business and strategic terms rather than martial terms. Wing also gives extensive commentary on applying these principles to one's daily life and work life.

Sun-Tzu's work has 13 sections, summarized below:

1. Calculations. Is the goal reasonable and worthwhile? If so then the time to act is now. If not, wait and build strength until it is reasonable. Self-reflection (e.g., benchmarking and audits) are necessary to understand one's own strength as well as that of the enemy. Both the self and the enemy must be understood.

2. Estimating the cost. Contingencies and alternatives must be weighed. Once the costs have been estimated, then commit to an unrelenting forward momentum. There must be unbelievably swift and decisive action, with no looking back.

3. Plan of attack. Is the thing isolated or integral? If the objective is isolated, then a forward attack is appropriate. If it is integral, then a broader, less obvious attack route may be necessary.

4. Positioning. Remove elements around you that allow backsliding. One must have a supportive environment. You cannot cross a river by going across halfway and then returning to shore to rest for the next half.

---

[2]Like the immune system, which can enter an autoimmune phase and destroy the very body it was protecting, so too can Quality Management consume its own as it metamorphoses a company via continuous change and quality improvement, making the new business very different from the old business.

5. Directing—Positioning your foe. Isolate your foe, make plans that do not include it; enlist the help of others. Act spontaneously.

6. Illusion and reality. The opponent must not be allowed to rest. When weak, feign strength. When strong, feign weakness.

7. Maneuvers and tactics. Indirect tactics such as logistics are essential to success. Indirect actions which lead to direct effect are the best. Competition within your own forces is dangerous. The focus must be on the enemy. Give your opponents the ability to flee. Damage the overtaken as little as possible. They are tomorrow's customers.

8. Spontaneity in the field. Sun-Tzu warns against being overly cautious, reckless, angry, fastidious, and attached to the organization or the status quo.

9. Moving the force—Confrontation. Self-control and discipline will grow from a determined challenge.

10. Situational positioning. Challenge only when certain. Put yourself between your foe and his support system, as disorder brings defeat.

11. The nine situations. Awareness of the situation is vital. Contradictory situations call for contradictory action.

12. The fiery attack. Do not prolong the confrontation. Enlist help from the outside and make every action count.

13. The use of intelligence. Information is the essence of success or defeat.

AESOP

Aesop was (perhaps) a Greek slave who lived in the 6th century B.C. While not providing the integrated framework of Sun-Tzu, Aesop's simple fables or parables provide profound insight into the nature of people's conduct in society. What is so instructive is that little has changed in human nature since the beginnings of civilization. Aesop has permeated Western literature for some 500 years, as browsing through a quotation dictionary will readily confirm. A small sample of Quality Management related morals is presented:

United we stand, divided we fall. **The four oxen and the lion.** A direct appeal to the use of self-managing teams, quality circles, and other uses of teamwork and shared vision.

Please all, and you please none. **The man, the boy, and the donkey.** A vision that has been defined in scope and is attainable is essential. Serving some segments of customers may mean ignoring others.

A liar will not be believed, even when he speaks the truth (cry wolf). **The shepard boy.** Credibility in management is regained everyday, and when lost, it may be lost until that management is replaced, and quite likely even longer. Trust is not bought by salary, or enforced by memos. Without trust, it is difficult to achieve the union required to succeed.

We often despise what is most useful to us. **The hart and the hunter.** Exempt, nonexempt, suits, Joe Punchclock, . . . the list is long that we use to demean the workers on whom we rely. As Martin Luther King, Jr., said, there is dignity in all work. Failure to understand this leads to strikes, lock-outs and boycotts, and management by fear.

You may share the labor of the great, but not the spoils. **The lion's share.** Rewards must be given out on a fair and regular basis, and it must be understood how one can achieve these rewards.

Perseverance wins the race. **The hare and the tortoise.** Aesop agrees with Sun-Tzu regarding careful calculations, plotting, and provisioning. Ambition and excitable energy are no substitute for wisdom and cunning.

SOCRATES

Socrates, the philosophizing stonemason, lived from 470 B.C. to 399 B.C., in Athens, during the Golden Age of Pericles. He is known to us from the works of Plato as a one of the greatest philosophers. A number of Quality Management concepts can be found in the various Socratic dialogues.

Root cause analysis. Socrates was always asking Why?, and even when told why, would keep probing until he reached the bottommost cause.

This rumination was meant to sweep away preconceptions and prejudices to reach the real reasons or truths of whatever subject was before him.

Empowering the worker. Democracy lends voice to all. Coupled with critical self-examination, this leads to an empowering force of continuous self-improvement.

Vision and ethics. "The unexamined life is not worth living" is a quote of Plato and Socrates. Constantly examining our beliefs and motives requires building a solid framework of morals and ethics.

Management by walking around. "Socratic Dialogue" refers to a technique of asking questions so that those answering them will be led to the answer. It is difficult to craft such questions, but it is highly rewarding for teacher and student alike. Socrates considered himself a gadfly, always goading others into thinking about their actions and beliefs.

BIBLIOGRAPHY

Aesop. *Fables*. (Numerous translations available.)

Clemens, John K., and Douglas F. Mayer. *The Classic Touch: Leadership Lessons from Homer to Hemingway.* Homewood, Ill.: Dow Jones-Irwin, 1987.

Griffin, Gerald R. *Machiavelli on Management*. New York: Praeger Publishers, 1991.

Juran, J. M. "China's Ancient History of Managing for Quality." *Quality Progress* 23, no. 7 (July 1990), pp. 31–35.

Sun-Tzu. *The Art of War*. trans. Samuel B. Griffith. London: Oxford University Press, 1963.

Sun-Tzu. *The Art of Strategy.* trans. R. L. Wing. New York: Dolphin/Doubleday, 1988.

Plato. *The Republic*. (Numerous translations available.)

PART 3

MANAGEMENT DYNAMICS

This section collects major components of organizational dynamics research. You could spend an entire career researching in just one of the subchapter areas. Therefore the coverage given here is necessarily introductory. The point is to provide beleaguered managers a means of diagnosing what problems may be encountered and where to turn. Invariably in implementing a Quality Management program, stumbling blocks are encountered such as: resistance to change, performance problems, open and unfocused anger, dalliance, and increased stress. It is important to be aware of these other subdisciplines of the management art when they arise. Tools alone are not enough. Successful Quality Management implementation requires leadership and an understanding of management dynamics.

Three tiers of management dynamics are presented. Companywide dynamics are those practices and techniques which act upon the entire organizational entity. Supervisory dynamics explores the relationship between supervisor and the supervised, although activities such as leadership and motivation can be instituted by anyone in any organization. Individual performance enhancement techniques are also included, and are of relevance to all levels of management.

There are, perhaps, as many approaches to management as there are managers, but some major trends can be designated. These are summarized (and simplified) in Exhibit 1. Probably no management school has advocated a pure approach, such as examining every problem with a team of operations researchers, and completely ignored the human element. The approaches have often been blamed for management problems: management science and operations research is often ridiculed in the United States. However management in Japan and Germany use such decision theory approaches with fervor and great success. Perhaps the problem is not with the tools, but the skill of those using them.

EXHIBIT 1
Management Approaches

Traditional Approach. Management by simple numerical goals and objectives, without regard to people. People are things to be overcome and used.

Empirical Approach. It is important to study other managers, irrespective of the nature of the work. Management skills are completely transferable.

Behavioral Approach. Management must consider the psychosocial context. People are part of the solution, not problem, and are the key to solving the problem.

Decision Theory Approach. A rational approach based upon mathematical models and processes.

Systems Engineering Approach. Management approach based upon input, output, processing, and constraints.

Integrative (Quality Management) Approach. Something of value is offered from each of the approaches. It is up to management to become aware of these approaches and select the most appropriate of them for the organization for maximum benefit.

Quality Management seeks to integrate the best of each school into a balanced approach to the myriad problems that face managers on a daily basis. Thus there are numerical tools from the decision theory, systems engineering and the rational schools; human-oriented tools and management dynamics from the behavioral school; commonsense approaches from the empirical and the traditional schools; and tools and dynamics that could be considered hybrids of more than one approach.

Management theories abound because of the vast number of researchers, the changing environment under which managers manage, and the increased awareness of management's impact on employee performance. Another, darker, reason is that new fads sell books and training programs. This has led some companies to look foolish as they seem (to their workers at least) to embrace the newest fad from the currently fashionable pop psychologist. Unfortunately, jaded employees may see TQM or Quality Management as another pretender, something to nod their heads to until the next management rage comes along. Fortunately, Quality Management is based upon solid research from the past 100 years of investigation into organizational behavior and management science. Quality Management is here to stay, although it will grow and change as new findings emerge, and as the business and societal climate changes.

The history of management (other than military leadership) as a discipline worthy or capable of being studied began in the first part of this century with a system by Frederick Taylor. Taylor termed his theory as *scientific management*. That is, work processes could be repeated with a degree of accuracy and could be taught, as opposed to an art, which is often totally unrepeatable, and to crafts, which are often, but not necessarily, repeatable. Today, it is more commonly known as Theory X, and is associated with the Hobbesian view that workers are slothful and must be tricked, prodded, and punished into good work habits. As a result, Taylor has something of a sinister reputation among management theorists today. Taylor was probably more interested in the idea of measurement of human performance than any behavioral or ethical models. There are also situations that may demand a Theory X approach.[1]

Most management researchers prefer the thesis termed *Theory Y,* which assumes that people like to work, and can be readily motivated, if one understands their situation as it relates to the organizational environment. (There is also a *Theory Z,* which has elements of the so-called Japanese style

---

[1]A crib tool manager observed that flashlights were being "lost" at an alarming rate. When she purchased inexpensive, day-glo pink flashlights, the "loss" rate fell to zero. No rebuking memos or other disciplinary actions were necessary. This avoided management by fear, but also got the point across.

of management. It is left to the reader as an exercise to develop management Theory AA.) The chapters in Part 3 are rooted in the Theory Y concepts. The validity of this approach is due to the nature of the worker today. Baby boomers are rather far along awareness and behavioral scales such as Maslow's Hierarchy of Needs (Chapter 3) and are in a position to demand humanistic treatment. The transformation from an autocratic environment of the Edwardian era to the self-managing, cross-functional teams of today is sketched in Exhibit 2.

Change and its attendant stress and conflict play a key role in transforming an organization. Organizations must be lead through these changes, rather than managed. Successful leadership requires understanding of human motivation.

BIBLIOGRAPHY

Bessinger, R. C., and W. Suojanen. *Management and the Brain*. Atlanta: Georgia State Universities Business Publications, 1983.

Blake, R. R., and J. S. Mouton. *Corporate Excellence through Grid Organization and Development*. Houston, Tex.: Gulf Publishing, 1968.

Ginzberg, Eli, and George Vojta. *Beyond Human Scale: The Large Corporation at Risk*. New York: Basic Books, 1985, pp. 218–19.

Levinson, H. *The Exceptional Executive*. Cambridge, Mass.: Harvard University Press, 1968; New York: The New American Library, 1971.

Likert, Rensis. *New Patterns of Management*. New York: McGraw-Hill, 1961.

McGregor, Douglas. *The Human Side of Organizations*. New York: McGraw-Hill, 1960.

Ouchi, William G. *Theory Z: How American Business Can Meet the Japanese Challenge*. New York: Avon Books, 1982.

Taylor, Frederick W. *The Principle of Scientific Management*. New York: Harper & Row, 1911.

EXHIBIT 2
Transformation from Theory X to TQM

	Theory X (Fear State)	*Classic Management by Objectives*	*Early TQM (Participation/ Involvement)*	*Advanced TQM (Empowerment/ Commitment)*
Organizational structure	Strict hierarchy	Hierarchy	Ad hoc teams	Autonomous teams
Improvement efforts	No changes allowed	Management determined	Suggestions, quality circles	Worker teams improve process
Mistakes/defects	Suicidal	Coverup, move on quickly	Valuable learning opportunities	Treasure
Decision power	Top	Top-down	Shared	Lowest level possible
Management focus	Dictatorial	Supervision	Coaching	Leadership
Tradition	Determines work methods	Preserved	Respects, changes as necessary	Respected, revised as needed
Motivation	Fear	Persuasion, fear	Humanistic	Humanistic and cross-cultural
Barriers	Creates them, crisis causing	Resources must be fought for, crisis oriented	Rearrange obstacles	Proactively removes and dissolves barriers
Locus	Self	Department oriented	Division, plant oriented	Total system (including vendors)

CHAPTER 13

COMPANYWIDE DYNAMICS

Key to the establishment of Quality Management is the integration of its vision throughout the entire company—all divisions at all levels. This is accomplished by understanding how systematic changes take place in organizations. Organizations take on behaviors and have a persona just as individuals do. Initially, the corporate culture or personality must be understood as it currently exists. Then a vision must be created of the new organization, with realistic expectations.

For example, a construction firm might have as part of its vision a safety conscious work force. Traditionally, construction workers consider themselves too virile to worry about safety. Indeed, if a construction worker was worried overmuch about safety, they would choose another line of work. Thus, a simple training session or the occasional videotape will probably prove ineffective. Significant emotional events, such as seeing a friend seriously hurt, would probably change behavior, but at enormous expense. Change will likely require a multiyear plan involving training, feedback, and a dedicated safety supervisor or inspector who understands the role that must be played (at least initially) as the *bad guy* enforcing new safety precautions.

The relationship of the four sections, "Corporate Culture," "Leadership," "Strategic Planning," and "Change Management," follows a Plan-Do-Check-Act cycle (Chapter 17) as illustrated in Figure 13–1. As organizational culture is revised, leadership is required to initiate changes carried out with strategic planning, resulting in a change management process. Changes in the business atmosphere may require changes in culture, and so on. Changes in any of the four are bound to result in the need for the other three. This process is an iterative and ongoing one.

CORPORATE CULTURE

The employer generally gets the employees he deserves.

Walter Gilbey

FIGURE 13–1
Companywide Dynamics and the Plan-Do-Check-Act Cycle

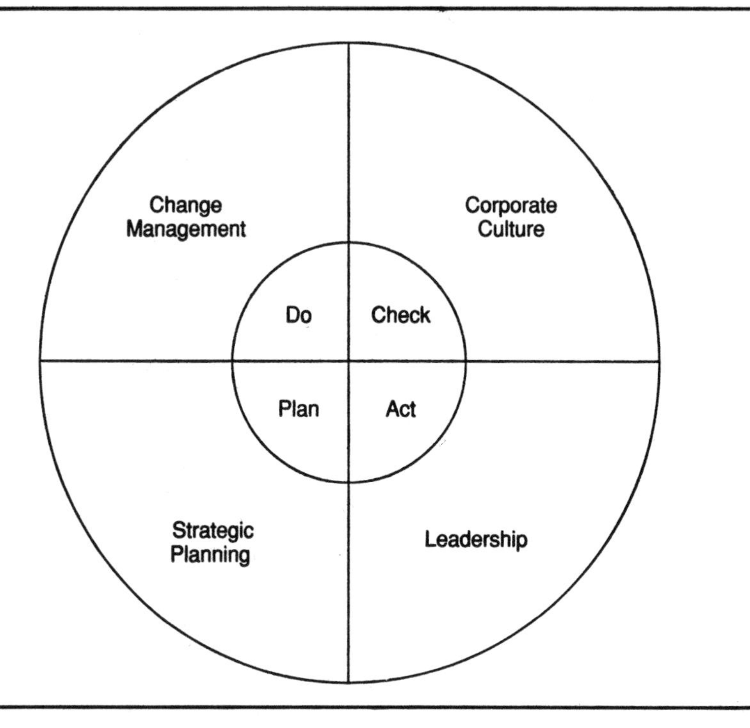

Aliases

Organizational behavior, belief system, and organizational culture.

Brief Description and Purpose

Just as any tribe or clan has certain rituals, behaviors, taboos, and preferences, so too does any organizational entity. This culture may be highly elaborate and conscious, or it may develop organically from within. All organizations have a culture, and it is up to top management to recognize it, analyze it, and, if desired or necessary, control and direct it. A sense of unity and purpose is forged from a diverse work force if workers and management are actively involved in continually shaping the organizational culture according to sincerely believed fundamental corporate principles. This culture is composed of the beliefs, values, and behaviors a group shares in an

FIGURE 13–2
Cultural Elements

1. Myth	5. Symbols
2. Ritual	6. Goals
3. Beliefs	7. History
4. Values	

environment (Figure 13–2), and is demonstrated in actions. Values left undemonstrated are not values.

Discussion

The term *culture* has been applied to the study of organizations for many years, but is still far from an exact science. Culture was defined by Margaret Mead as "a body of learned behavior, a collection of beliefs, habits, practices, and traditions, shared by a group of people (a society) and successively learned by new members who enter the society." This definition is helpful in considering why people in an organization resist change. As with scientists studying primitive societies, managers, as an advocate of change, "must be aware that you are dealing with a pattern of human habits, beliefs and traditions which may differ from yours and which may, therefore, view this change in a way totally different from your view," says Juran.

Corporate culture is derived from beliefs that are set forth by guiding principles, which may be external to the organization (e.g., the Ten Commandments), and are demonstrated by daily beliefs or actions. Actions set corporate culture in motion, not memoranda. Corporate culture cannot be changed by dictum. Management must sincerely believe in the corporate culture and demonstrate the values on a daily basis for it to pervade the company. Corporate culture must be continually reevaluated to account for rapidly changing perceptions about job expectations, and increasing work force diversity. Ways should be found to integrate the quality effort into all existing organizational activities, including strategic and financial planning. Reward systems must reinforce the corporate values. Changes must be implemented deliberately and publicly.

Buck Rodgers, former vice president of marketing at IBM, describes his personal guiding values (Figure 13–3) in his book *Getting the Most Out of Yourself and Others*. He attempts to demonstrate these values daily. The list is short enough to be easily remembered, simple enough to be able to be accomplished with skills almost anyone possesses, yet laudable and worthy to society and company alike. These values are not likely to be changed over

FIGURE 13–3
An Example Set of Values

1. Life is meant to be enjoyed.
2. Work is a blessing to be enjoyed.
3. One's work can be one's play.
4. Unrealistic expectations are handicaps.
5. Use all your talents.
6. Believe in yourself—create opportunity.
7. Make a difference.
8. Have a sense of humor.
9. Believe that people are good.
10. Pay your civic dues.

Adapted from Buck Rodgers, *Getting the Most Out of Yourself and Others* (New York: Harper & Row, 1987).

a period of time, and they are broad enough to allow flexibility in demonstration of these beliefs, yet specific enough for anyone to readily cite an example of how such a belief can be demonstrated.

Relation to Other Concepts and Tools

Changing corporate culture should be understood prior to instituting change and employee involvement. Culture is often the thing we are trying to change via leadership, employee involvement or empowerment. Tools useful in defining culture include auditing and benchmarking. Strategic planning tools include cause effect diagrams, work flow analysis, force field analysis, and goal setting. Quality function deployment is well suited in determining missing elements of corporate culture and daily beliefs.

Process

1. Define the current organizational quality culture. This involves asking managers, supervisors, employees, vendors, and customers such questions as:

- What is it like to work here?
- What styles of management are most commonly found in the organization?

- How well do work groups function together?
- Do current organizational practices tend to help or hinder the organization's effectiveness?
- What is the nature of our business? Is the final product commodity-oriented or value-added oriented?

Organizations which view themselves as being in a commodity area often value lowest cost over all other attributes. Value-added organizations search for distinctions from their competitors and proclaim these differences to their customers. Hybrids are also possible, such as in the personal computer market, where lowest prices and outstanding service (from the same company) are possible. Any company capable of doing this while maintaining a positive balance is sure to be a market leader.

In formal organizations, where everything is required in writing, the ability to respond to a customer's needs can be severely diminished. The effectiveness of individuals, and ultimately the entire organization, is impacted by the policies related to the general quality of work life. Compare the legendary performance of Federal Express employees to your organization. Why do the Federal Express employees perform differently? How are employees who use imagination and resourcefulness to solve a problem treated versus those who dogmatically follow a rule book? How would customers respond to these two choices? Are decisions of immediate importance to customers decided at a managerial level, or with the person who makes contact with the customer?

Management must go to the source and ask the employees what quality problems they face, what keeps them from doing the best job possible, and what helps and hinders their performance. Bringing the employees into the change process in the definition phase is the first step toward gaining their trust and cooperation. Recruit vendors to describe the company culture.

Surveys can be undertaken to define the current culture. Statements could be made with which the respondent can agree, strongly agree, disagree, strongly disagree, or have no opinion. An example of some types of questions are listed in Figure 13–4. Open-ended questions may also be appropriate, and some examples are provided in Figure 13–5. The problem with surveys is that employees are usually very adept at guessing the answers they think management would like to hear. Surveys should be bolstered with interviews and observations of the behavior reported.

If it is too difficult to describe the current culture with descriptive words and phrases, then take a situational approach. Describe a situation such as an employee who has an idea to improve a process in a different department. How do they proceed, if at all? Situations are also useful in that beliefs of

FIGURE 13–4
Culture/Goal Matrix

Cultural Value: Dedication *Observed values:*	*1*	*2*	*3*	*4*
Overtime is done willingly (but paid)	+			+
Employees work hard	+			
Bull sessions common	–	+ (?)	– (?)	+
Consistently do more than expected	+			+
New tasks taken on willingly	+	+		+
Benefits are minimally adequate, if single				–
Benefits are poor, if married				–

Selected Organizational Goals:
1. Fast response time.
2. Innovation.
3. Accuracy.
4. Low turnover.

Interactions
+ Positive, reinforcing interaction.
– Negative interaction.
? Requires further investigation.

FIGURE 13–5
Example Open-Ended Questions

How is one promoted in this organization?
What motto or phrase would describe this organization?
What tactics should be employed to initiate a new idea or process?
What qualities are observed in successful managers in this organization?
What behavior or task could you do that you think would be rewarded?
Would you want to be a supplier to this organization? Why or why not?

differing organizational layers can be compared. This may reveal that certain beliefs held at the managerial level (the policy manual says that "improvement suggestions are welcome from all employees") may not be held at other levels (a supervisor might actually discourage such actions). Examples may need to be elicited, rather than a hypothetical reaction, as the levels may simply parrot what they perceive to be the guiding beliefs (and what management wants to hear) rather than the demonstrated belief.

2. Match current values with desired values. Determine how the current culture meshes with strategic goals, and what beliefs need to be modified or introduced to be aligned with these goals. An example of part of such an analysis is provided in Figure 13–4. A matrix tool has been used to list current daily beliefs in the rows (in this case, factors related to dedication were used). The columns are corporate goals. The matrix is then used to determine which daily beliefs support these goals, and which hinder their attainment. Extensions to this matrix could be made, showing the compatibility of the goals and benchmarking criteria. A "house of quality" could be developed (Chapter 17).

3. Involve everyone. Provide information to everyone in the organization. Everyone in the organization must have a clear understanding of what values the company holds. Conversely, management must understand when the values need to be changed. This requires a comprehensive feedback system between all levels. Encourage employee participation at all stages (see Employee Commitment in Chapter 14).

4. Adopt a change management process. Utilize a systematic, formalized approach to change (Change Management, Chapter 13).

5. Management must lead the culture change. Rosabeth Moss Kanter, in *The Change Masters,* said "the more profound, comprehensive, and widespread the proposed change, the more absolute is the need for deep understanding and active leadership by the top managers." A constancy of purpose must be created by taking a long-term view, establishing meaningful goals, and ensuring that behavior is consistent with the goals. Leadership style and structure (see the section Leadership in this chapter) may need to be assessed and changed to be aligned with the new values.

6. Expand culture to suppliers and customers. Any organization is part of a larger culture. Export your culture to the organizations directly involved. If successful, these values will quickly become part of the industry. Mail-order sales, for example, had suffered from a bad reputation of poor delivery, no service, and questionable business practices. The major mail-order houses of today provide next day service at nominal cost and on-site support, accept returns graciously, and keep the customer informed on any items that have to be backordered, and promptly refund for items that are unavailable. This culture, once established by a few major leaders, quickly led to everyone following these policies as customers came to expect this level of service.

Some suggestions on involving suppliers:

- Simplify the acquisition process.
- Involve suppliers at the decision/planning stage.
- Use suppliers in problem-solving sessions.
- Minimize the number of vendors.
- Help suppliers improve. Establish training to key vendors.
- Reward vendors who improve or meet quality criteria with more business, reduced paperwork, or some appropriate incentive.

When involving customers:

- Make them aware of policy changes.
- Allow them easy access to feedback mechanisms. Suggestions boxes are a start, but involve only a self-selected group. What do the other customers think?
- Benchmark customer reactions to your values and those of your competitors.

Dos and Don'ts

• Don't let a big nonevent occur. Memos, meetings, picnics, and the like do not ensure changes in daily beliefs. Personal changes occur in personal settings, not in companywide rallies. Allow each team or group to discuss current and needed cultural beliefs and behaviors, and let them formulate a response to the needed changes.

• An organization must direct cultural values or the daily, practiced beliefs will, and these values may be those of the lowest common denominator. One poor example left unchecked in an organization can wreak havoc on other employee's behavior. Don't let the tail wag the dog.

• People learn by personal example. The next time you're strolling through the plant and spy a coffee cup next to an operating machine, throw it away yourself, rather than barking at the offender.

• Don't overemphasize personal values. Your crusade with the Save-the-Box-Elder-Bug Society may not appeal to everyone. Don't impose inappropriate values on the work force. Some crusades are best when kept private.

• The customer is often a driving force in creating a culture. If you try to change your culture to one that is incompatible with your customer, prepare to find new customers, if this wasn't the goal to begin with.

• Alter history. There is no reason to carry a burden of past history when change is necessary for corporate survival. The Walt Disney Company, producer of classic, high-quality children's movies for decades, would have never produced a live action motion picture with an R rating. To survive at the box office, however, it was necessary to produce such movies, since they could support the incredibly expensive animated productions. To preserve its image, the new movies were produced under the name Touchstone. Thus, history was respected, and change was implemented.

LEADERSHIP

Managers do things right—Leaders do the right things.

Warren Bennis

Leadership is intelligence, credibility, humanity, courage, and discipline.

Sun-Tzu

Aliases

The term *leadership* is often accompanied by an adjective, such as transformational, charismatic, and practical.

Brief Description and Purpose

Managers can perform the tasks as well or better than their subordinates. Leaders must direct job functions that they personally may not be able to do. With rapid technological and social change the norm, leadership has taken on a prominent role in management ability.

Discussion

Leadership, outside of a military context, as a thing that can be analyzed and repeated, has only been studied in the last 50 or more years. It has received so much attention lately because of the changing nature of the way in which we must conduct work. We simply cannot know or be able to do all of our workers' jobs. Instead we must lead them. The research concerning leadership is now beginning to put together metamodels (Figure 13–6), weaving leadership styles, managerial behavior, and end goals. The best news would seem to be that leadership can be learned. It is not necessarily innate. It does

FIGURE 13–6
Leadership Metamodel

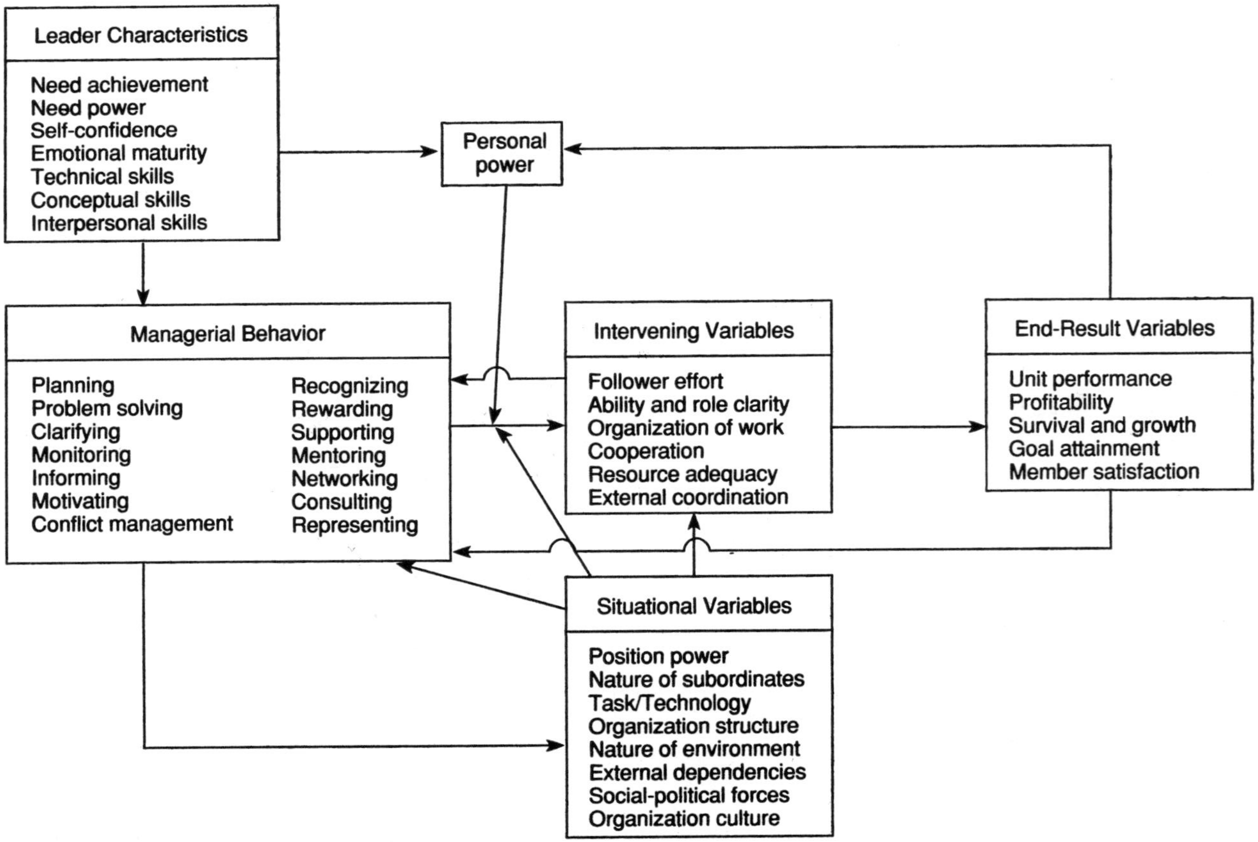

Leader Characteristics

Need achievement
Need power
Self-confidence
Emotional maturity
Technical skills
Conceptual skills
Interpersonal skills

Personal power

Managerial Behavior

Planning
Problem solving
Clarifying
Monitoring
Informing
Motivating
Conflict management
Recognizing
Rewarding
Supporting
Mentoring
Networking
Consulting
Representing

Intervening Variables

Follower effort
Ability and role clarity
Organization of work
Cooperation
Resource adequacy
External coordination

End-Result Variables

Unit performance
Profitability
Survival and growth
Goal attainment
Member satisfaction

Situational Variables

Position power
Nature of subordinates
Task/Technology
Organization structure
Nature of environment
External dependencies
Social-political forces
Organization culture

not belong to any one gender, ethnic group, or age-group, although maturity is a critical factor in attaining leadership.

Some common models of the structure of leadership are:

- Authoritarian.
- Charismatic.
- Paternalistic.
- Participative.
- Laissez-faire.

Authoritarian structures are well known to all of us, and need little discussion. That they are failing rapidly is also self-evident. They treat humans as expendable commodities, and in times of recession, *downsize*—fire large amounts of people in a desperate ploy for survival. This has led to a tremendous dissatisfaction with the workplace—polls show that more than 75 percent of employers are unhappy with their work ("CBS News," May 6, 1990). Downsizing creates a terrible work burden on those remaining, as they soak up "other duties as assigned" and wait for the ax to swing in their direction.

Charismatic leadership works well for religious and political institutions, but offers little for the business entity, which must survive its charismatic leader. Paternalistic management almost sounds like patronizing management, which it often is. Participative management engages workers to be leaders as well, and encourages employees to take an active, involved, and committed view of employment. Laissez-faire (hands off) management works with highly autonomous individuals who have little direct accountability to the parent organization (e.g., professors who earn a salary from grant money), and would not seem to be an appropriate model for most businesses. The authoritarian, participative, and laissez-faire styles are illustrated in Figure 13–7.

Leadership is also often characterized into a bestiary of styles. One such zoo of leadership styles shown in Figure 13–8 presents a synthesis of some of these styles. Too much can probably be made of these styles, but they are of interest in ferreting out potential role models for a given business situation.

Relation to Other Concepts and Tools

An examination of leadership styles is necessary prior to launching change management. Kouzes and Posner in their works on leadership have found that leaders must:

FIGURE 13–7
Leadership Structures

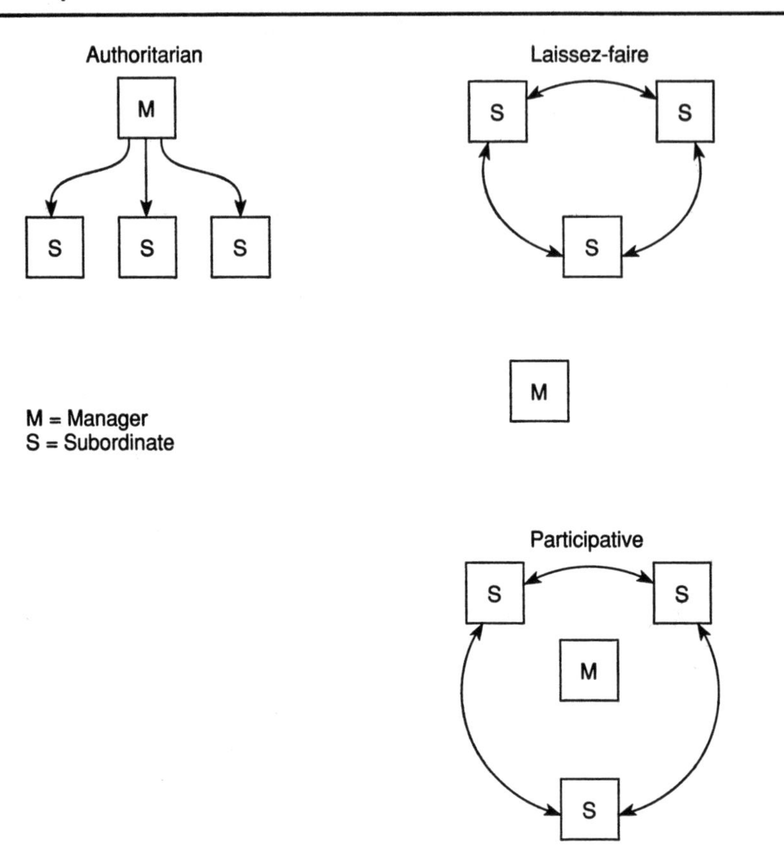

- Challenge the process.
- Inspire a shared vision.
- Enable others to act.
- Model the way.
- Encourage the heart.

The most important upshot of leadership studies for business managers is that leadership should not be equated to professional sports leaders—the rare upper 1/10 of 1 percent who take home Super Bowl rings or gold medals. Leadership comes from organization, clarity of vision, and deliberateness of execution (see Figure 13–9). Sun-tzu (Chapter 12) describes a leader as one who possesses intelligence, credibility, humanity, courage, and discipline.

FIGURE 13–8
Leadership Styles

Meditative	
Strengths:	Weaknesses:
Specific goals	Resists other assignments
Single focus	Dislikes multiple projects
Highly innovative	May delegate too much
Logical	May resist hunches
Challengers	Dislikes rules, policies
Highly rational	May appear blunt

Meditative and Intuitive	
Strengths:	Weaknesses:
Specific goals	Inconsistent
A few focuses	Slow moving
Innovatively intuitive	Relies on intuition for proof
Reasoning	Holes in reasoning
Investigators	May analyze too much
Moderately rational	Not rational enough

Meditative and Negotiative	
Strengths:	Weaknesses:
Balances goals of all	Indecisive alone
Flexible focus	Unclear
Rational and personal	Seen as inconsistent
Practical and empathetic	May move too slow
Open minded to feelings	Softhearted
Involves people in decisions	Too slow making decisions

Meditative and Directive	
Strengths:	Weaknesses:
Project specific	Resists change
Assignment focused	Overfocused
Moderately innovative	May not delegate too much
Practical	Resists hunches and guesses
Implementers	Likes rules, policies too much
Systematic and factual	May seem insensitive

Examples	
Meditative:	Meditative/Negotiative:
F. Lee Bailey	John F. Kennedy
Sigmund Freud	Lee Iacocca
Jawaharlal Nehru	Abraham Lincoln
Thomas Jefferson	Mikhail Gorbachev
Robert Kennedy	Gen. Omar Bradley
Prometheus	
Meditative/Intuitive:	Meditative/Directive:
Albert Einstein	Oliver Cromwell
Margaret Thatcher	Anthony Eden
John Sculley	Charles Schwab
Jimmy Carter	Charles Darwin
Harry Truman	Karl Marx

FIGURE 13–8 *(continued)*

Intuitive

Strengths:	Weaknesses:
Goals multiple	Resists assignments
Multiple focuses	Dislikes single projects
Highly intuitive	May not delegate
Imaginative	May rely on hunches
Explorers	Dislikes rules, policies
Highly original	May appear bizarre

Intuitive and Meditative

Strengths:	Weaknesses:
Goals general	Resists other assignments
Many focuses	Dislikes multiple projects
Intuitively innovative	May delegate too much
Inventive	May resist hunches
Excursionists	Dislikes rules, policies
Moderately original	May appear blunt

Intuitive and Negotiative

Strengths:	Weaknesses:
Goals creative and group	Resists working alone
Many group focuses	Dislikes single projects
Moderately intuitive	May get too involved
Mildly imaginative	May resist hunches
Explorers	Dislikes rules, policies
Highly original	May appear flippant

Intuitive and Directive

Strengths:	Weaknesses:
Goals creative and do-able	Argument about their work
Many directed focuses	Seemingly scattered
Systematically intuitive	May be too organized
Imaginative	May resist hunches
Creative implementors	Dislike rules, policies
Specifically original	Appears argumentative

Examples

Intuitive:	Intuitive/Meditative:
Clarence Darrow	Winston Churchill
Anwar Sadat	Gen. Robert E. Lee
Steve Jobs	Ted Turner
Fred Smith	Andrew Carnegie
Thomas Edison	Napoleon
Apollo	

Intuitive/Directive:	Intuitive/Negotiative:
Frank Lloyd Wright	St. Francis of Assisi
Atilla the Hun	Vincent Van Gogh
Gen. George Patton	Abraham Lincoln
Moses	Marc Anthony
Henry Ford	Corazon Aquino

FIGURE 13–8 *(continued)*

Negotiative

Strengths:	Weaknesses:
Group goal oriented	Will not think alone
Varying focuses	Erratic
Highly personal	Too personal
Friendly	Not serious enough
Joiners	Too involved with others
Very empathetic/sympathetic	Too sensitive

Negotiative and Meditative

Strengths:	Weaknesses:
Group goal oriented with facts	Become argumentative
Variety of focuses	Inconsistent
Mildly personal	Too personal
Caring	Can't make decisions
Consensus type leader	Indecisive alone
Mildly empathetic/sympathetic	Listens too much

Negotiative and Intuitive

Strengths:	Weaknesses:
Group goal oriented	Often won't think alone
Mixed focuses	Not consistent
Personal and individual	Aloof at times
Friendly and creative	Disruptive to group thinking
Selective joiners	Often overly involved
Mildly empathetic/sympathetic	Often too sensitive

Negotiative and Directive

Strengths:	Weaknesses:
Group goal oriented	Will not think alone
Varying focuses	Erratic
Highly personal	Too personal
Friendly	Not serious enough
Joiners	Too involved with others
Highly empathetic/sympathetic	Too sensitive

Examples

Negotiative:
- Mother Theresa
- Carl Rogers
- Norman Vincent Peale
- Toulouse-Lautrec
- Hans Christian Andersen
- Dionyslus

Negotiative/Intuitive:
- Nero
- Mark Twain
- Albert Schweitzer
- George Gershwin
- Jane Adams

Negotiative/Meditative:
- Mahatma Gandhi
- Florence Nightingale
- Peter Uberroth
- Steve Wozniak
- Martin Luther King, Jr.

Negotiative/Directive:
- George Bush
- Pope John XXIII
- Robert Schuller
- Roy Wilkins
- Edward Manet

FIGURE 13–8 *(concluded)*

Directive

Strengths:	Weaknesses:
Goal controlled	Resist other assignments
Single focused	Dislike multiple projects
Highly innovative	May delegate too much
Logical	May resist hunches
Challengers	Dislike rules, policies
Highly rational	May appear blunt

Directive and Meditative

Strengths:	Weaknesses:
Goal ordered	Resist other assignments
Single focused	Dislike multiple projects
Highly innovative	May delegate too much
Logical	May resist hunches
Challengers	Dislike rules, policies
Highly rational	May appear blunt

Directive and Intuitive

Strengths:	Weaknesses:
Goal controlled	Resist other assignments
Single focused	Dislike multiple projects
Highly innovative	May delegate too much
Logical	May resist hunches
Challengers	Dislike rules, policies
Highly rational	May appear blunt

Directive and Negotiative

Strengths:	Weaknesses:
Goal controlled and friendly	Mildly too sensitive
Single group focused	Dislike multiple projects
Highly systematic in groups	May organize too much
Practical and personal	May become ambiguous
Managers	Torn between rule and people
People organizer	May become too involved

Examples

Directive:	Directive/Intuitive:
Adolf Hitler	Moshe Dayan
Henry Ford II	Teddy Roosevelt
Billy Martin	Roy Disney, Jr.
Khomeini	Pope Julius
William Jennings Bryant	Horace Greeley
Epimetheus	

Directive/Meditative:	Directive/Negotiative:
Charles DeGaulle	Franklin Delano Roosevelt
Dwight D. Eisenhower	Ronald Reagan
Donald Peterson	Pope Paul VI
Moamar Khadafy	King Arthur
Herman Goering	Mao Tse-Tung

Source: R. A. Black, "Fact, Creativity, Teamwork, and Rules: Understanding Leadership Styles," *Industrial Management,* September–October 1990, pp. 17–20.

FIGURE 13–9
Leadership Skills

Task-Oriented Characteristics
1. Need for achievement.
2. Drive for responsibility.
3. Responsible.
4. Enterprise initiative.
5. Task orientation.

Social-Oriented Characteristics
1. Ability to enlist cooperation.
2. Administrative ability.
3. Cooperativeness.
4. Attractiveness.
5. Nurturance.
6. Popularity, prestige.
7. Sociability, interpersonal skills.
8. Social participation.
9. Tact, diplomacy.

Adapted from Ralph Stogdill, *Handbook of Leadership: A Survey of Theory and Research* (New York: Free Press, 1974).

Dos and Don'ts

• Experiment with leadership styles. With whatever style you choose, infect your workers. Let them in turn become leaders, not followers.

• Determining a style of leadership based upon your personality is easy. But it may not be the best style for the tasks at hand. If a style seems ill-suited, but necessary, tell yourself you'll act the part. By acting like something, and looking like something, we may become that something. Studying biographies of great people may be inspiring, but be vigilant to the one dimensional biography. Most people are far more complex than a simple biography can detail.

• Leadership in a charismatic sense may be overvalued as an asset in establishing Quality Management. The leader who demonstrates commitment to the guiding principles and knows when to get out of the way is far more important to the health of an ongoing organization than a magical, charismatic leader.

STRATEGIC PLANNING

What people say you cannot do, you try and find that you can.

Henry David Thoreau

Aliases

Hoshin planning, defining a new vision.

Brief Description and Purpose

Strategic planning defines the strengths, weaknesses, opportunities, and threats during the next one to five years, and initiates a strategy that can be implemented into daily operations.

Discussion

Without a mechanism to watch for upcoming threats and opportunities, an organization may be blindsided when faced with a threat or opportunity. Strategic planning formalizes the looking ahead process. How far ahead? Matsushita claims a 100-year plan, as does Royal Dutch Shell (who thinks that gasoline will not be needed 100 years hence). Most authors favor looking ahead two to three years. The Hoshin planning system described in the "Process" subsection includes a long-range vision (five years) and a shorter range plan (one year). This is short enough to be implementation oriented, but farsighted enough to observe upcoming impacts.

Planning should be integrated with goal setting, and individual initiative. This is in direct contrast to planning in autocratic organizations, which dictate plans and goals. These organizations are ignoring the talent within. Planning, to be effective, must be blended with daily operations.

Planning requires that one eye be cast to improvement, and one eye toward the status quo. Organizations must do the things that they regularly do well, as well as the new improvements mandated by the planning.

Process

Robert King describes the Hoshin planning system illustrated in Figure 13–10. The three internal actions are plan, audit, and execute, which could be recast into plan, do, check, and act, (Chapter 17, section "Deming Cycle") with the action being a revision of the planning stages. Note the heavy emphasis on feedback mechanisms, such as the monthly audit. This audit is performed by the individual(s) implementing the plan.

FIGURE 13–10
Hoshin Planning

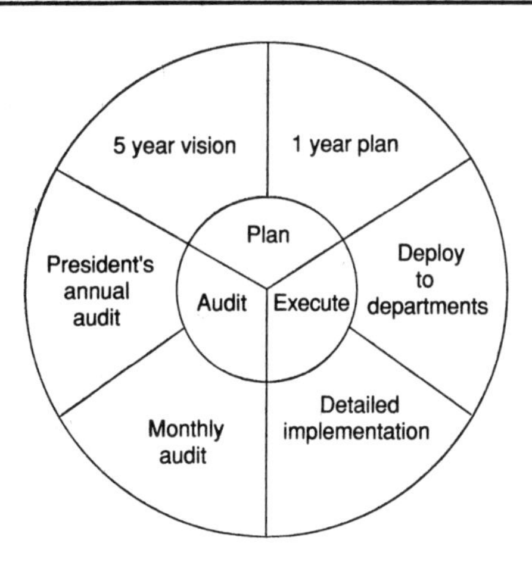

Source: Adapted from Robert King, "Hoshin Planning: The Foundation of Total Quality Management," *ASQC Quality Congress Transactions*, 1989.

1. Formulate the Plan. In formulating the plan, a force field analysis may be used to collect information regarding the present and projected business climate. This force field would analyze the strengths and weaknesses of the organization and/or its products, and identify threats and opportunities. A quality function deployment diagram (Chapter 17) may be necessary. Research required into this planning certainly will be ongoing and may be continuous.

2. Deploy to departments. The plan must now be communicated to all departments involved. Departmental goals and objectives must be set in accordance with the plan and individual strategies must be established. Feedback at this stage may be desirable to refine the plan prior to its final initiation.

3. Implementation. Strategies must be implemented. A well-designed quality function deployment effort smoothly translates planning goals into production techniques.

4. Audits. Feedback must be given regularly to evaluate the progress and make midcourse corrections. Each individual, as well as each department, must conduct a self assessment. Results of the audit should be communicated to the work force and reflected in the updated plan (see the tools in Chapter 18, Self Examination).

Dos and Don'ts

• Employee commitment is essential. Without communicating the need for improvement, employees cannot effectively implement the change. Planning must have the attention of the highest level of management.

• Examine the planning schemes of companies unrelated to yours. What threats are they bracing for? What new opportunities do they foresee?

• Don't be surprised by technology. Assign a staff member to periodically research and assess new technologies that appear in the scientific and technical literature. A dedicated technology assessment officer is useful to separate the feasible from the possible from the improbable.

• Study your best competition. What directions are they moving in? What niches are they leaving unfulfilled? What niches is your company ignoring?

CHANGE MANAGEMENT

Change is debilitating when done to us, but exhilarating when done by us.

Rosabeth Moss Kanter

No more good must be attempted than the people can bear.

Thomas Jefferson

Aliases

Transformational leadership.

Brief Description and Purpose

The ambition of the art of change management is to transform the organization with as few shock waves as possible. Change is the fulcrum on which the success of a Quality Management program rests. This art demands great attention to the individual psyche, for without careful guidance, anger and resentment are the inevitable result. Change is not brought about by directives from top management, but by direct communication. Some studies suggest that after translation from the vice president, director or division head, manager or branch chief, and supervisor, the employee receives about 20 percent of the information. Therefore change management, to be effective, must be brought about organically, on an individual or nearly individual basis.

Discussion

Change is probably so difficult to accomplish because it is against our nature as a species to change. Humans existed for tens of thousands of years in much the same way. It was not until large, fortified, stable cities were invented that continuous, riotous change became the norm. Then change was necessary to resolve overcrowding, famines, and armed conflicts. The current business climate changes quickly, and the rate of change increases with time. In the computer industry, a new "generation" evolved every 5 to 10 years in the beginning. Then this period shortened to two to five years, then one to two, and conventional wisdom now places this interval at about 10 to 12 months. The consumer electronic industry has new generations of features and equipment about every three to nine months. Perhaps the ground that the United States lost to Japan in this area was not due to deficits in technical proficiency or productivity, but rather to reluctance to implement change as an ongoing, and manageable process. Another (and rather facetious) indication of U.S. inability to cope with change as a predictable, workable process is the bumper sticker declaring "Stuff Happens." We need to change this attitude to "I Make Stuff Happen."

Psychologists and organizational behavior specialists are beginning to unravel the change knot, but at present there are a myriad of theories to select from. However almost all of these have elements in common. We will examine one of these theories which seems typical, and then propose a generic, implementable process. One model may not fit any situation, and hybrid models may be necessary to fit a given organization. Figure 13–11 compares three (and by no means all) models. The Buckly/Perkins models and the Lewin model will be discussed in detail. Juran's model is self-explanatory from the titles of the processes.

Buckly/Perkins Model

Karen W. Buckly and Dani Perkins make a distinction between transformation and change as they discuss a process to accomplish "transformative change." They define transformation as "a profound fundamental change in thought and action which creates an irreversible discontinuity in the experience of a system." Change, on the other hand, is defined as "the modification of beliefs, behaviors, and attitudes. Change is moving to another location on the same floor. Transformation is moving up a floor." Transformation involves large numbers of people and significant segments of an organization. These authors present a seven-stage model through which an organization must move to achieve successful transformative change. They acknowledge that movement through the model may not be smooth, that some regression to a previous step may occur and that the time required to traverse the model will vary widely. The seven stages are shown in Figure 13–11.

In the *unconscious stage* the organization is beginning to experience some discomfort with the way things are. Something is not right. The initial

FIGURE 13–11
Change Process Models

Juran	*Lewin*	*Buckly/Perkins*
Breakthrough in attitude	Unfreeze	Unconscious
↓	↓	↓
Pareto analysis	Change	Awakening
↓	↓	↓
Mobilize for breakthrough in knowledge	Refreeze	Reordering
↓		↓
Establish steering arm		Translation
↓		↓
Establish diagnostic arm		Commitment
↓		↓
Resistance to change		Embodiment
↓		↓
Breakthrough in performance		Integration
↓		
Transition to new level		

warning signs are being observed but not connected. As the organization becomes aware of the need for a change, it enters the second stage awakening.

In the *awakening stage* the need for change crystallizes. Satisfaction with the status quo is disrupted. The rate of awakening varies with conditions in the environment, the extent of the acceptance of the need for change within the organization and the willingness of the people to move to the next stage, reordering.

The *reordering stage* begins the analysis process of collecting information on the gap between where the organization is currently and where it should be. If the need is for a transformative change, involvement of the employees likely to be impacted is crucial.

The *translation stage* encompasses the integration of the information, analysis, and concepts developed in the first three stages. This integration leads to a vision of what the organization wants to achieve, the organizing and setting directions toward that goal.

The *commitment stage* occurs when the organization and the individual stakeholders decide whether or not to sign up for the initiative. If people don't sign up (i.e., commit themselves to the need for change), the organization will not move and the initiative will languish and die. At this stage, management must assess the current likelihood of success of the change effort. If it is determined that the organization and the individuals involved are prepared to proceed, the next stage is appropriate. If they are not yet prepared, then further groundwork is required.

If the decision is to press on, the *embodiment stage* is entered. This stage includes the challenge of managing the interaction of three elements: (1) consciousness shifts—attitudes, beliefs, and assumptions; (2) structural changes—patterns of organization; and (3) behavioral changes—actions, relationships, and communications. This stage is filled with experimentation, trial and error, and a start/stop tempo without which lasting results will not be achieved. This is the transition period. It could also be characterized as the implementation period when the plans of the first five stages are made operational and refined into workable form.

The final stage is *integration*. "A supportive environment of trust, cooperation, and openness is developed." The turmoil of the embodiment stage is calmed and a new peak level of performance is achieved. All elements of the formal and informal organization are essentially working together. Fine tuning of the new organization is continuing but the change has become the accepted mode of operation. This completes the transformation process and the organization is again entering the *unconscious stage*.

The authors characterize this process as one of fixing something defective or supplying something that is missing.

Lewin Model

Kurt Lewin views change as a three-step process: unfreezing, changing, and refreezing.

The objective of *unfreezing* is to prepare the work force to accept a major organizational change. Unfreezing clarifies the need for change. It opens up the organization to critical review, introducing the concept of continuous process improvement as a matter of organizational policy. Once involved in the process of self-analysis, people within the organization begin to feel that they own part of the problem. They then become less resistant to change. The unfreezing phase provides alternatives for improvement.

The next step is to effect the *change*. The cornerstone for a successful change was laid when employees throughout the organization participated in defining where the changes should take place. Now the specific changes must be communicated to the entire work force, understood, and accepted. The change is often initiated through a new policy on quality, which addresses many of the issues uncovered in the unfreezing process. Once change is established, the process itself consists of a series of initiatives to make the new policy a reality. Management must ensure that all employees are afforded ample opportunity for training, education, and individual development, an opportunity to participate in control of the process, and a chance to excel.

With the new organizational culture in place, the change becomes part of the daily routine. The new behavior—continuous process improvement in the case of Quality Management—is internalized by the work force. Productivity, efficiency, and effectiveness are at a higher level than before the change took place. Managers must now create and maintain an environment which constantly reinforces the change, or *refreeze*.

Summary

While no management model will perfectly fit your organization, each provides a structure to guide you through the change effort. They enable you to ask the right questions and to obtain answers as to how to proceed. Note that all of these processes assume that change is to be brought about as a result of commitment rather than compliance. Organizations today cannot afford to alienate their employees by demanding compliance (often colloquially referred to as *golden handcuffs*). These handcuffs tend not to work on the baby boom generation, and they certainly do not work on the baby buster (born after 1965) generation. Figure 13–12 sketches the two approaches.

Relation to Other Concepts and Tools

Introducing any sort of change, whether or not it is dramatic to the supervisor, may have a sizable impact on the employee. Changing desks, for example, may be trivial to the supervisor, but not to the employee. It is a mark of status, and changing prestige is always a dramatic change. Thus it requires employee involvement (Why should the manager care where they sit?), a knowledge of organizational culture, leadership ability, planning, and may encompass virtually all of the tools in Part 3.

Process

The process presented here is a hybrid model of Lewin's concepts and Kirkham's 24-step process.

Three roles are involved in this process for managing change:

Sponsors, who are the individuals or group who authorizes or are empowered with the change.

Implementors, who set the changes in motion.

Users, who are directly impacted by the change and/or are expected to achieve the desired change.

The pace at which change can take place can be categorized in four ways:

All-at-once. Both social and technological changes are brought about immediately. This method induces an enormous amount of stress. Liker

FIGURE 13–12
Compliance versus Commitment

Change by Compliance	*Change by Commitment*
Focus on the employee	Focus on the improvement desired
Responsibility is on the employee's supervisor	Responsibility is on the user of the changes
Communication is at an employee	Communication is with an employee
Management is authoritarian	Management is collaborative

describes an example of this type of chaotic change at Basic Industries, where even some of the key players were so disturbed that they left.

Technical systems first. In this scheme technical systems are changed and then the human resources aspect is changed.

Social systems first. Here the human resources aspects are changed first. Zygmont discovered that when Deere & Company implemented flexible manufacturing systems, changing social systems first obviated some changes in technical systems. Since changing human resources is often less expensive than technological change, the advantages to this scheme are obvious.

Staged approach. Change is implemented gradually. In computer-integrated manufacturing, for example, "islands of automation" (separately automated work cells with no links to other work cells) were common during the mid- to late-1980s. These islands are now becoming integrated as sophisticated software and standards emerge. Due to the extraordinary complexity of integration, it is best to focus on the integration aspects alone, without worrying unduly about each station operating properly—this has already been done during the island stage.

Sponsorship Phase (Unfreeze Phase)

1. Identify the need for a change. Establish credibility of the sponsors. Sponsors must be keenly aware of past experiences, the success or failure of similar changes in the company, and the attendant reasons for success and failure. The sponsors must also eliminate all indecision.

2. Define the what-how-why-when-and who. Determine how long the change will take. Decide how you will know the difference after the changes will take place and how you can measure success. Identify implementers and users. In developing a strategy for planned change, the following should be considered:

Environmental factors: These may include economic, financial, social, and related factors.

Climate: This includes the way in which the organization functions, the basic communication process, decision making, and goal-setting processes, motivation, control, and reward.

Management characteristics: The concern here is with the style of management used to operate the organization. Whether managers are participative, autocratic, or situational can have a tremendous impact on the success of a Quality Management implementation. Flexibility is the key phrase here—but blind acceptance should not be expected. It is healthy to question change in order to understand and accept it.

Inter-work group relationships: This refers to the specific ways in which the groups within the organization work together, how effectively they function, and types of conflict that occur.

Organizational practices: This looks at the factors that tend to help or hinder effectiveness. Specifically, it requires a look at policies related to the human organization, such as promotion policies, transfer policies, developmental policies, and the like.

Once the current organizational culture is identified and the ultimate goals of Quality Management are established, it is essential to determine and implement methods to measure whether or not the change intervention is working. A series of measurements over an extended period of time is needed.

Implementation Phase (Move Phase)

3. Consider how to announce the changes. Be careful of rumors bouncing about. The individuals directly involved should know as soon as possible. Ideally the users should be the sponsors.

4. Promote correct interpretation of changes. Teaching foreign nationals and small children has taught us that we assume a tremendous amount of background knowledge. Care must be taken to ensure appropriate interpretation.

5. Determine if the users are able to implement the changes. Do they have the requisite knowledge, tools and skills? If not, how will they obtain them?

6. Decide if the users are prepared psychologically for the change. This is a time of low stability and high emotion, often unfocused. To counteract this, lines of communication must be open, and they must be honest, even if the answer is "I don't know." An answer of I don't know followed by "Let's find out" starts a powerful process. Use Socratic dialogue (Chapter 12) to draw out fears and hidden reservations.

7. Prepare for and overcome pockets of resistance. There are individuals in every organization who harbor corrosive feelings about the organization. Defuse them as early as possible. Make them a sponsor, if practicable. Use peer pressure to reinforce changes. Some of these individuals may be used as early warning devices for problems in the change process. In any event, don't discount them or ignore them, unless you intend to terminate them.

Some strategy and tactics on conflict management will help (see also Chapter 15, the section on "Stress Management"):

- Focus on one issue at a time.
- Choose the "combat zone" carefully.
- Don't overreact to unintentional remarks, or take remarks out of context.
- Don't be too quick to judge or react. Settling disputes before going to bed may be good advice for married couples, but it may be too simple for business decisions affecting many people's livelihoods.
- Don't corner someone. Leaving a business opponent or employee with nowhere to turn creates a devastated employee who may be bent on revenge.
- Agree to disagree. There may be more than one way to accomplish the same task.
- Maintain a sense of humor and perspective. A colleague of ours is fond of saying that there is no long-term solution—in the long term we're all dead. In this dark humor lies a sense of perspective. Remember that a job is just a part of life, perhaps the least part, when compared to friends and family. Carrying a grudge against a work colleague is silly and self-destructive.

Usage Phase (Refreeze Phase)

The new order of business is now the established order. Assess the impact of the change. Prepare a lessons learned memo so that future action groups can learn from your experience.

Dos and Don'ts

Figure 13–13 contrasts the vision with reality. While at times it is unavoidable for visions to degenerate, knowing what they can degenerate into can be a useful lesson in avoidance.

- Any change effort in which changes in individual behavior are required, regardless of initial focus, must include means for ensuring that such changes will in fact occur. Saying that it will be done, does not make it so.
- Organizational change is more likely to be met with success when key management people initiate and support the change process. It requires a hands on approach.
- Organizational change is best accomplished when persons likely to be affected by the change are brought into the process as soon as possible.
- Successful change is not likely to occur following the single application of any technique.
- Successful change programs must rely upon informed and motivated persons within the organization if the results are to be maintained.
- No single technique or approach is optimal for all organizational problems, contexts, and objectives. Diagnosis is essential.
- We aren't all winners. Sometimes implementing change requires that some people lose prestige, status, control or even their position entirely. Termina-

FIGURE 13–13
Vision versus Reality in Change Management

Beautiful Vision	*Brutal Reality*
"No one will lose their job."	"We'll help you with your outplacement status."
Training for everyone.	Only one class held—during lunch.
Work force has vision statements emblazoned on laminated wallet cards.	Work force uses the card for a bookmark.
Profit sharing implemented.	What can you buy for 10 cents?
Wealth of benefits to select from.	Benefits book and form require a law degree to read.
Increased communication with the employees.	More useless meetings.
Management communication by walking around.	Management by getting in the way.
Best consultants money can buy.	Consultants don't have to live with the mess they create.

FIGURE 13–14
Reasons for Resistance to Change

Knowledge and skills could become obsolete.
Fear of job or other economic loss.
Individual or group negatively impacted by the proposed change.
Ego defensiveness (a change may imply prior failure).
Comfort with the status quo.
Peer pressure.
Lack of information.
Limited information/perspective.
Too little time to adapt.

tions brought about by change should be done as simultaneously as possible. Assure those not terminated that when the firing stops, it is not their turn. Nobody said Quality Management was easy, or that implementing discipline in a modern environment was simple.

• Don't forget what was done right previously. Change can be exciting, but don't lose sight of procedures that worked well before. Are those doing things the 'old way' left unsupported?

• Cultures are changed and transformed, not removed. If a culture is destroyed, and one is not put in its place, an undesirable culture may form.

• Some reasons for resistance to change are outlined in Figure 13–14.

• Start change with small wins. implement changes in a "lead-the-fleet" department or division. Once the bugs have been worked out, begin exporting this model elsewhere. Rely on inter-plant or division rivalry: "If the guys in Yuba City can do it, we can do it too."

• Sometimes the best of visions and plans does not work out in reality. Feedback throughout the process is critical for a successful outcome. It may be necessary to make course changes midstream.

BIBLIOGRAPHY

Corporate Culture

Davis, Stanley M. *Managing Corporate Culture*. New York: Harper & Row, 1984.

Deal, T. E., and A. A. Kennedy. *Corporate Cultures*. Reading, Mass.: Addison-Wesley Publishing, 1982.

Kanter, Rosabeth Moss. *The Change Masters*. New York: Simon & Schuster, 1983.

Rodgers, Buck. *Getting the Most Out of Yourself and Others*. New York: Harper & Row, 1987.

Sashkin, Marshall. "A Theory of Organizational Leadership: Vision, Culture and Charisma." *Proceedings of Symposium on Charismatic Leadership in Management*. Montreal, Que.: McGill University, 1987.

Schein, Edgar. *Organizational Culture and Leadership*. San Francisco: Jossey-Bass, 1985.

Uttal, B. "The Corporate Culture Vultures." *Fortune* 108, no. 8 (October 17, 1983), pp. 66–72.

Webster, Cynthia. "Toward the Measurement of the Marketing Culture of a Service Firm." *Journal of Business Research* 21, no. 4 (December 1990), pp. 345–62.

Leadership

Bennis, Warren, and Burt Nanus. *Leaders: The Strategies for Taking Charge*. New York: Harper & Row, 1985.

Black, R. A. "Fact, Creativity, Teamwork, and Rules: Understanding Leadership Styles." *Industrial Management,* September/October 1990, pp. 17–20.

Bothwell, L. *The Art of Leadership*. Englewood Cliffs, N.J.: Prentice Hall, 1983.

Hersey, Paul. *The Situational Leader.* New York: Warner Books, 1984.

Kiechel, W. "Visionary Leadership and Beyond." *Fortune*. July 21, 1986, pp. 127–28.

Kouzes, James, and Barry Posner. *The Leadership Challenge: How to Get Extraordinary Things Done in Organizations*. San Francisco: Jossey-Bass, 1987.

Levinson, H., and S. Rosenthal. *CEO: Corporate Leadership in Action*. New York: Basic Books, 1984.

Manz, Charles, and H. P. Sims. "Leading Workers to Lead Themselves: The External Leadership of Self-Managing Work Teams." *Administrative Science Quarterly* 32, no. 1 (1990), pp. 106–29.

McGregor, Douglas. *Leadership and Motivation*. Cambridge, Mass.: MIT Press, 1983.

Schein, Edgar. *Organizational Culture and Leadership*. San Francisco: Jossey-Bass, 1985.

Stogdill, Ralph. *Handbook of Leadership: A Survey of Theory and Research*. New York: Free Press, 1974.

Tannenbaum, Robert, and Warren H. Schmidt. "How to Choose a Leadership Pattern." *Harvard Business Review Classic*, May/June 1973, pp. 162–80.

Trice, Harrison M., and Janice Beyer. "Cultural Leadership in Organizations." *Organization Science* 2, no. 2 (1991), pp. 149–69.

Westley, Frances, and Henry Mintzburg. "Visionary Leadership and Strategic Management," *Strategic Management Journal* 10, no. 3 (1989), pp. 17–32.

Strategic Planning

Albert, K. J. *The Strategic Management Handbook*. New York: McGraw-Hill, 1983.

Gale, Bradley T., and Robert D. Buzzell. "Market Perceived Quality: Key Strategic Concept, Planning Review." *International Society for Planning and Strategic Management,* March/April 1989.

Juran, J. M., and Frank M. Gryna, Jr. *Quality Planning and Analysis*. New York: McGraw-Hill, 1980.

Juran, J. M. *Juran on Planning for Quality*. New York: Free Press, 1988.

King, Robert E. "Hoshin Planning, The Foundation of Total Quality Management." *ASQC Quality Congress Transactions*, 1989.

Thompson, P.; DeSouza, G.; and B. T. Gale. "The Strategic Management of Service Quality." *PIMSLETTER, No. 33*, Cambridge, Mass.: Strategic Planning Institute, 1985.

Willborn, Walter. *Quality Management System: A Planning and Auditing Guide*. Milwaukee, Wis.: ASQC Quality Press, 1989.

Change Management

Ackoff, Russell. "The Management of Change and the Change It Requires of Management." *Systems Practice* 3, no. 5 (1990), pp. 427–40.

Bennis, Warren. *The Planning of Change*. New York: Holt, Rinehart & Winston, 1976.

Buckly, Karen W., and Dani Perkins. *Managing Complexity of Organizational Transformation*. Alexandria, Va.: Miles River Press, 1984.

Hersey, Paul, and K. H. Blanchard. "The Management of Change." *Training and Development Journal* 26, no. 1 (January 1972); 26, no. 2 (February 1972); 26, no. 3 (March 1972).

Kanter, Rosabeth Moss. *The Change Masters*. New York: Simon & Schuster, 1983.

Kanter, Rosabeth Moss. "Managing the Human Side of Change." *Management Review*, April 1985, pp. 52–56.

Kirkham, Roger L. "How to Manage Changes Coming from Tight Times." *Federal Manager's Quarterly*, January 1987, pp. 30–37.

Lewin, K. "Group Decision and Social Change." *Readings in Social Psychology*. New York: Holt, Rinehart & Winston, 1958, pp. 197–211.

Liker, Jeffrey K.; David B. Roitman; and Ethel Roskies. "Changing Everything All at Once: Work Life and Technological Change." *Sloan Management Review*, Summer 1987, pp. 29–47.

Pondy, L. R.; R. J. Boland, Jr.; and H. Thomas. *Managing Ambiguity and Change*. New York: John Wiley & Sons, 1988.

Recardo, Ronald. "The What, Why and How of Change Management." *Manufacturing Systems*, May 1991, pp. 52–58.

Schultz, Louis E. *The Role of Top Management in Effecting Change to Improve Quality and Productivity*. Minneapolis, Minn.: Process Management Institute, 1985, pp. 1–8.

J. Zygmont. "Flexible Manufacturing Systems: Curing the Cure-All." *High Technology*, October 1966, pp. 23–24.

CHAPTER 14

SUPERVISORY LEVEL DYNAMICS

The beatific and occasionally amorphous visions of upper management must be transformed into work-a-day goals, plans, specifications, and guidance. This is left to the middle and supervisory levels of management. Supervisors are caught between satiating the needs of both top-level management and their employees, often pleasing no one. The practices described in this chapter help the supervisors guide their employees to their peak level of performance for the organization. This requires a long-term commitment and focus. Payback may not be immediate, and any immediate payback that is seen may not last (experiments have shown that introducing *anything* new to the work force may briefly enhance productivity).

Enhancing productivity means more than producing more widgets per hour. It is allowing the work force to determine if these widgets are worth making, at any price, and if so, how should they be made, what their design is, and who should use them and how. This view of the employee imbues them with intelligence and the ability to participate meaningfully. There are many examples of companies doing precisely this in the United States and around the globe.

MOTIVATION

Capacities clamor to be used.

Abraham Maslow

Aliases

Performance enhancement, productivity improvement.

Brief Description and Purpose

Motivation is a challenge common to all supervisors as well as staff. What is motivation? It can be viewed as a force that moves one toward an objective. The key elements, then, are the *force* and the "objective." In terms of Quality Management, it is crucial that the motivation force come from within each individual of the organization. If an objective is defined in a manner that makes it important to an individual, that individual will want to accomplish the objective.

Discussion

There are many sources and theories for employee and self-motivation. Much has been written on the subject due to the perceived bottom-line impacts of improved staff productivity. To successfully motivate an individual is to understand something about that person; each employee is unique, therefore it is important to structure motivation techniques to the individual's situation and personal goals.

Motivation can be viewed as an attitude. Attitudes are often self-fulfilling prophecies. Therefore, we must be careful with the force part of motivation. Remember, the force should be from within for long-term success. Successful motivation results in a positive work attitude; abuse (or fear of it) is not a successful management technique. Management by fear techniques may work over the short term, but the long-term impacts can be fatal to an organization. All people need to feel valued and valuable. Value does not necessarily imply money (though money is one indicator of an employee's value to the organization), but rather the concept that your ideas are important and that you make a difference.

Successful organizational changes are brought about by individuals. These individuals must clearly understand their organizational role in the change process, as well as the dedication to accomplish the organization's goals. This dedication correlates to motivation. Employees (at all levels) need to know "What's in it for me?" and "Why is this important?" in order to be spurred into action that leads to change.

Relation to Other Concepts and Tools

Motivation requires an inquiry into the current factors affecting performance (both positive and negative). Some sort of self or group reflection might be useful in this regard, as behavior may only be a symptom or manifestation of a root cause (weight loss counselors have a saying: "It's not what you're eating—it's what is eating you''). Performance management is needed to ensure that goals have been met, and to set new goals. Brainstorming and a subsequent cause-effect analysis, root cause analysis and/or Pareto analysis may be apropos. Traditional behavior modification techniques for food and drug addictions are also related to employee motivation, and such ''12 step'' therapies may be modified for office use for curing serious behavioral problems.

Process

1. Determine the characteristics of the job. Some jobs operate on commission. Some are salaried with a bonus tied to performance. Some jobs have much travel. Each job has a set of characteristics that it requires of the person filling it. Recognizing these qualities is the first step in selecting the right person for the job, or for correcting problems. A convenient way to analyze a job is to scale it relative to two factors: cognitive and affective. The more cognitive the job, the less people contact it has, and the more mental skill and knowledge it may require. Affective jobs require people skills, including cold calling, face-to-face meetings with clients, and public speaking. Some require a combination of these skills, such as teaching.

2. Determine the motivational context of the employee. A match must be made between the job characteristics and those of the employee. The employee factors may change throughout their career. While employees have young children, they are unlikely to desire a great deal of travel. Employees with high school age children may be unwilling to sever the ties their children have made at this fragile time. Divorce dramatically and unpredictably alters employee behavior and motivations. Each situation is different. An overall corporate culture provides a motivational basis and flavor, but when examining individual performance, a case-by-case analysis is needed.

3. Identify discrepancies between the job, the individual, and management. The influence of supervision needs to be examined as well. Are the wrong messages being sent (i.e., the boss often saunters in late but censures subordinates who do so)? Is the culture inappropriate to the reality? A checklist for determining various motivational factors is provided in the ''Dos and Don'ts'' section.

4. Create and implement a strategy for motivation. Strategies for behavior modification take on one of four characteristics:

1. Positive reinforcement.
2. Negative reinforcement.
3. Punishment.
4. Extinction.

Positive and negative reinforcement seek to increase the frequency of the behavior it is trying to modify. Punishment seeks to decrease the frequency. Extinction, or termination, decreases the frequency by eliminating the employee. Punishment is usually not effective (witness the high recidivism rate among prisoners in the United States). Negative reinforcement is also not as effective as positive reinforcement. Any strategy other than positive reinforcement may be viewed as management by fear. Philip Grant's work is an excellent starting point for developing a detailed exploration into motivation causes and cures.

An alternative is to radically alter the way work is organized. Establishing self-managing teams for example, where problems such as tardiness, who sits where, who does what, and who can play what radio station are determined by the team, leaving management free for planning and aiding the team in unusual problems. Self-managing teams allow peer pressure and the human need for conformity to motivate employees, eliminating or reducing the corrosive Us versus Them mentality.

EMPLOYEE COMMITMENT

Tell me and I forget
Show me and I remember
Involve me and I understand.

Confucius

Aliases

Employee involvement, participative management, team building, employee motivation, sense of ownership.

Brief Description and Purpose

The finest tools and equipment are useless without good people. Participation or involvement is crucial to fostering a sense of ownership in a job or task. With ownership, pride of workmanship nearly always follows. Involvement and participation lead to the commitment keystone of Quality Management and continuous improvement. Without employee commitment, all of the tools and techniques will not bear fruit, and a stifling status quo will reign.

The focal point for quality is not found in the traditional quality functions. The responsibility and accountability for quality in every process, for each product and service, is shared by employees throughout the organization. The role of quality professionals becomes one of mentor and adviser, chartered to ensure that everyone understands what is expected of him or her in order to meet the total quality challenge.

Do List[1]

Skills

- ☐ Does the employee have the skills and training necessary for the job, including affective domain and cognitive domain skills?
- ☐ Have these skills been updated or regularly honed?
- ☐ Do employees help establish their performance goals?
- ☐ Is the organizational environment conducive to providing help and assistance when needed?

Self-esteem

- ☐ Does the employee understand their importance to the organization?
- ☐ Are facilities adequate?
- ☐ Is enough authority to do the job given?
- ☐ Are feedback and evaluation provided?
- ☐ Is performance appraisal based on individual improvement, not comparison with others?

Reinforcement

- ☐ Rewards are known a priori, and could be attained by anyone.
- ☐ Employees are not socially isolated.
- ☐ Reward for specific acts.
- ☐ Allow discretionary ("free") time. Increase it if the time is used productively.
- ☐ Reward only for work-related acts.
- ☐ Employees set standards for rewards.
- ☐ Devise rewards other than money.
- ☐ Praise in public, censure in private.
- ☐ Employees must perceive that awards are fair and worthwhile.
- ☐ Unacceptably low performance is dealt with swiftly and decisively.

Support structure

- ☐ When possible, allow employees to adapt tasks to personal energy rhythms.
- ☐ Provide incentives to exercise and watch their dietary needs.
- ☐ The cafeteria provides healthy choices at a reasonable cost.
- ☐ Adequate vacation time is provided, and employees take adequately long vacations (at least a week).
- ☐ Ensure the work environment is as safe as possible and as comfortable as needed.
- ☐ Periodically alter the pace of work.
- ☐ Cultivate a social system that will promote the venting of frustrations.
- ☐ Value worker mistakes as much as possible.
- ☐ Rotate jobs and/or change the environment periodically.

Minimize outside distractions

- ☐ Discourage excessive, time-consuming and unrelated hobbies, civic activities, and outside business interests.
- ☐ Provide employees with help dealing with family problems.
- ☐ Discourage excessive involvement in nonwork-related clubs and associations.
- ☐ Help workers locate within a reasonable commuting distance of their place of work.

- ☐ Avoid hiring workers with excessive interest in community politics.
- ☐ Prevent work goals, and the number of different tasks assigned the employee, from mushrooming.
- ☐ Design work areas to minimize socializing.

Don't List[1]

- ☐ Daily deadlines without explanation or planning.
- ☐ Changes in schedule without rationale.
- ☐ Impromptu meetings without subject disclosure.
- ☐ Public reprimands.
- ☐ Lack of public praise.
- ☐ Surveillance of daily activities.
- ☐ Short impractical milestones.
- ☐ Have no assigned priorities.
- ☐ Give short, unnecessary suspenses.
- ☐ Give work only to the best people, undertask others.
- ☐ Blame upper management.
- ☐ Do employee's work, instead of delegating or managing.
- ☐ Leave actions until suspenses are too late.
- ☐ Be impersonable and discourteous.
- ☐ Let support staff delay final product completion.

[1]Adapted from an unpublished list by Jim Wasson, U.S. Army, Huntsville, Alabama.

Discussion

A quip quickly illustrates the difference between involvement and commitment. A chicken proposed to the pig that they give the farmer her favorite meal for her birthday: ham and eggs. The pig replied that while the chicken was involved, total commitment was required by the pig.

While involvement is often espoused and would appear to be an end goal, it is really an intermediate step, with the real end goal being commitment from the work force. This commitment comes through empowerment, a natural evolution of the involvement process.

Relation to Other Concepts and Tools

Employee involvement is closely related to managing change, and multifunction teams. It requires motivation and performance management to continue to involve employees at their highest levels of accomplishment.

Process

Implementation of employee involvement programs often invokes one or more of the following elements:

Senior management involvement. Top management is the only management layer that can really empower workers. All authority comes from this level and, unless it is delegated properly, employee involvement will be a hollow phrase.

Ownership/Stakeholders. For some this concept means literal ownership, such as employee stock option plans (ESOP). What can be as effective, and not require such complex instruments, is to allow an employee to *own* a task, problem, or function. In a mail-order house, this may mean that the customer service representative can solve the problem without asking the supervisor. In the factory, this may mean allowing each cell to determine what constitutes acceptable quality, when to perform preventive maintenance, and when to order new or repair parts. Ownership requires trust on the part of management, and competence on the part of the employee. If only one concept is used from this list to empower employees, use this one.

Empowerment. Empowerment is a rather fancy word for trust, which is too scary a word for most managers to use. Without trust, employees are *us* and managers are *them* and the two become so involved in outmaneuvering and outscoring each other over trivial issues, that both lose sight that they work for the same company, presumably working under the same strategic vision. Empowerment also seeks to actively identify and remove barriers to doing a better job.

Rewards. We all like to get rewards, even when they are unmerited (unmerited rewards or forgiveness are termed *grace* by Christians, *wa* by the Japanese). Rewards should be in proportion for the work done, and they should be available to anyone in the company. Clear criteria for awards must be established, or else rewards will be regarded as "something for the apple polishers." Rewards need not always be monetary: trips to a conference, a course, and time to pursue a pet project are all ways to reward without a direct monetary bonus. They also indicate thought (don't you prefer getting presents over money?). Consider forgiveness as well. Destroy personnel records over three years old. Why should something somebody did 15 years ago when they were new have a bearing on their promotion now? If it was a devastating incident, either someone will remember, or it was not as devastating as originally thought.

Training, Cross-Training, Job Rotation. Cross-training and job rotation allow us to walk a mile in the other guy's shoes. It provides a superb background for all employees. This knowledge leads to empowerment by employees knowing why work rules exist (or why they shouldn't). Some companies (e.g., Johnsonville Sausage) have changed to a pay system whereby the more you train, the more you earn. Obviously the employee that can do two or more jobs is considerably more valuable than the one that can do only one. Do not exempt staff employees (including the CEO) from "getting their hands dirty" in production work. A few days on the line will give them an experience they will recall for quite some time. Harvey McKay relates how he regularly and unannounced, works at a station in his envelope factory. Even though the event doesn't get posted in the company newsletter, word of this always spreads quickly through the work force and makes a loud statement, nonetheless.

The United States Airforce has a reliability program devoted to this sort of activity called *Blue Two Visits*. A design engineer goes out on the flight line as a maintenance mechanic airman (who is usually of two-stripe rank, hence the name). This program has been responsible for many real-world lessons in design, and has resulted in numerous design improvements.

Communications. Empowerment requires knowledge, and no knowledge is gained without communication and experimentation. Since most workers are not originating knowledge, communication is the sole arbiter of knowledge. Give employees more information than you think they need. When they start asking questions such as "What does the financial report have to do with me?" tell them.

The extent to which employee involvement is practical and effective depends entirely on the following conditions:

- An open system of communication—formal and informal, both vertical and horizontal.
- Management style that is flexible and appropriate to specific situations—situational leadership.
- Organizational, work unit, and individual goals that are explicit and aligned, and progress that is reviewed and evaluated regularly.
- Clearly defined areas of responsibility for each employee.
- Minimal restrictive policies, procedures, work rules, and regulations—an effective system to remove barriers to efficiency and effectiveness must be implemented.
- Participation by everyone in goal setting, decision making, problem solving, setting-up procedures, defining responsibilities, and so on.
- Encouragement of creativity and initiative at all levels and appropriate rewards.

- An organizational environment characterized by flexibility and adaptability to change.
- Emphasis on individual worth and pride in workmanship, and work group relationships characterized by a high degree of trust and interdependence.
- Individuals responsible for their own behavior.
- Fair and equitable reward and recognition systems closely related to individual or work group performance.
- Emphasis on group cooperation and team building, and small group loyalties integrated with the greater goals and purposes of the organization.
- Freedom to experiment, to make mistakes, and to take risks.
- Management systems developed through collaboration and participation, people using the systems who understand and know how to use them; a feeling that management systems can be changed when they are not working or need improvement.

Dos and Don'ts

• Encourage "cheating." Our educational system, kindergarten through graduate school, goes to extraordinary lengths to undermine human cooperative spirit. Cooperative effort is almost never allowed. It is often labeled *cheating,* with serious penalty. It will take a great deal of training to undo the damage, especially since you are working with adults, who are slow to change. It could be beneficial to offer team building sessions at the elementary school level. Too few children of today's busy set take part in organized activities devoted to enhance team skills.

• Avoid commitment without involvement and participation. The U.S. school system requires a commitment from every parent and taxpayer, yet involvement is not encouraged, and is often actively discouraged. That this can result in a discordant situation is obvious from any U.S. news magazine article on education.

• Receive feedback, unvarnished, from employees concerning the employee involvement process. Management learning to trust employees is the hardest part of implementing Quality Management, and is at the heart of employee involvement.

• When apportioning rewards, be sure to include everyone involved. This means the people who make sure we have a clean and safe work area, and the vendors who juggled schedules to make us look good.

TEAM BUILDING

Two heads are better than one.

John Heywood

Aliases

Multifunction teams, cross-functional teams, process action teams, and quality circles.

Brief Description and Purpose

"Two heads are better than one" is an old phrase that has new life and auspicion in modern management practice. Teams and group techniques are proving to be a simple, highly effective tonic to organizational creativity, quality improvement, and quality of work life. Organizational development studies have shown the optimal size of teams to be about four to eight members. Multifunction or cross-functional teams are teams whose members cut across functional boundaries, such as a team composed of a member from accounting, purchasing, engineering, and production. Team members are often internal customers of one another.

Discussion

The initial group selected should include decision makers from key groups of stakeholders—those identifiable groups that will be impacted by the change and/or can impact upon the success of the change effort. These groups may include employees, peers, senior managers, customers, suppliers and staff support elements. This is not to imply that every possible group is represented. The selection process must be thoughtful, as this team is central to a fully successful change effort.

A key characteristic of the individuals selected should be that they are able and willing to make decisions/commitments for their respective groups. Further, these commitments are likely to be supported by actions necessary to make them happen. Many meetings are held where representatives are sent in lieu of the decision maker or attendees are invited because of misplaced protocol concerns. In both cases, they are not productive participants. In many cases they thwart progress due to their unwillingness or inability to commit their organization to a course of action. If you are seeking participation from outside organizations, be sure that the group member has sufficient authority to represent their organization.

A companion issue in building your planning team is to seek members who will actively participate. Showing up for the planning sessions is not enough. Each individual should have something to contribute and be willing to do so. Members should not be designated simply because they happen to have the available time, they should be selected based upon what they can bring to the group to achieve the objective. If the change is important enough to pursue, then consider adjusting other work load to ensure that the right people participate when needed in achieving the objective.

At various points in the change process, individuals may be added to the original planning group or additional subteams may be formed. It is important in change efforts that take a long time to complete, and to maintain some continuity in the planning and execution oversight group. Otherwise the organization can experience considerable disruption through constant shifting of direction.

The success of teams can be found documented in a variety of trade and business magazines. For illustration, one issue of *Fortune* revealed that:

- A Federal Express team found and fixed a $2.1 million billing problem.
- Aetna reduced the ratio of middle managers from 1:7 to 1:30.
- A General Mills plant in Lodi, California, increased productivity 40 percent by using self-managing teams.

Figure 14–1 illustrates some different types of teams, describing whether the primary action is in reaction to a situation, or proactive in outlook, who has membership on the particular team, whether the focus is long or short term, and an example is cited. Although many variations of team types exist, they all require communication, conflict reduction, and increased cohesion and commitment among group members.

Relation to Other Concepts and Tools

Team building in the United States requires a change process. Working in groups is almost totally alien to the educational experience in the United States, and 12 to 18 years of reward for individual performance (only) cannot be metamorphosed instantly. Leadership by example, employee involvement and participation, and team building pilot projects should be attempted first. Chapter 19 explores teams used as productivity tools.

FIGURE 14–1
Some Types of Teams

Corrective Action/Tiger Team:
 Reactive.
 Members primarily internal to company or division.
 Short-term, single-task focus.
 Example: Determine location of problem in systems integration project.

Quality Improvement/Quality Circle:
 Reactive and proactive.
 Members primarily internal to company or division.
 Long-term, multiple-task focus.
 Example: Examine methods improvements in purchasing department.

Focus Group:
 Reactive.
 Members primarily external to company or division.
 Short-term, single-task (possibly multiple).
 Example: Determine effectiveness of a new TV commercial.

Self-Managing:
 Reactive and proactive.
 Members internal to company or division.
 Daily operations.
 Example: Assembly unit in a manufacturing plant.

Process Action Team:
 Proactive.
 Members primarily internal to company or division.
 Short term, but perhaps regularly formed.
 Example: Determine automation needs for the next 12 to 18 months.

Process

1. Select the facilitator. The facilitator should not be the expert on the subject. In fact they may know little about the topic under study. This individual should know a good deal about organizational behavior. Selecting a team leader can be disastrous without adequate training. As a society we eschew group activity, and leaders (chairpersons) are often *stuck* with handling the entire task. Avoid picking a leader if you can. If you must, pick a leader that is not generally thought to be a *natural* leader: the quiet worker that nobody knows well, for example. These people are often full of surprises. Their modesty may be masking talent waiting to be unleashed.

2. Select the team. Ideally membership should be voluntary, but this may not be practical as teams first begin in your organization. Sometimes assignments must be made. Be sure that all team members are aware of the purpose and role. Be realistic. Time must be made for the team. It cannot be part of "other duties as assigned."

The team size should be in the range of four to eight members. It may be wise to avoid mixing beginners and superstars (depending upon the superstar). Then again, how do beginners grow into superstars without seeing first hand how the superstars act? Reserve this sort of mixing until a team mentality has been well established.

3. Work together. Teams must have ownership of the problem, believe it is important to solve the problem, and have the authority to do so. Otherwise the team will fail. Set ground rules early. Distribute the minutes, putting action plans and the like in writing. If possible, have a special room where items the team leaves will be undisturbed. The meeting room should have a blackboard and/or flip chart, and/or an overhead projector.

Anticipate the team ending and determine how the team will decide when it is done before they begin. Assess team performance by process criteria, not results criteria. Examine how well they function while a team. Do they resolve conflicts, work creatively, and trust each other? Some rules for the team to follow are suggested in Figure 14–2. Figure 14–3 establishes some ground rules for supervisors who organize and/or manage team performance. Characteristics of effective (and ineffective) teams are sketched in Figure 14–4. Figure 14–5 lists some common barriers to effective team

FIGURE 14–2
Ground Rules for the Team

Insist on careful time control. Start on time and end on time. Enough time allowed to get the work done and no more.

Be sure committee members are sensitive to each other's needs and expressions. People listen to and respect others' opinions.

Foster an informal relaxed atmosphere, rather than a formal exchange.

Ensure good preparation on the part of the chairman and committee members. Materials should be prepared and available.

Members are all qualified and interested. They want to be a part of the committee.

Interruptions are avoided or held to a minimum.

Good minutes or records are kept, so that decisions are not lost.

Periodically, the committee stops and assesses its own performance. Necessary improvements are worked out.

FIGURE 14–3
Ground Rules for the Supervisor

Do not assume that requisite skills exist in the group.
Give early and positive feedback on team performance.
Appraise the group as such, not individual members.
Encourage unplanned, informal meetings.
Be aware of divisiveness in the group.
Do not barge in on group meetings. Be invited or ask permission.
Let the team decision stand. Trust your employees.
Clearly define the role, purpose, objectives, and goals of the committee.

FIGURE 14–4
Effective and Ineffective Team Characteristics

Effective	*Ineffective*
Innovative and creative.	Complaints, confusion.
Commitment.	"I'll have to clear it" attitude.
Members highly independent.	Isolation of members.
Resolve conflicts within group.	Conflict avoidance.
High levels of trust.	Lack of trust.
High enthusiasm, interest.	Low enthusiasm.

building. Research on group behavior is an energetic field of research, and managers should keep tabs on new research in this area.

Dos and Don'ts

- Committee members must feel they are given some kind of reward for their committee efforts. Recognition and appreciation are given, so that they feel they are really making a contribution. Adopting the recommendations graciously may be considered a reward.
- The work of the committee must be accepted and used, and seems to make a contribution to the organization.
- Avoid psychobabble when introducing team techniques. At least translate such phrases as "conflict avoidance mechanisms" to "getting along." Group technique studies have been prone to faddish research, but the basic group

FIGURE 14–5
Barriers to Team Building

Barrier	*Barrier Removal*
Role conflicts	Define scope at the outset. Have the group determine where it envisions itself in the solution of the problem or process at hand.
Unclear objectives	Define budget, time, and other constraints at the outset. Allow members access to find out other parameters as the need arises.
Team member is a no starter, wet blanket	The team must be empowered as a team to remove a member when it becomes necessary.
Team leadership conflict	Ideally, there should be no team leader, only a facilitator. Tasks such as note taking, and liaison, should be rotated among group members.
Team personnel selection	Workers need to know what the objectives are, what their authority is in obtaining the objectives. They must understand their role, and why they were selected. Don't be afraid to reappoint the entire team. Don't be afraid to omit the "subject matter experts."
Poor or irrelevant results	Was authority commensurate with responsibility?

concept has been around for millennia (hunting a woolly mammoth requires no small coordination of effort), and will continue to be.

PERFORMANCE MANAGEMENT

> *The best executive is the one who has sense enough to pick good men to do what he wants done, and self-restraint enough to keep from meddling with them while they do it.*
>
> Theodore Roosevelt

Aliases

Evaluation of employees and employee performance.

Brief Description and Purpose

Employee performance is managed, not appraised. Deming and others advocate the elimination of annual performance ratings. Performance management is dynamic, ongoing, and process oriented, rather than passive, clock watching, and results oriented.

Discussion

The changes that Quality Management require in dealing with employees and work processes also require changes in the way performance of employees is measured. Skills can be measured by purely objective means, such as the number of widgets made in a day. This would be the same as management by numbers, or management by fear as Deming calls it.

Even without Quality Management emphasis, the changing demographics of the work force are demanding a change in the way appraisals are made. With downsizing and flatter corporate organizational structures, the need for middle managers has dropped considerably. This creates large numbers of workers who are plateaued at their current organizational level. The organizational challenge is how to make these workers feel fulfilled by their work. Systems that were created in days of large numbers of fast trackers will need to be revised.

Performance appraisal systems (PAS) are passive, one-way "communications" of the employees' performance. Perhaps they can respond on a section of the form, but those who do so invite reprisal. These forms are dutifully forwarded to a "human resource department" that "examines" them and files them. The PAS fails for several reasons:

Supervisor bias. Everyone is subjective when rating another person, and these ratings may be colored by jealousy, incompetence, and prejudice. It is unreasonable for someone to recreate an individual's performance over a one-year period in one or two hours. Quite likely all we remember is what annoyed us about their performance.

Performance standards. If you expect that everyone should meet or exceeds certain production quota then you are either ignorant of what average means, or your quota is too low. Half of your employees will be below the institutional average, half will be above. Even if you belong to an exceptional organization, there is still an institutional average. To berate an employee because they are typical (plus or minus one standard deviation from the mean) is demeaning, and damages their work ethic.

The performance criteria becomes worse if there is nothing simple like widgets produced per hour, and the work is of a clerical, administrative, or professional nature.

They're never thrown away. An incident from 17 years ago can be as fresh as your rating from last year.

PAS welcomes mediocrity. Too bad a performance appraisal reflects badly on the supervisor as well as the employee. Too good, and the employee may bypass the supervisor. This situation leads to system that "almost always" (a favorite PAS phrase) leads to an achieving of the management by objectives (MBO) plan, but not much more, or less, surprisingly. This abuse has nearly led to the ruin of the fine, original concept of MBO.

PAS squelches teamwork. Because "merit" systems emphasize individuals, this leads to a stifling of teamwork. A modern organization cannot long retain its status by being a collection of Lone Rangers. A complex, global network of teams and groups are required, sensitive to vagaries of the marketplace and organizational needs.

Relation to Other Concepts and Tools

Performance management requires skilled communication. Tools that enhance communication are: cause-effect diagrams, brainstorming, flow charting, Delphi technique, and an awareness of how to negotiate and set objectives.

Process

1. Eliminate the Annual Performance Appraisal. A year is far too long to let poor behavior continue or good behavior go unrewarded. Substitute the performance appraisal with an ongoing assessment. Unless you're the CEO, you probably can't eliminate the harmful exercise of annual reviews, but their negative effect can be mitigated. If you are the CEO, eliminate the annual performance appraisal. It's really an easy thing do, although the personnel department will give you many reasons for keeping it. All of their objections can be overcome—their primary concern is that poor performance be documented should questions arise from a dismissal. Poor performance should be documented for that purpose, but the feedback should be rendered immediately to allow the employees to correct their behavior now, not months later at an annual review.

2. Don't eliminate procedures concerning terminating an employee. Most companies have a detailed process for filing grievances, giving reprimands, appeal procedures, and the like. These components are a necessity to protect the employee's right to due process. Poor performance that leads to termination must be documented, and the employee must be made aware of this in writing.

3. Substitute ongoing assessment. Ongoing assessment requires communication and example. Each task must be prioritized, expectations should be negotiated and made clear, and performance goals must be agreed upon. It should be easy to recognize a ''winner'' of a product, service, or performance for both employee and supervisor. The supervisor must act as a coach, and set an example for her staff. She cannot take the occasional two-hour lunch and expect her staff not to, as well.

Utilize peer pressure as much as possible to enhance performance. Docking an employee a half-hour of pay for being 10 minutes late results in an employee who will more than take the half-hour payback in surly service or shoddy workmanship, costing the business customers. Peer pressure, however, can bring employees in line with the corporate culture. (It may be that the corporate culture is 10 minutes late. The authors know of one company where the employees prided themselves on being late for work and leaving early. Sounds irresponsible, until you realize that the 10 to 15 minutes saved each day, at the most, was more than made up by frequent extended travel and taking work home, which was uncompensated.)

A list of essential ingredients in performance management is listed in Figure 14–6. A performance management cycle model (Figure 14–7) developed by Baird, Beatty, and Schneir relies on goal setting (Chapter 17), feedback, and coaching. This model requires an ongoing commitment from the supervisor (or self-managing team) to evaluate performance. In a supervisory setting, input from peers of those being performance managed should be considered, as conditions may have changed considerably since the supervisor was ''in the trenches'' or was familiarized with the climate in which the employees operate.

If you must provide a performance appraisal anyway (as many government agencies are required to), at least consider some of the following:

- Have the employee write their own review. You may be surprised at how hard they can be on themselves, and surprised at what they considered to be their achievements and shortcomings.

FIGURE 14–6
Performance Management Ingredients

1. Immediate feedback.
2. Negotiate goals.
3. Delegate strategy.
4. Coach by example.
5. Experiment with performance measures.
6. Analyze performance by more than one measure.
7. Peer input.
8. Provide training and tutoring.

- Have their peer group write their appraisal. This may seem risky, but peers do it informally and unceasingly. Peer standing may be incongruous with the supervisor's ranking, and some rethinking of either the individual's performance or performance measures may be needed.
- Ensure participation of goal setting for the PAS. This could be done either individually, or better still, in groups. The peer group or work may yield some insights as to what measures need to be improved.
- Review performance more frequently than the PAS specifies.
- Make sure that the rating is not a surprise.
- Force the human resource department (or whatever the guardian of the PAS is called) into an advisory, rather than enforcing role. Find out ways that they can assist the employee in improving performance.

Dos and Don'ts

- "Praise in public, reprimand in private" is always appropriate.
- Performance management is closely linked to motivation; review that chapter in conjunction with this one.
- Base performance criteria on group goals, not individual ones.
- Performance goals and criteria must be known by all affected employees, and should be established by these employees.
- Abstain from numerical goals as much as possible. They are easy to manipulate, and may actually cripple performance by having been set too low.
- Train, train, train. This includes formal courses, conferences, trade shows, skull sessions, and reading trade journals and books.
- Measure performance in several different ways.

FIGURE 14–7
Performance Management Cycle

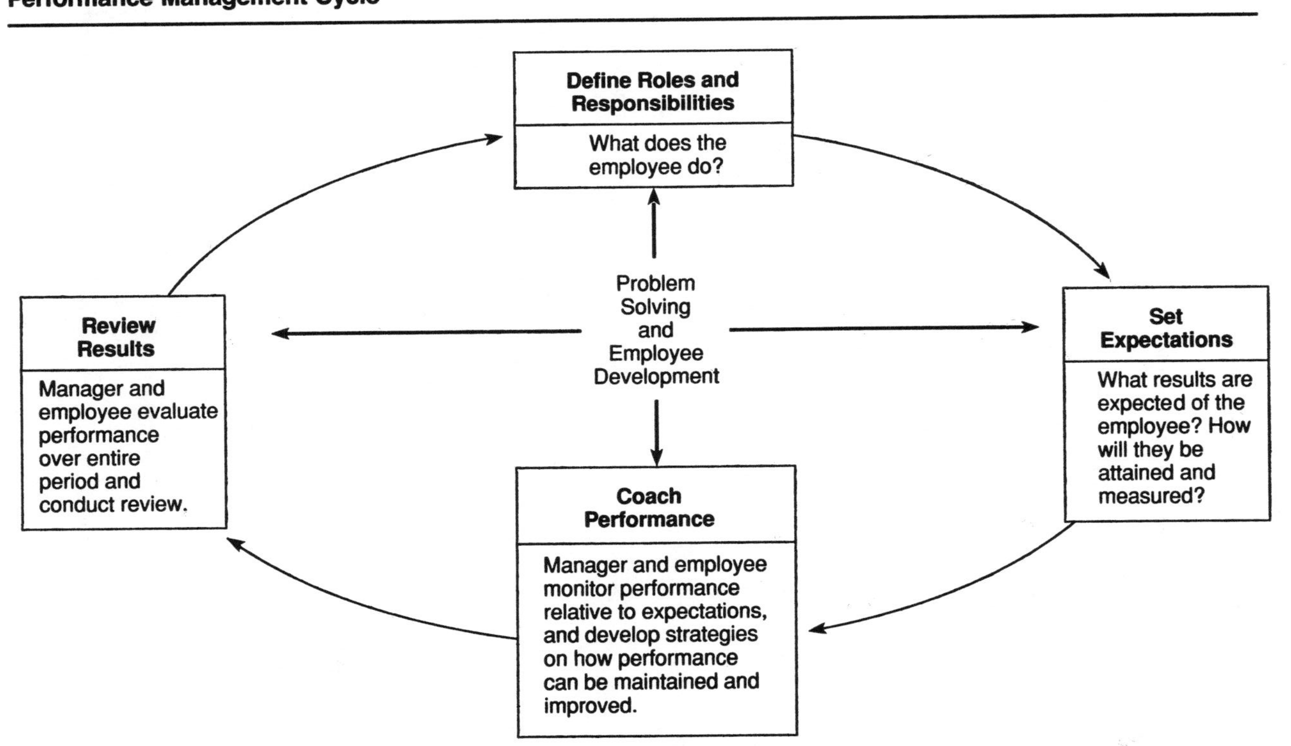

Source: Lloyd Baird; Richard W. Beatty; and Craig E. Schneier, "What Performance Management Can Do for TQI," *Quality Progress* 21, no. 3 (1988), pp. 28–32.

BIBLIOGRAPHY

Motivation

Grant, Philip C. *The Effort-Net Return Model of Employee Motivation*. Westport, Conn.: Quorum; 1990.

Kast, D. "The Motivational Basis of Organizational Behavior." *Behavioral Science* 9, no. 2 (1964), pp. 131–43.

Kazdin, A. E. *Behavior Modification in Applied Settings*. Homewood, Ill.: Richard D. Irwin, 1980.

Lewin, Kurt. *The Conceptual Representation and the Measurement of Psychological Forces*. Durham, N.C.: Duke University Press, 1938.

Luthans, Fred. *Organizational Behavior*. New York: McGraw-Hill, 1973.

Luthans, Fred, and R. Kreitner. *Organizational Behavior Modification*. Glenview, Ill.: Scott, Foresman, 1978.

Maslow, Abraham H. *Motivation and Personality*. New York: Harper & Row, 1954.

Massey, C. "What You Are Now Is What You Were When." Videotape.

McGregor, Douglas. *Leadership and Motivation*. Cambridge, Mass.: MIT Press, 1983.

Vroom, V. *Work and Motivation*. New York: John Wiley & Sons, 1964.

Employee Commitment

Dyer, Constance. *Canon Production System: Creative Involvement of the Total Workforce*. Cambridge, Mass.: Productivity Press, 1984.

Lawler, E. E. III. *High Involvement Management: Participative Strategies for Improving Organizational Performance*. San Francisco: Jossey-Bass, 1986.

Pierce, Richard J. *Involvement Engineering: Engaging Employees in Quality and Productivity*. Milwaukee, Wis.: ASQC Press, 1986.

Team Building

Aubrey, Charles A. II, and Patricia Felkins. *Teamwork: Involving People in Quality and Productivity Improvement*. Milwaukee, Wis.: ASQC Press, 1988.

Bolivar, J. "Management, Workers Unite to Boost Productivity." *Production Engineering*, July 1986, pp. 14–16.

Cole, R. E. *Work, Mobility, and Participation: A Comparative Study of American and Japanese Industry*. Berkeley: University of California Press, 1979.

Doering, Robert D. "An Approach toward Improving the Creative Output of Scientific Task Teams, *IEEE Transactions on Engineering Management* 20, no. 1 (February 1973), pp. 29–31.

Dumaine, Brian. "Who Needs a Boss?" *Fortune,* May 7, 1990.

Flaherty, Gerald S. "The Total Quality Team." *Proceedings of IMPRO 88*. Wilton, Conn.: Juran Institute, 1988.

Fried, Louis. "Don't Smother Your Project in People." *Management Advisor* 9, no. 3 (March 1972), pp. 46–49.

Gryna, F. *Quality Circles: A Team Approach to Problem Solving*. New York: American Management Association (AMACOM), 1982.

Hardaker, Maurice, and Bryan K. Ward. "Getting Things Done, How to Make a Team Work." *Harvard Business Review,* November 1987.

Jones, Louis, and Ronald McBride. *An Introduction to Team-Approach Problem Solving*. Milwaukee, Wis.: ASQC Press, 1990.

Jones, Steven D.; Randy Powell; and Scott Roberts. "Comprehensive Measurement to Improve Assembly-Line Work Group Effectiveness." *National Productivity Review* 10, no. 1 (Winter 1990/91), pp. 45–55.

Lewin, Kurt. *The Conceptual Representation and the Measurement of Psychological Forces*. Durham, N.C.: Duke University Press, 1938.

Lewin, Kurt. "Frontiers in Group Dynamics." *Human Relations* 1, no. 1 (1947).

Luthans, Fred. *Organizational Behavior*. New York: McGraw-Hill, 1973.

Manz, Charles, and H. P. Sims. "Leading Workers to Lead Themselves: The External Leadership of Self-Managing Work Teams." *Administrative Science Quarterly* 32, no. 1 (1990), pp. 106–29.

Persico, John Jr. "Team up for Quality Improvement." *Quality Progress,* January 1989.

Rubinstein, Sidney P. *Participative Systems at Work: Creating Quality and Employment Security.* Milwaukee, Wis.: ASQC Press, 1987.

Scholtes, Peter R. *The Team Handbook*. Madison, Wis.: Joiner Associates, 1988.

Thamhain, H. J., and D. L. Wilemon. "The Effective Management of Conflict in Project-Oriented Work Environments." *Defense Management Journal,* 11, no. 3 (July 1972), p. 975.

Performance Management

Baird, Lloyd S.; Richard W. Beatty; and Craig E. Schneier. "What Performance Management Can Do for TQI." *Quality Progress* 21, no. 3 (1988), pp. 28–32.

Camp, Robert. "Benchmarking: The Search for the Best Practices That Lead to Superior Performance." Parts 1–5. *Quality Progress,* January–May 1989.

Daniels, Aubrey C. *Performance Management: Improving Quality Productivity through Positive Reinforcement*. 3d ed., Tucker, Ga.: Performance Management Publications, 1990.

Davis, J. H. *Group Performance*. Reading, Mass.: Addison-Wesley Publishing, 1969.

Scholtes, Peter. *An Elaboration of Deming's Teachings on Performance Appraisal*. Madison, Wis.: Joiner Associates, 1987.

CHAPTER 15

MAXIMIZING INDIVIDUAL PERFORMANCE

Maximizing individual performance requires an awareness of one's own capabilities and limitations. Stress management and time management allow an understanding for the finite capacity of any individual, and pursues avenues that ameliorate feelings of inadequacy, and frustration that lead to procrastination and failure. Stress and time management are the obverse and reverse of the same coin. Managing time well reduces stress, and reducing stress leads to better time management. This may not mean greater quantity of output, but better quality. They are skills that the supervisor can little influence, other than to make available appropriate training, furnish resources, and identify and remove organizational roadblocks. Removing individual roadblocks may well require professional psychological counseling and therapy.

Creativity and innovation imbues a work environment with vitality and a sense of forward momentum. That there can even be a "process" to be "creative" seems an oxymoron, but research into invention suggest that there are indeed steps that can be taken to cultivate our ingenuity.

Motivation could have been included in this chapter, but the tight coupling of employee motivation to supervisor compels it to be placed in supervisory dynamics. Individuals should be aware of motivational factors, however, as modifying behavior rests upon self-knowledge of motivating causes.

Self-management is another term for maximizing individual performance, and there have recently been a number of books written about self-management techniques.

STRESS MANAGEMENT[1]

Stress is the spice of life.

Hans Selye

[1]This chapter was written by Emily Schlenker.

Aliases

Organizational stress, catalytic management.

Brief Description and Purpose

In any complete managerial skills checklist can be found items dealing with promoting organizational effectiveness and managing change. In an era of rapidly advancing technology and escalating complexity, change is surely a given, and just as surely complicates the task of fostering organizational effectiveness. Since stress occurs as a response to change, this section will focus on the manager's role in managing organizational stress.

Discussion

Most managers are aware that the impact of rapid and drastic change results in increased employee stress. Change means loss—loss of status quo, loss of comfortable habits and patterns, loss of old modes of thinking and operating, and loss of "the way we've always done it." Loss involves grief, and loss always means challenge and stress.

Managers have to be cognizant of stress levels in their organizations, and of the effect on employees and organizational effectiveness. But they must also develop a particular awareness of their own role as organizational catalysts. The effective executive assumes the important managerial task of stimulating creative change and helping employees channel the energy of the human stress response to manage change. In the role of catalyst, managers not only have stress, they cause stress. Converting that stress from a negative, defensive reaction to a positive, creative force in the workplace is a crucial function of management.

A quick inventory (see Figure 15–1, Organizational Stress Checklist) will be a useful tool in gaining an idea of the amount of stress circulating in a given organization. The checklist items are either known causative factors of stress, or known stress symptoms. Some of the causative items refer to environmental stressors which can exist subliminally. When combined with other causative agents involving greater or lesser degrees of change to which employees must adjust, they can be damaging.

The classic definition of stress is the response to a demand to adapt to change. Thus stress can be seen to be connected to physical moves, reorganizations, reductions in force or refocusing of tasks. Presence of these factors in the organizational milieu is a straw in the wind, telling the manager that stress is active. Increases in employee tardiness, daydreaming, use of sick

FIGURE 15–1
Organizational Stress Checklist

Are any of the following items applicable to your organization, either currently or at any time in the last 12 months? If so, circle the number following those items, and ONLY those items, which are applicable. If an item is not applicable, do not circle the number.

Item	Points
1. Physical move of office location	80
2. Poor environmental conditions (noise, temperature, lighting)	40
3. Reorganization in chain of command	90
4. Change in organizational task, focus, or priority	100
5. Change in supervision or management	90
6. Poor vertical and/or horizontal communication	40
7. Reduction in work force	100
8. Increase in work force	30
9. Voluntary employee terminations	40
10. Over- or underqualified employees	20
11. Increase in number of committees and/or meetings	30
12. Increase in employee use of sick leave	20
13. Increase in employee tardiness	10
14. Employee daydreaming	10
15. Increase in reprimands or disciplinary actions	40
16. Presence of unresolved conflict	30
17. Unmet task deadlines	60
18. Increase in customer service complaints	60
19. Unclear supervisory expectations of employees	40
20. Recurring crises	70

The total score measures the amount of stress present in the organization. A score of 150 or less is considered within normal range. Scores of 150 to 250 indicate moderate organizational stress. Scores of 250 to 350 indicate severe organizational stress. A score above 350 indicates a dangerous and probably disruptive level of stress in the organization.

leave, flight from the organization, or a rise in the need for disciplinary activity on the part of the manager are symptomatic of higher stress levels, leading to recurrent crises, unaccomplished tasks and customer dissatisfaction.

Although the checklist is obviously not all-inclusive, these items constitute a valid general set of organizational stress indicators. Point totals below 150 can be considered *normal* levels of organizational stress. Scores of 150 to 250 denote a moderate level of stress which, while probably not disruptive of day-to-day function, would bear watching. Scores of 250 or above indicate severe stress and call for specific intervention if the organization is to remain productive.

Process

Having totaled the organizational score on the checklist, how can the manager intervene to alter the course of the stress response? The following steps shepherd our path.

1. Recognizing the givens. The manager must recognize and communicate to employees the fact that change is a constant in the workplace and the world. Since change surrounds us all of the time, it will happen whether or not we like it or plan for it. It is necessary, then, to internalize the idea that change is needed rather than imposed. Change can be internally rather than externally driven. Workers can be helped to accept change as potentially positive—an opportunity to learn, stretch, and grow in many dimensions. If change is perceived as negative, threatening, and destructive, the result will be a threatened, rigid, resistant work force lacking the ability to move forward. A kind of self-fulfilling prophecy occurs which hinders problem solving and drags down the organization. By contrast, if the manager is able to convey the idea that change can be interesting, exciting, and desirable, employees will feel less negative stress at the prospect of its occurrence. It's more pleasant and productive to enjoy and participate in the change process.

2. Shedding the fears. Too often managers and subordinates view each other with adversarial suspicion and distrust. The outcome is a workplace darkened by the fear that one side will take advantage of, or find ways to subvert the ideas of, the other. This phobic atmosphere is nonproductive at best. At worst, it causes a deep gulf between upper and lower levels of the organization. Managers need to trust employees and to verbalize their own confident expectations of high worker motivation, good performance, and participation in the change process. Since people tend to do what they are expected to do, trusted workers become trusting workers. They are more likely to buy into the goals of the organization. Communication flows freely in both directions and effectiveness increases.

3. Promoting laughter. In any stressful situation one of the intensifying factors is panic, and panic is a stressor in itself. Laughter serves as a block to panic and helps immunize both manager and staff against stress. It will be wise, however, to keep in mind one caveat. Negative humor, such as sarcasm, ridicule, angry jokes, sick jokes, jokes at another's expense, is dysfunctional and stress producing. Positive humor, by contrast, aerates the lungs, produces endorphins, decreases tension, creates energy, maintains perspective, and opens the individual to new ideas. If the manager is able to promote positive humor, employees will be helped to view challenges as fun. They will see problems as solvable, gain confidence in their own abilities, and begin to enjoy the change process.

4. Removing risk from creativity. The concept of risk carries negative connotations for most people. In a change situation it is a human reaction to try to minimize risk and maximize safety, and therein lies a problem. To progress, risk is necessary. Problem solving and change management require the participants to take risks. Managers can promote creative problem solving on the part of the team by encouraging new ideas, experimentation, and learning. While this action does not remove all risk, it will eliminate the perceived risk of managerial displeasure as employees gather and try out possible methods and techniques. In addition, by displaying openness to novel ideas and insights, the manager will, by example, foster creativity in the work force.

5. Allowing control. The issue of perceived control is critical to managing personal stress, and has a serious impact on organizational stress as well. The individual who sees himself as helpless in a situation usually adapts poorly to change, becoming passive and immobile while the process surges around him. If the manager exerts too much control over employees, this becomes a group phenomenon and the entire work force may remain in a state of paralysis. Conversely, the individual who perceives himself able to exert even minimal control over his situation will respond far more adaptively to change. The manager who is able to relinquish control, allowing subordinates to pick up that prerogative, will be rewarded with an adaptive organizational climate.

6. Fertilizing and facilitating. Experimentation, risk taking, creativity, and productive change management do not occur in a sterile or nonsupportive environment. Even in the face of budgetary restrictions, managers must provide psychological and physical resources needed by employees to accomplish tasks. Certainly it is necessary to operate within fiscal constraints, however attitudinal resources do not require heavy financial outlays. Subordinates need to be aware that their manager is willing to support them in their endeavors, in words and actions. Often, too, simple things such as adequate computer time, library time, paper supplies, and clerical support are make-or-break factors in getting the job done. A manager's encouragement and receptivity to requests is a high-priority item.

7. Providing time to gel. In the most supportive, most creative of environments, ideas need time to mature before they are ready for final realization. The wise manager will recognize that periods when employees seem to be standing still or making only slight forward movement are not necessarily periods of stagnation. A hands-off policy at such times will allow for a gelling of ideas into workable form, whereupon progress ensues.

8. Communicating the positives. Last but by no means least important, the good manager will realize that there will be failures in the process. These should be seen neither as dead ends, nor as defects in the attitudes or abilities of the work force. It is not a Pollyanna technique to find positive aspects in failure, since having the freedom to fail yields fewer failures. The manager who is able to emphasize positive points in disaster situations helps subordinates to maintain adaptive attitudes, pick up the pieces, and move on to successful methods of problem solving and productive change.

Dos and Don'ts

- Becoming aware of organizational stress levels and taking to heart the foregoing recommendations will not enable the manager to eliminate stress from the organization. The eradication of all stress is neither feasible nor desirable. The manager will, however, gain the ability to replace negative stress with positive stress, yielding high-energy levels and associated increases in creativity and organizational effectiveness.
- Additional techniques can be incorporated in the organizational stress management plan. These have to do with lowering personal stress levels.
- Help employees create supportive physical and interpersonal environments. An individual can exert some control over negative stress levels by structuring a calm physical setting and cultivating a support network of people who will help solve problems.
- Ration the stressors active at any one time. Since stress is cumulative, it makes sense not to allow all stressful activities to occur at once. If the individual has a difficult evaluation in the morning, followed by a lengthy staff meeting in the afternoon, that is an inappropriate date to schedule the root canal after work.
- Encourage employees to practice personal de-stressors. Healthy diets, physical exercise, occasional breaks, deep breathing, and guided imagery exercises are all effective in managing individual stress.

TIME MANAGEMENT

Work expands so as to fill the time available for its completion.

C. Northcote Parkinson

Aliases

Managing time, efficiency.

Brief Description and Purpose

Managing our own time is a such a basic skill that is often ignored. It is quite easy to dictate to others how to conduct their time; it is another matter entirely to manage our own time. Thus the techniques in this section are intended primarily for individuals.

Discussion

Time is the elusive "fourth dimension." As humans, we are ill-equipped to comprehend the passage of time. Our memories are so vivid that the passage of decades seems to be short. Our diurnal rhythms are amiss as well. Recent research indicates that we operate on a 25-hour clock (this may vary from one person to the next) which is reset by external clues daily. Almost any business leader has mastered managing time. Most seem to have enough time to run the business, exercise, have a family, and engage in a social life. Many follow the process outlined to take advantage of all the time they are given.

The problem of time management increases with management responsibility and autonomy. It is difficult to limit access of associates, problems crop up at the wrong time. In addition, we often have the freedom to schedule our own time, to plan what we think is important. This can lead to procrastination (workaholism and procrastination are different sides of the same coin), starting a downward spiral in self-esteem and productivity.

At the heart of the time management issue is control. This control is a scary thing; we often would prefer to let some else take control—as Sartre said, we are "condemned" to be free. However, by selecting the tools and tips in this section that are suitable to an individual personality we can regain control over our intangible companion, time.

Relation to Other Tools

Diligent management of time reduces stress, and can increase our motivation. Goal setting may aid in managing time.

Process

Not all of the following techniques work for all people. Try the techniques, and revise them to suit your situation and tastes.

1. Time Logs

• First catalog how you spend your day. Keep a very meticulous log for at least one day, preferably for about two weeks. Don't worry if the week selected is "unusual," all weeks are unusual. Keep the entry simple, perhaps just using two columns: start time and task. The ending time should correspond to the starting time of the next task on the list. Be sure and include things such as phone calls and other interruptions.

• Do a Pareto analysis (Chapter 20) of how much time you spent on the phone, in meetings, in bull sessions, waiting in line, sleeping, and so forth. Is this how you would choose to allocate your time?

• Can you eliminate or otherwise manage time robbers (Figure 15–2)? Can you employ a phone answering machine, or establish "office hours" in which you are generally available? Can you delegate? Be careful you don't fall into the do-it-yourself trap. Your time is valuable—all of it! Spend it on what you enjoy. If lawn care or house cleaning are relaxing, then do them. If these are things you hate, hire a service so you can spend time elsewhere. Expensive? How much is your time worth per hour? Balance this against the cost of the service.

• Is there anything you can do to change time robbers into time well spent? While waiting can you read, make To-do lists, or catch up on correspondence?

• Can you say no? To be "nice guys" we often do not refuse a request from a friend or associate. "Nice guys" however, also manage their time wisely and budget it accordingly. Your friends should respect your decision (and perhaps your newfound assertiveness as well) if you decline invitations with good grace.

2. Calendars

Calendars are excellent ways to see the "big picture" of your goals. Some people prefer the "Week-at-a-Glance" variety or other proprietary methods such as Day Timers™. Automated calendars can be easily updated and printed out as needed. One can also take a task list and provide it in a daily, weekly, monthly, quarterly, or yearly format.

3. To-Do Lists

If you adopt only one of the techniques mentioned here, make it the to-do list. It can be a simple, handwritten list, or it can be automated. To-do lists

FIGURE 15–2
Time Robbers

Telephone calls.
Failure to delegate.
Poor communication.
Incomplete work.
Delayed decisions.
Social visits.
Broken or poorly operating equipment.

can be arranged by time, by type of task, or some combination. They should also employ some sort of priority notation, such as using an A to denote primary tasks which must be done, B to denote tasks desired to be done, but not essential, and C to tasks which would be "nice" to do, but can be delayed for some time to no detriment. Use the completed list as a work log or diary.

Establishing realistic, achievable goals is essential to good time management. Conversely, good time management is essential to achieving these goals. Goals are discussed at length in Chapter 17.

4. Determine What Driver Works Best for You

People can be motivated in many ways. Following are some examples.

Time driven. Let the clock be your taskmaster. Doggedly change into the next task when the clock tells you. If you do this for a while you may be surprised at the amount of work you're able to accomplish, particularly if you parcel out tasks into increments no longer than one hour (adult attention spans are about 45 minutes long).

Task driven. Pursue a task from point A to point B with little regard to how long it takes. This may require an unswerving dedication that we may not possess. It may be better to approach the task using the MacKenzie's Swiss cheese theory: Poke holes in the task by doing small portions of it at a time.

5. Stop Procrastination

Procrastination results in more work. The workaholic loves this situation (perhaps subconsciously) because the work is never done. Therefore workaholics are often procrastinators and vice versa. James Sherman, in his work *Stop Procrastinating,* identifies a number of reasons for procrastination and suggests some techniques to overcome the habit:

Be happy. Author Norman Cousins closeted himself in a hotel room and indulged in devouring all manner of humorous books. His mortal cancer went into remission, to the astonishment of the medical establishment. Humor would seem to have a link to physiological as well as psychological well-being. Realize how miraculous it is that any one of us is alive, and how incredibly fortunate we are to live in an age of technological convenience, and comparative peace. Dale Carnegie has built a training empire based on this simple (but far from simplistic) premise. If nothing makes you genuinely happy, then act. Don't procrastinate because you have low self-esteem or lack proper perspective. Seek help. If you look good, sound good, act good, then fairly soon you will *feel* good as well.

Know yourself. Know your limitations as well as your abilities. When necessary, delegate, organize, and plan better. Use time management techniques to increase your time for relaxation. Analyze your reasons for procrastinating. All of the causes listed in Figure 15–3 are surmountable. Is there a secret fear of failure or success? Is there an inner voice holding you back, telling you to do what's safe? Conquer negative fantasies that tell you how horrible something might be if you do or don't do it? Philosophers call this technique *reductio ad absurdum*. Laymen prefer to call it, *What's the worst that can happen?* Will you go to jail? Die? Contract a disfiguring disease? Probably the worst thing is far less worrisome than these possibilities. And probably your worst fears will *not* be realized.

Visualize. Successful athletes have used this technique for decades, and in the 90s it seems to be catching on for health improvement as well. You might call this technique focused daydreaming or self-hypnosis. The technique is to imagine (visualize) your task completed. Imagine also the steps required to lead to the finish. What problems might arise? Imagine dealing with them. Medical studies have shown that visualization is a powerful tool in increasing self-esteem, and "psyching up" to accomplish a task effectively.

Do a leading task. A leading task is some small, easy to do or fun way of starting a project. Really start the project, don't simply shuffle papers or engage in busy work. This will require breaking the task into smaller parts, which in itself may decrease apprehension at doing the project. This divide and conquer trick is one of the most useful ways to get started.

FIGURE 15–3
Causes of Procrastination

Confusion.
Lack of priorities.
Fear of risk.
Avoiding unpleasant tasks.
Depression.
Obsessive/compulsive behavior.
Boredom.
Fatigue.
Manipulation by others.
Manipulation by yourself.
Physical disabilities.

Control your environment. Create a physical atmosphere conducive to task completion. Establish office hours. Get an answering machine. Take control. Other will adapt. Your coworkers will not disparage you. Rather the reverse will occur. This is a step one must do as an individual. No one can do it for you. Make a decision. Stick with it.

Develop alternate activities. Alternate positive stress is sometimes highly beneficial. Develop an engaging hobby; one that requires as much (possibly more) mental concentration as your current job. Be sure to get enough exercise. At least 20 minutes (even 10 minutes twice a day is sufficient exercise, if the heart rate rises to 70 percent of its maximum) three times a week, ideally one hour daily. Studies conflict as to the minimal amount of time required for cardiovascular improvement. To enjoy a sport, however, probably an hour per set is needed to gain proficiency, thereby enjoying the sport more. This modest investment will pay off in increased output in other tasks.

Develop a broader perspective. Believe in something larger than yourself. Realize your place and role in the sweep of history. Pray, meditate, attend worship services. If you have a distaste for organized religion or are not theologically inclined, develop interests in the humanities. Realizing the struggles of our ancestors by studying history, archaeology, or even art and music helps achieve a broader vista.

Dos and Don'ts

- ***It really isn't urgent.*** Unless it *really is* on fire or you're going to the emergency room, it can be scheduled. These "urgent" tasks seduce us into thinking we are accomplishing as much as we can in the time we have, when in fact they rob us of efficiency.

- ***Be aware of what is not important.*** "Writer's block" is a well-known procrastination technique that diverts your attention from what you are supposed to be doing, so that instead of deciding what to do with the problem employee in human resources, it seems very important to sketch out a new company letterhead. This sort of deception is very insidious, particularly when we have meetings or otherwise have people waiting for us.

- ***Ask your peers how you waste your time (and maybe theirs).*** We may waste time in ways we don't even realize. Ask a friend how they perceive you are wasting time, and let them know why you are asking. Don't defend yourself to them!

• ***Avoid the "As busy as I am I must be very productive" syndrome.*** Being busy is not the same thing as being productive. Scientists, engineers, strategic managers, and others can be their most "productive" when they stare at the wall. Be sure that your busyness is moving in a forward direction.

• ***Establish realistic deadlines.*** Pad the due dates. Always underpromise and overdeliver.

• ***Be wary of meetings.*** Start all meetings on time. Establish an end time. Do not degenerate into discussion on sports. Allow no interruptions.

• ***Avoid memos.*** Use the phone. Use the FAX machine. Handwrite on the original letter and return it (uncopied). It may not seem professional to do this, but failing to answer as fast as possible is far less professional. Do anything to avoid assistant prepared, perpetually filed, never read memos.

• ***Practice calculated neglect.*** Let small things slide. If they re-occur, do them.

• ***Throw it away.*** Some studies suggest that 80 percent of all information filed is never used. Be ruthless in throwing it away. However, be aware of how long records must be kept for internal audits, the IRS, and so on.

• ***Delegate.*** How much do you make an hour? Keep this figure in mind as you are scheduling tasks. Will that time spent pay for itself? Should you photocopy that document yourself, have a subordinate do it, or buy inexpensive desktop photocopy machine. For instance, if you make $25 an hour, does it really make sense to spend two hours to mow and trim your lawn when a service will do it for $35? If you need the exercise, though, reconsider this delegation!

• ***Give it time to work.*** Behavioral scientists have determined that establishing new habits (or breaking old ones) takes about 21 to 30 days. Thus, if you were to rise every day and do 10 minutes of stretching exercises, on the 22nd day you would be very disturbed (you may possibly feel physical effects) if you did not do the stretching routine.

• ***Prepare for backsliding.*** It is common, perhaps even necessary, for occasional backsliding to occur. As long as this is temporary and does not permanently disrupt your routine, there is usually little harm done. Exceptions to this are behaviors that are extraordinarily hazardous, such as drug addiction.

CREATIVITY AND INNOVATION

A hundred thousand men were led
By one calf near three centuries dead.
They followed still his crooked way,
And lost one hundred years a day;
For thus such reverence is lent
To well-established precedent.

Sam Walter Foss

Aliases

Breakthrough engineering.

Brief Description and Purpose

Creativity and innovation at the personal level are essential to the survival of a Quality Management effort. The planning, organizing, and group tools of Part 4 all rely heavily on creative thinking. Breakthrough thinking is need to overcome the status quo mentality infecting most businesses.

Discussion

"Creativity" should not be equated with trying to become Einstein junior. It means seeing new ways to perform and enjoy our work. It involves eliminating barriers to our naturally innovative minds. Creativity allows richer convergent thinking and admits divergent thinking. Convergent thinking focuses on the immediate task at hand. Divergent or lateral thinking looks for relationships that may not be obvious, or tries to merge two seemingly unrelated concepts. For example, a researcher in the early 1980s might have kept a data base on small cards. Mulling over improvements in such a system would have a convergent thinker considering the size of the cards, sortation methods, cross-referencing indexes, and the like. The divergent thinker might have explored the capabilities of the newly emerging desktop computer. Lateral thinking encourages shaking up the status quo, looking for radical solutions.

While managers cannot make an employee "more creative," they can eliminate the barriers and behavior that stifle creativity in the work force. Creativity doesn't have to result in a page one discovery. Finding a better way to enter information on a form, to handle exceptions, or to satisfy a complaint is probably of more immediate use to the organization.

Process

1. Remove organization-imposed barriers. This requires an examination of the corporate culture. How does the company react to new ideas? How do new ways of doing business come about? Some examples of cultural barriers to creativity are:

- *Is there a suggestion box?* Suggestions should be solicited constantly. To have to write out a suggestion and slip it into a locked box is absurd and insulting. Employees deserve a better hearing than that.
- *Is some new idea greeted with "I'll have to pass it up the chain" or some other limp excuse?* Develop a process to allow ideas to be passed along to someone with authority to implement them. Better still, delegate authority to the lowest possible level, allowing those in direct contact with a problem to be allowed to deal with it creatively.
- *Are employee (or customer) problems always resolved by management?* Let the employees resolve the problem. They have the most direct stake.
- *Have you heard the phrase "Because that's the way we do things around here" lately?* Even unruly children warrant a more cogent response. Status quo thinking is the most dangerous barrier to the implementation of Quality Management.
- *Does everyone wear a uniform (even though there is no compelling reason to do so) everyday?* Have a play clothes or costume day. Do something else that may require creativity—have a company play or variety show for example. Do *not* coerce an employee into participating, or even attending.

2. Help open the mental locks. The first step in opening the personal psychological barriers to creativity is to know of their existence. Roger van Oech calls these barriers "mental locks." Figure 15–4 describes a number of these locks. The opening of these locks will increase self-esteem, creativity, and may boost efficiency and morale as well. These locks cannot open, however, unless the organizational culture allows creativity to bloom.

3. Use tools that encourage creative thinking. Tools such as brainstorming, Delphi, nominal group technique, quality circles, process decision program chart, foolproofing, who-what-why-where-when-how (all are in Part 4, "Tools and Techniques") are all dependent upon creativity for them to work.

FIGURE 15–4
Mental Barriers to Creativity

1. There Is One Right Way to Do Things.
   There is always more than one right way to solve a problem.
2. That's Not Practical.
   This dampens creativity, and doesn't allow building from an intriguing, but "impractical" idea to a similar, but "practical" idea.
3. That's the Way We Do It / Follow the Rules.
   Status quo is the mark of someone who is retired in place (RIP). People with this attitude spill their bile on everyone around them. Slavishly following rules or behavior patterns is boring, and eventually fatal to a business. Considering the rapid advances of technology, consider the following rule: ***Every Rule Can Be Broken Except This One.***
4. That Mistake Is Going to Cost You.
   We have to value mistakes as opportunities. Never making mistakes leads to fear of failure—we have never learned to cope with making mistakes.
5. That's Not Our Problem.
   Dodging a problem doesn't make it go away. This attitude is one that is built up over the years in a rigid, bureaucratic environment, and points out a cultural flaw.
6. Don't Be Silly.
   Foolishness is a key to being creative. Court jesters were used in the ancient Roman and medieval eras to remind rulers of their morality and their shortcomings. Use the fool inside of you to arrive at new ways to look at a problem.
7. Grunts Don't Think.
   This repressive view is shared by management and "grunts" alike. Working on an assembly line may not require an advanced degree, but it does require intellect, and an expertise is developed. Management may have never worked on the line, or worked on it so long ago they have forgotten what it was all about. Let those directly involved have a voice in decisions regarding the line. It took nature four billion years to create humans—to waste such a hard won resource as the human mind is beyond belief.

4. Install an idea process. Design an idea process model such as the following:

- Idea origination—can come from anyone.
- Idea screening—the idea originator shares the idea with peers and the supervisor, refining and fleshing out the idea.
- Peer review—workers in the effected area review the idea, further refining it. Judgment of the idea must be suspended at this stage.
- Seek a sponsor or champion—someone with authority takes responsibility for the idea and sees the idea through to implementation.

Dos and Don'ts

• Creativity training and sessions to increase awareness to creativity are a good place to start, but without a concerted change in organizational culture the sessions will be viewed by the work force as something designed to "entertain the troops."

• Watch for creativity killer phrases as closely as you would discriminatory ethnic and gender comments. The result of the disparaging remarks is the same: lowered worker esteem and morale, leading to reduced productivity, if not outright sabotage of work efforts. Discard such phrases as: "We tried it before and it didn't work," "My boss won't go for it," "We don't pay you to . . .," "When did you have time to come up with that idea?"

• Capitalize on the "free spirit" cultural value of American workers. Entrepreneurship and innovation has been a hallmark of the "American Way" for over 150 years. There is no reason not to expect employees to be as creative as the founder. Combined with unified goals and direction (something the Japanese have excelled in), an unstoppable business entity is sure to evolve.

BIBLIOGRAPHY

Stress Management

Cooper, Kenneth. *Aerobics*. New York: M. Evans, 1967.

Cousins, Norman. *Anatomy of an Illness*. New York: Bantam Books, 1981.

DeBono, Edward. *Tactics: The Art and Science of Success*. Boston: Little, Brown, 1984.

Friedman, M., and D. Ulmer. *Treating Type A Behavior and Your Heart*. New York: Alfred A. Knopf, 1984.

Ghiselin, B. ed. *The Creative Process*. New York: Doubleday, 1966.

Gray, Jeffrey. *The Psychology of Fear and Stress*. New York: McGraw-Hill, 1971.

Kiechel, Walter, III. "Executives Ought to Be Funnier." *Fortune*, December 12, 1983, p. 208.

Kirschmann, John D. *The Nutrition Almanac*. New York: McGraw-Hill, 1984.

Loehr, Dr. James E., and Peter J. McLaughlin. *Mentally Tough*. New York: M. Evans, 1986.

Loehr, Dr. James E., and Dr. Jeffrey Migdow. *Take a Deep Breath*. New York: Villard Books, 1986.

Maddi, Salvatore R., and Suzanne C. Kobasa. *The Hardy Executive: Health under Stress*. Homewood, Ill.: Richard D. Irwin, 1984.

McGrath, J. *Social and Psychological Factors in Stress*. New York: Holt, Rinehart & Winston, 1970.

Mindess, Harvey. *Laughter and Liberation*. Los Angeles: Nash, 1971.

Samuels, Mike, and Nancy Samuels. *Seeing with the Mind's Eye*. New York: Random House, 1975.

Schlenker, Emily C. *An Organizational Stress Inventory*. Rock Island, Ill.: Emily Schlenker, 1991.

Selye, H. *Stress without Distress*. Philadelphia: Lippincott, 1974.

Time Management

Cousins, Norman. *Anatomy of an Illness*. New York: Bantam Books, 1981.

Kendrick, John W., and John B. Kendrick. *Personal Productivity*. Armonk, N.Y.: M. E. Sharpe, 1988.

MacKenzie, R. Alex. *The Time Trap*. New York: McGraw-Hill, 1972.

Creativity and Innovation

DeBono, E. *Lateral Thinking for Management*. New York: McGraw-Hill, 1971.

Von Oech, Roger. *A Whack on the Side of the Head: How to Unlock Your Mind for Innovation*. New York: Warner Books, 1983.

Weisberg, R. W. *Creativity: Genius and Other Myths*. New York: Freeman, 1986.

West, Michael A., and James L. Farr. *Innovation and Creativity at Work*. New York: John Wiley & Sons, 1990.

PART 4

TOOLS AND TECHNIQUES

- Quality Circles
- Service Quality

20. Statistical Tools
    - Statistical Measures and Sampling
    - Control Charts
    - Design of Experiments
    - Evolutionary Operation
    - Pareto Analysis
21. Specialized Techniques

Quality Management tools are analogous to a carpenter's tools. Use the right tool for the job, but don't blame the tool for incompetent implementation. When using the tools, experiment with the format and process so that it fits in with the organization's culture and vision. There is nothing magic about the precise processes presented here, although the approaches are the most common. Some tools may require more training than can be allowed here. This is especially true with the statistical and numerical tools in Chapter 20. Self-examination tools (Chapter 18) are rapidly expanding, and multivolume checklists are becoming available for self-reflection.

Other ways of earmarking the tools can be found; dividing them into the Seven Tools of Ishikawa (Chapter 8)—seven new tools, other tools, and the like—Exhibit 1 gives some different ways to associate the tools. The present allocation is based on what stage of development the tool may be useful. Of course, some may be a planning tool and an organizing tool, such as force field analysis. Some may be primarily a planning tool, such as quality function deployment, but require self-examination. The statistical tools are collected in Chapter 20 because of the background required to implement them.

Deming has strongly advocated the use of statistical tools, almost to the exclusion of any other tools. While statistical quality control is extraordinarily useful (and still underused in the United States) in moderate to high-volume production, it is inadequate in analyzing problems in the one-at-a-time production or job shop. Besides, statistical tools generally indicate the existence of a problem, and not necessarily its solution. Other tools must be brought to bear to determine the cause and a solution.

New tools are being invented all the time for continuous improvement. Therefore journals and trade magazines such as *Quality Progress*,

EXHIBIT 1
Alternative Means of Categorizing Quality Management Tools

Graphical Tools:
- Cause-effect diagrams
- Control charts
- Data presentation
- Flowcharts
- Pareto analysis
- Process decision program chart
- Quality function deployment
- Systematic diagram
- Work flow analysis

Companywide Techniques:
- Auditing
- Benchmarking
- Deming cycle
- Goal setting
- Quality circles
- Quality costs
- Quality function deployment
- Service quality

Data Analysis:
- Check sheets
- Control charts
- Design of experiments
- Evolutionary operation
- Force field analysis
- Pareto analysis
- Sampling

Problem Identification:
- Brainstorming
- Cause-effect diagrams
- Check sheets
- Control charts
- Data presentation
- Delphi technique
- Force field analysis
- Nominal group techniques
- Pareto analysis
- Quality circles
- Quality function deployment
- Root cause analysis

Decision-Making Tools:
- Auditing
- Benchmarking
- Force field analysis
- Nominal group techniques

Modeling Tools:
- Benchmarking
- Flowcharts
- Quality function deployment
- Work flow analysis

Preventive Tools:
- Control charts
- Design of experiments
- Evolutionary operation
- FMECA
- Foolproofing
- Goal setting
- Pareto analysis
- Process decision program chart

Creativity Tools:
- Brainstorming
- Evolutionary operation
- Foolproofing
- Nominal group techniques
- Process decision program chart
- Quality circles
- Who-what-when-where-why-how

Quality Engineering, *Quality*, and others should be perused on a regular basis to keep up to date, and to monitor implementation issues and controversies.

BIBLIOGRAPHY

Burr, John T. *SPC Tools for Operators*. Milwaukee, Wis.: ASQC Press, 1989.

Ciampa, D. *Manufacturing's New Mandate: The Tools for Leadership*. New York: John Wiley & Sons, 1988.

Gitlow, Howard; Shelly Gitlow; Alan Oppenheim; and Rosa Oppenheim. *Tools and Methods for the Improvement of Quality*. Homewood, Ill., Richard D. Irwin, 1989.

Kane, Victor E. *Defect Prevention: Use of Simple Statistical Tools*. Milwaukee, Wis.: ASQC Press, 1989.

Karatsu, Hajime. *Mastering the Tools of QC*. Tokyo: JUSE, 1987.

Mizuno, Shigeru, ed. *Management for Quality Improvement: The 7 New QC Tools*. Cambridge, Mass.: Productivity Press, 1988.

Ozeki, Kazuo, and Asaka Tesuichi. *Handbook of Quality Tools*. Cambridge, Mass.: Productivity Press, 1990.

"The Tools of Quality." *Quality Progress* (June through December), 1990.

CHAPTER 16

ORGANIZING

A little neglect may breed mischief: for want of a nail the shoe was lost; for want of a shoe the horse was lost; for want of a horse the rider was lost.

Benjamin Franklin

Organizational tools aid in collecting, classifying, and presenting information. Organizational tools force looking at a problem from several viewpoints, particularly when coupled with the group techniques. Brainstorming, for example, may be used as a data collection technique regarding problems in the manufacture of a product. Cause-effect diagrams may then be used to identify relationships. Data presentation techniques and/or Pareto analysis might then be used to determine the root cause. As a result of this analysis, the control charts may be revised or a new goal may be established.

The tools in this section work with a minimal amount of complexity on a small scale at the departmental level. Problems that require a company-wide scope necessitate that the tools be layered and partitioned. One master cause-effect diagram or context flowchart may represent the company as a whole. This diagram must then be partitioned into subdiagrams that may reflect divisions or product lines or some other type of categorical breakdown. Cause-effect diagrams and flowcharts partition readily, and can be used as a simplified representation of a complex puzzle.

CAUSE-EFFECT DIAGRAMS

Aliases

Cause-effect diagrams, Ishikawa diagram, fishbone diagram, enumeration diagram.

Brief Description and Purpose

Cause-effect diagrams provide a graphical representation (they look like a fishbone) and categorize causes, stimuli, or factors that impact on an effect or outcome. They are useful in sorting out the results of a brainstorming session (Chapter 19). Kaoru Ishikawa introduced cause-effect diagrams in 1943.

Discussion

The cause-effect diagram is a graphic representation of causes and effects for a particular problem. The diagram can identify problems early in data collection and analysis. When analyzing various alternatives, it may be used to identify various influences the solution may have on the problem (process) if implemented. The diagram may be constructed by an individual, but most often the construction of the diagram is a group effort.

The major categories often considered are the "6 Ms": Money, Machines, Material, Methods, (Wo)Manpower, and Management. Causes are then sorted into these categories, or others, and the diagram begins to take on a fishbonelike appearance. Causes may be subdivided as appropriate. A blank cause-effect diagram with the 6 M's is shown in Figure 16–1. Figure 16–2 displays the causes inputs to the major causes.

Relation to Other Concepts and Tools

Inputs to the diagram can be made by brainstorming. Pareto analysis may reveal which causes are the most dominant. Cause-effect diagrams are also

FIGURE 16–1
Generic Cause-Effect Diagram

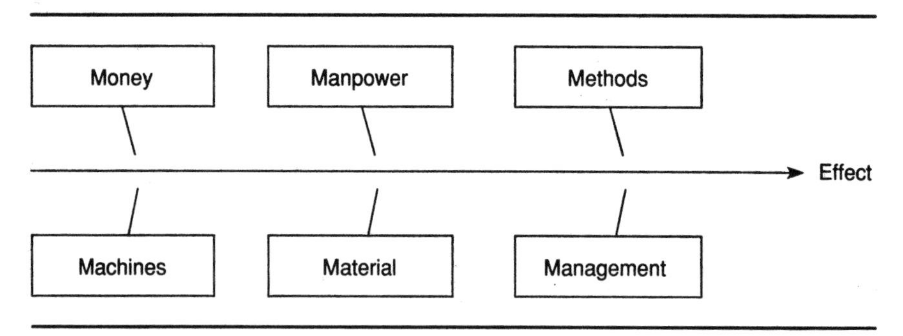

FIGURE 16–2
Minor Causes Inputs to the Major Cause

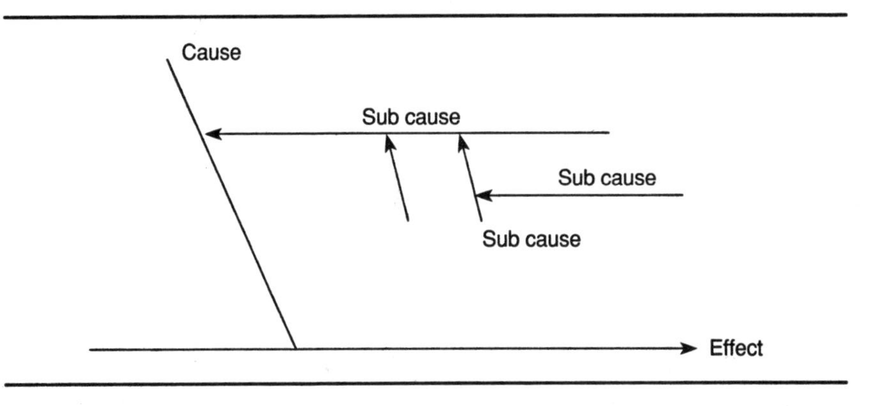

useful in analyzing statistical process control (SPC) problems, as SPC detects an obstacle, but can pose no solution.

Process

1. Identify effect. Identify the effect and begin the diagram. The effect may also be a desired goal, rather than a specific problem.

2. Identify major cause categories. The causes branch out from the center portion of the diagram. Use the 6 Ms or some other way of considering different influences on the effect. Not all of the 6 Ms will apply to every problem. These major causes are typically generic, rather than specific.

3. Identify specific minor causes. Inputs are identified and each positioned on the diagram as inputs to the associated major cause or category. The minor causes should be specific, and minor causes can have subsets. Each of the major inputs (causes) may in turn be treated as an effect, and the analysis may be performed at a lower level and additional divisions may be made to whatever level of detail is desired. Concentrating on who-what-when-where-why-how can help in associating a minor cause with a major category. The categorization process that occurs during brainstorming can be useful in assigning categories.

4. Analysis. Once the cause-effect diagram has been completed with the causes identified, those causes which may lead to the greatest potential improvement in the situation must be determined. The most probable minor causes should be selected for additional investigation. Pareto analysis is also useful to isolate the most important causes of the problem.

Example

A software development team wanted to analyze the factors relating to its software inspection process. Initially, a brainstorming session was held, and then a cause-effect diagram was created by the same group. During the cause-effect diagramming, the group wanted to prune the categories, eliminating material and machines. Instead, they combined the two realizing that the hardware and software environment could play a vital role in software inspection. Note that some of the minor causes are repeated when appropriate. After a brainstorming session, the cause-effect diagram of Figure 16–3 was created by the group.

Dos and Don'ts

- Work toward long-term results rather than short-term patching of problems. This can be done by varying the effect. Instead of using as an effect "sources of customer complaints," try "complete customer satisfaction."
- Use as large a diagram as practicable; it can be reduced later.
- If a category becomes too large or complex, create a separate diagram.
- Do not brainstorm and create a cause-effect diagram simultaneously, this will add a great deal of confusion. However, do revise the list of causes should it warrant. Brainstorming may create a large number of difficult to organize ideas. Don't be afraid to prune.

CHECK SHEETS AND DATA COLLECTION

Aliases

Check sheets, data sheets, checklists, forms.

Brief Description and Purpose

Check sheets simplify the process of data collection by providing a well-designed form on which to enter data. Accurate data collection is vital to any statistical effort. Check sheets help to minimize errors and confusion.

FIGURE 16–3
Cause-Effect Diagram for Improving a Software Inspection Process

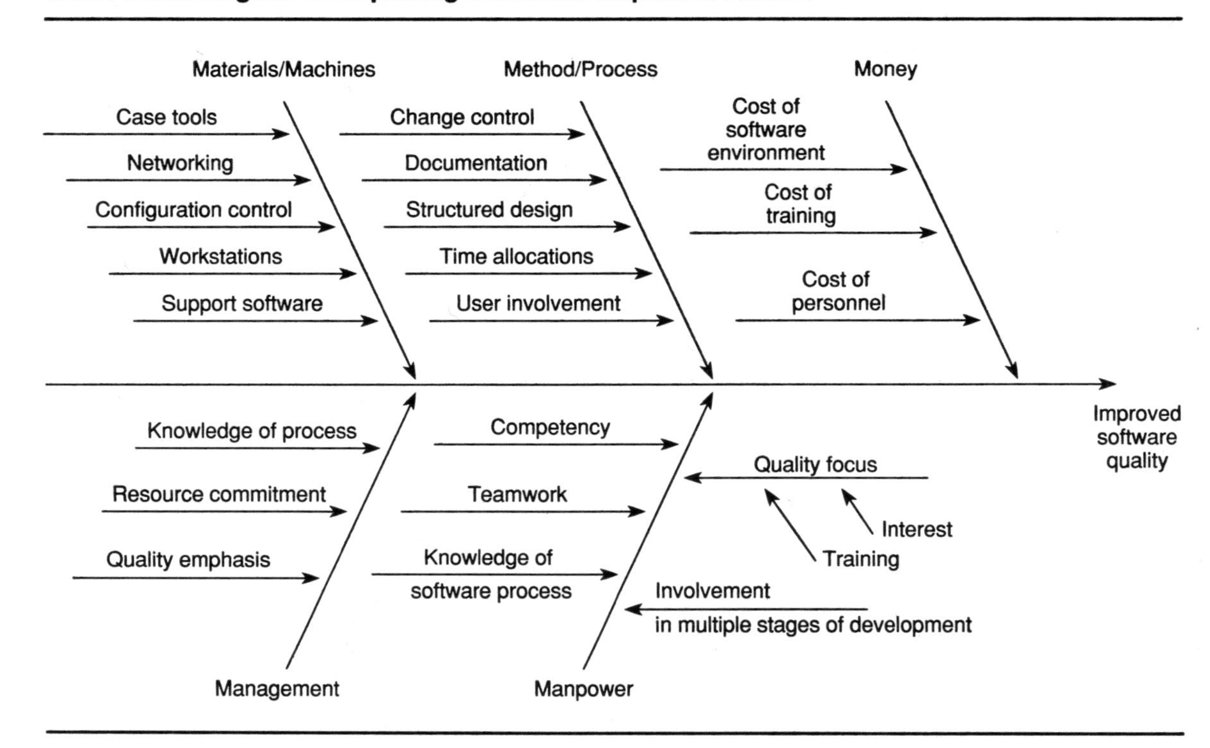

FIGURE 16–4
Check Sheet for Variables

Inner Diameter Measurements

Clean calipers after each use

Under 200 (rework)	𝍸 II	7
200–204.8	𝍸 𝍸 III	13
204.9–205.1	𝍸 𝍸 𝍸 𝍸 𝍸 𝍸 III	33
205.2–209.9	𝍸 𝍸 I	11
210 and over (scrap)	I	1

Examiner: ______________________ Date: ____________ Shift: _____

Calipers no.: ______________________

FIGURE 16–5
Check Sheet for Attributes

Envelope Defects by Attribute Check Sheet

Type of Defect	*7–9 A.M.*	*10–12 A.M.*	*12:30–2:30 P.M.*	*2:30–4:30 P.M.*	*Total*
Glue failure	II	IIII	II	III	11
Improper fold	𝍸	𝍸 III	𝍸 IIII	𝍸 II	29
Poor cut	II	III	II	IIII	11
Total	9	15	13	14	51

Discussion

Data collection may be obtained by several means: check sheets, checklists, and data sheets. Checklists and check sheets are generally suited for attribute data collection, and data sheets for variable data collection or measurements. However, check sheets can be used for variables as well as other uses (see Figures 16–4, 16–5, and 16–6). Check sheets are easy to complete forms that allow for rapid entry of information. They often allow for hash marking.

FIGURE 16–6
Check Sheet for Defect Location

Appliance Model 870

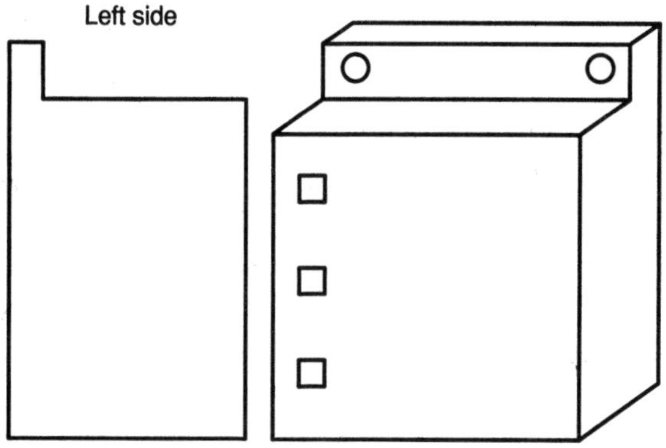

Left side

Examiner: ______________________ Date: ______________

Serial Number of Item: ______________________________

Remarks: ______________________________________

Checklists provide a comprehensive assessment. Both check sheets and checklists must be carefully laid out in advance, or data may be missed, although they should be dynamic and revised as necessary.

Data collection should not be limited to manufacturing processes. Services can also be measured. Caution must be employed in data collection of services—an average alone tells nothing. The range and dispersion (Chapter 20) must also be used to understand patterns. Minutiae should be avoided as well. Counting the number of keystrokes per 10 minute period may be possible with computers, but is it of value? Probably not. Because of the widespread misuse of service statistics, managers are often warned to avoid "management by numbers," which would seem to be at odds with the Quality Management message to "measure, measure, and measure." The point is to use the data as a source of strategic information, not as a preeminent performance management tool.

Relation to Other Concepts and Tools

Data collection is a first step to almost any statistical measure. Brainstorming, cause-effect diagrams, and small group techniques can be used to design the data collection form or activity. Flow charting and work flow analysis are also useful in assuring utility of the checklists, check sheets, and data sheets.

Process

1. Determine what needs to be collected. This input should come from a quality circle or other small group, using TQM tools. Determine the who-what-how-why-where-when. With multiple shifts and/or multiple vendors unusual situations may arise. What happens during the day or with one vendor may not happen at night or with a different vendor.

2. Draft the check sheet/checklist/data sheet. Ask those who will be filling out the form for their input. Practice with a few situations and revise the document.

3. Implement the check sheet/checklist/data sheet. Depending upon the complexity of the form and how much of a change it represents from business as usual, implementing a check sheet may require a change process (Chapter 13, "Change Management" section).

4. Regularly audit and revise the instrument. Guard against bias that may be introduced (see Chapter 20). Proactively determine how each of the different types of bias might effect the results, and revise the procedure or form.

Example

Figures 16–4 through 16–8 show example check sheets for variables, attributes, and defect locations, a checklist, and a data sheet.

Dos and Don'ts

- Constantly revise and update the forms as new ideas emerge.
- Compare forms from different companies, and different fields. Borrow ideas from unrelated sources. Analyze what makes one magazine or brochure easy to read.

FIGURE 16–7
Checklist Example

Print Shop Client Order Checklist

- ☐ Client name and business name: ____________________
- ☐ Daytime phone: ____________________
- ☐ Bill or ☐ Payment upon receipt of printing (Checks higher than 300 only)
- ☐ Type of item
  - ☐ Envelopes
  - ☐ Stationery
  - ☐ Brochure   Number of pages: ____
  - ☐ Business card
  - ☐ Photocopy, one side
  - ☐ Photocopy, two sides
  - ☐ Other: ____________________
- ☐ Quantity
- ☐ Ink color(s): ____________________
- ☐ Paper weight: ______
- ☐ Paper color: ______
- ☐ Paper remarks: ____________________
- ☐ Trim size: ____________
- ☐ Required delivery date: __________
- ☐ Pickup or ☐ Deliver
- ☐ Typesetting required or ☐ Camera ready copy
- ☐ Customer ok's cost estimate

Order taken by: __________ Date: __________ Time: __________

- Get input from the people who use the forms on a daily basis.
- Simplify the form. Is there a way to extract information from the form, or in simplifying record-keeping, rather than making calculations on the form (e.g., recording only starting times, assuming that one time period's start is the other's end reduces the end point entry, and allows the interval of time to be determined later)? Complex forms may get filled out later, possibly misrepresenting information.

FIGURE 16–8
Data Sheet Example

Patient Temperature

Nurse	*Date*	*Time*	*Temp° F.*

DATA PRESENTATION

Aliases

Graphical displays.

Brief Description and Purpose

As the Chinese proverb suggests, one picture is worth 10,000 words, since pictures are readily fathomed, while words may not be. Rapid pattern recognition and matching is what makes humans fundamentally different from most computers (experimental computers that are based upon pattern matching are being developed). Understanding that the mind more readily accepts visual images rather than a stream of numbers (which is what a computer prefers), we can use data presentation as a tool itself. With humans, seeing is believing.[1]

[1]This is not true of other species. Cats, for example, rely on smell more than seeing. Hence, if a plate of food looks awful but smells good, humans will avoid it, but a feline will lap it up.

Discussion

A wide number of options are available, such as the pie chart, bar chart, scatter diagram, figure charts (the newspaper *USA Today* employs a cavalcade of these), and so on. The two often referred to as quality tools are histograms and scatter diagrams (both are part of Ishikawa's Seven Tools. See Chapter 8). Another interesting presentation tool is the boxplot, which plots the high and low value, and draws a box around the values where 50 percent of the observations lie (Figure 16–12). This tool is very useful in plotting highly concentrated data that nonetheless has outliers (i.e., data points that are considerably removed from the rest of the data points), a common occurrence in quality control applications.

Process and Examples

With software for desktop computers, data can be input and then manipulated into graphical format very simply. Some commonly used graph types are:

Line graphs connect the data points that are plotted on a graph. Usually the variable (measurement) of interest is plotted on the vertical scale, and the independent variable (such as time or length in centimeters) is plotted on the horizontal axis.

Histograms are bar charts, as shown in Figure 16–10. Histograms are bar charts that show a cluster of data over a specified range. This clumping together of data makes presentations easier to interpret. Not surprisingly, histograms can also be used to smooth out or hide anomolous or unusual data points. Figure 16–9 presents a suggested guideline on how many intervals (or bars) to make given the number of data points. Notice that histograms integrate readily with check sheets. The check sheet of Figure 16–4 even looks like a histogram turned on its side.

Scatter or XY graphs are "raw" data plots, as shown in Figure 16–11. No interpretation is attempted in scatter diagrams, although the choice of

FIGURE 16–9
Histogram Interval Guidelines

Number of Data Points	*Number of Intervals*
Less than 50	5–7
50–99	6–10
100–249	7–12
Over 250	10–20

FIGURE 16–10
Histogram (defects per hour)

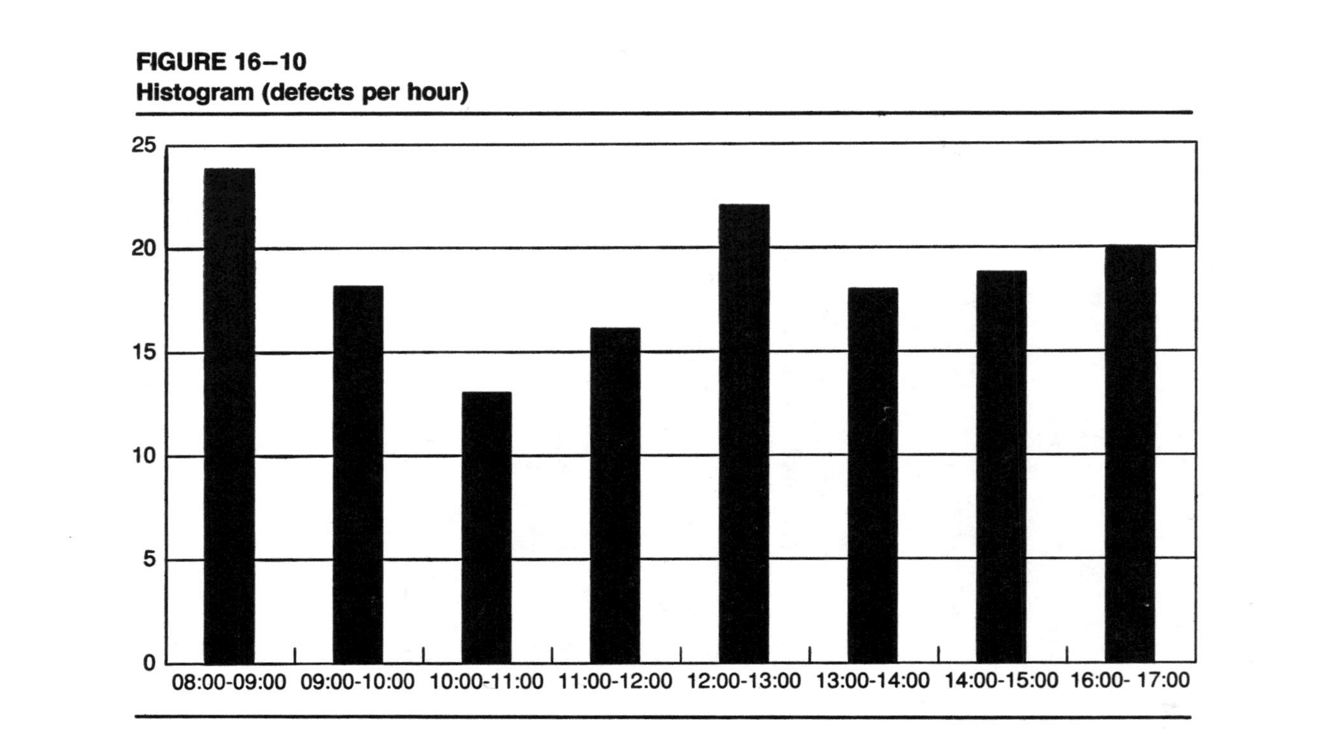

FIGURE 16–11
Scatter Plots

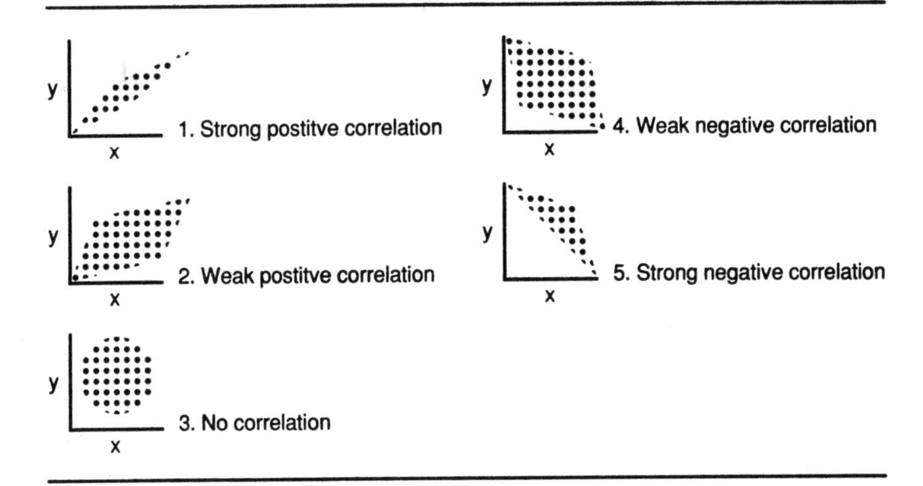

scale can distort the data, making it look as though there is a better (or worse) correlation to the data points than there is. A correlation is an association that can be inferred from the data. For example, a plot of blood pressure versus body weight indicates that the greater the body weight, the higher the blood pressure.

Pie charts are circular graphs useful in showing the percentage that each element occupies. Data that adds up to whole is amenable to being displayed in a pie chart, such as segments of a budget.

Column charts are the same as pie charts, except the data is presented in a columnar format.

High-Low Charts show the high and low for the data element. These charts are very useful in plotting stock prices, as they can show the high, low, and close.

Boxplots (or whisker and box) display the extremes (the whiskers) and put a box around 50 percent of the values. This sort of plot is useful when data are expected to be close to one another, but extremes or outliers are possible. Figure 16–12 shows the boxplot.

Dos and Don'ts

- Try more than one way to present the data.
- Try multiple presentations on the same graph: superimpose the scatter diagram on the histogram, for example.
- Produce a raw data scatter diagram, and then produce a histogram. Even scatter diagrams can be biased in their presentation, if the scale of the axis is

FIGURE 16–12
Boxplot

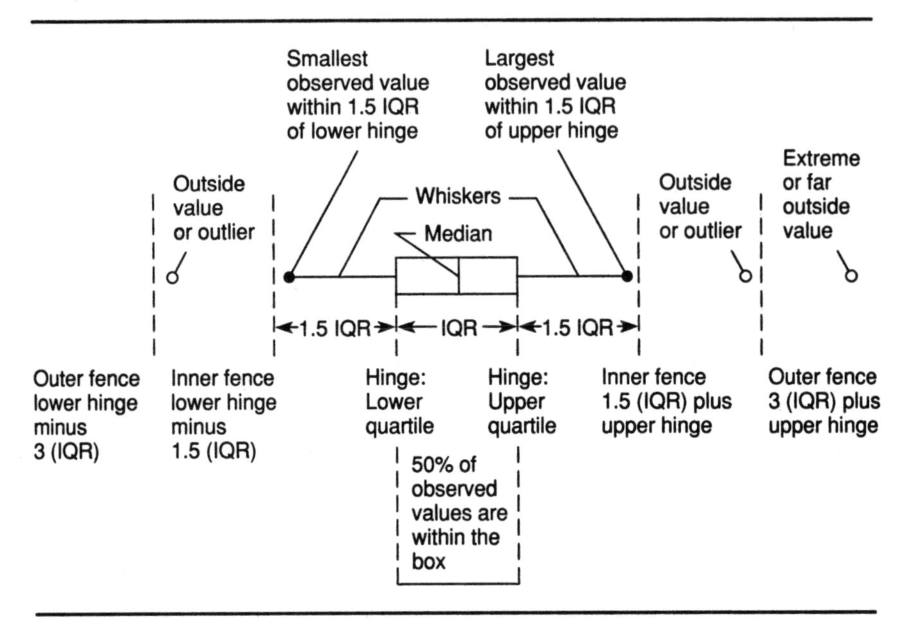

chosen poorly—too narrow or close on one axis and too broad on the other will distort the data. The adage "figures don't lie, but liars figure" certainly applies to the interpretation of graphs and drawing inferences without further investigation of the underlying metrics.

• Does it make sense to infer a correlation? Just because a correlation can be drawn doesn't mean there really is one. *Physics Today* once had a plot of the stork population in West Germany since World War II and superimposed on that was a plot of births since the war. As all children who are told that the stork brings the babies (a German folk tale), the nearly perfect correlation came as no surprise! Be willing to be suspicious of assumptions, and to perform experiments to determine and isolate the underlying cause.

FLOWCHARTS AND INPUT-OUTPUT ANALYSIS

Aliases

Hierarchical input-output analysis, relations diagram, Petri nets, and other specialized techniques similar to flowcharting.

Brief Description and Purpose

Flowcharts graphically establish the data and/or control sequence of events for a particular process. Objectives and inputs to the process are identified, as are outputs or outcomes, thus establishing the requirements of the process. Flowcharts, Petri nets, and input-output analysis are useful in coordinating multifunction or multidivision communication when integrating processes, or adding new processes to established ones. They can model a process as it is or as it could be.

Discussion and Examples

Input-Output Analysis

Input-output analysis is primarily concerned with enumerating the various inputs and outputs of a process, ignoring the inner workings of the process temporarily. By focusing on inputs and outputs, requirements can be detailed, without concern of the specifics of the process. The process can be detailed at a later time. An added utility of this tool is that managers can see how their staffs will be impacted, without being bogged down by process details, something usually left to the operational personnel. Cross-departmental coordination requirements may be easier to see using this method. An example is shown in Figure 16–13. Input-output may also be hierarchical, where the output from one process is the input to the next process.

Flowcharts

Flowcharts can model either control or data information. Control information is concerned with the if, ands, buts, whys, and wherefores of the sequence of events. Data flow is much laxer in that it simply models a transformation

FIGURE 16–13
Input-Output Analysis Example

Inputs	*Process*	*Outputs*
Corporate revenue trends		
Projected economic conditions		Five-year goals
Personnel outlook		
Payroll data		Revenue projections
Competitive benchmarking		

of data, and is generally unconcerned with the particulars of the transformation. Data flow models are useful in modeling strategic and planning elements, as they ignore process details, yet give an indication of the sequence of events. Control flowcharts express process details, and become muddled if they grow too large. Thus data models are usually constructed first, with operational details filled in using control flowcharts. We will use traditional nomenclature and call data flowcharts data flow diagrams, and control flowcharts simply flowcharts.

Data flow diagrams chart the sequence of data transformations. For example, hours worked per project are recorded on a time card, which is then given to the supervisor for verification. This verified time card is passed along to the accounting department, which also verifies the accuracy and posts the card, and produces a paycheck, which is (probably) verified by the employee. At each instance, the data was transformed: from hours worked to *verified* hours worked to a paycheck.

Data flow diagrams resemble simplified input-output charts which have been strung together. Details concerning the data (payroll rates, tax rates, social security, and so on) can be cataloged in a data dictionary. In Figure 16–14, the data transformation is noted by the process bubble or circle, and the data itself by the line. Data flow diagrams are so simple in execution that they need almost no explanation. Data flow diagrams can be *partitioned*—that is, one data transformation bubble can represent several bubbles "inside of it." That is, process bubble 2 could really consist of several more processes, perhaps internal to itself or that department. This feature makes data flow diagrams outstanding tools for providing both a high-level view for top- and mid-level management, so that different departments who process similar information understand what the previous and the next department to receive its output need, while not forgetting details for implementors. Partitioning is illustrated in Figure 16–15.

FIGURE 16–14
Data Flow Diagram Example

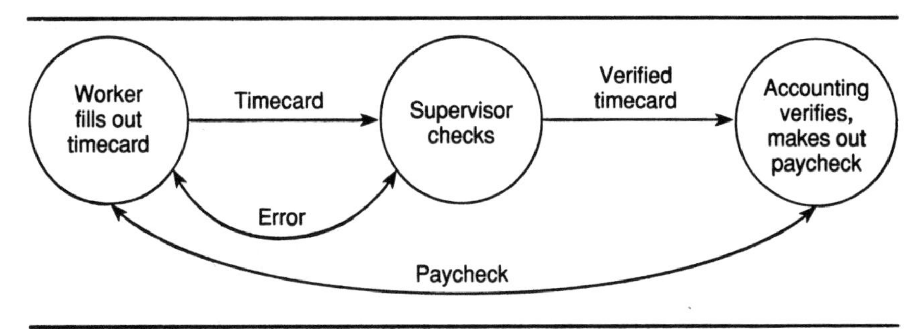

FIGURE 16–15
Partitioned Data Flow Diagram

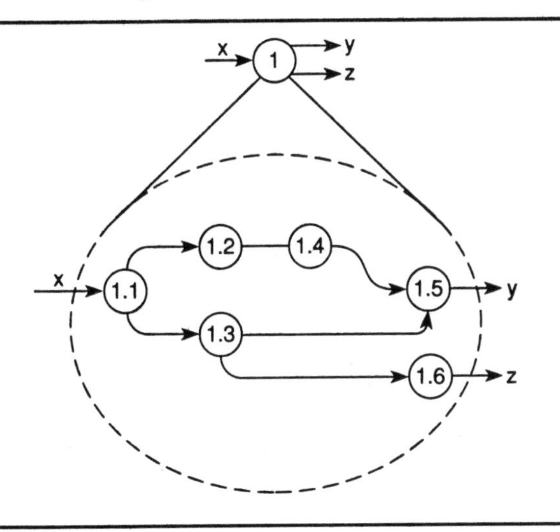

Flowcharts model the if, ands, and buts of a process, or the logic of a process. Traditional programming-like flowchart symbols are summarized in Figure 16–16, along with an example. Flowcharts can become cumbersome for complicated processes, so it is best to apply flowcharts to a portion of the problem. Ideally, the flowchart should not exceed four pages, and the optimal size is one page. Flowcharts can be used to model the process itself, after a data flow diagram has been made.

Relations Diagrams

The relations diagram is one of the "seven new tools" (Mizuno), and is a nonpartitioned chart showing the relationships between various cause-effect diagrams. The purpose of a relations diagram is to discover the root cause, or key processes in a web of causes.

This model is not terribly concerned about the data, it prefers to model what to do with it, and how to handle exceptions.

Relation to Other Concepts and Tools

Determining inputs, outputs, and process details can benefit from group techniques, as well as planning and organizing tools. Work flow analyses are flowcharts with a specific purpose in mind.

FIGURE 16–16
Flowchart Example

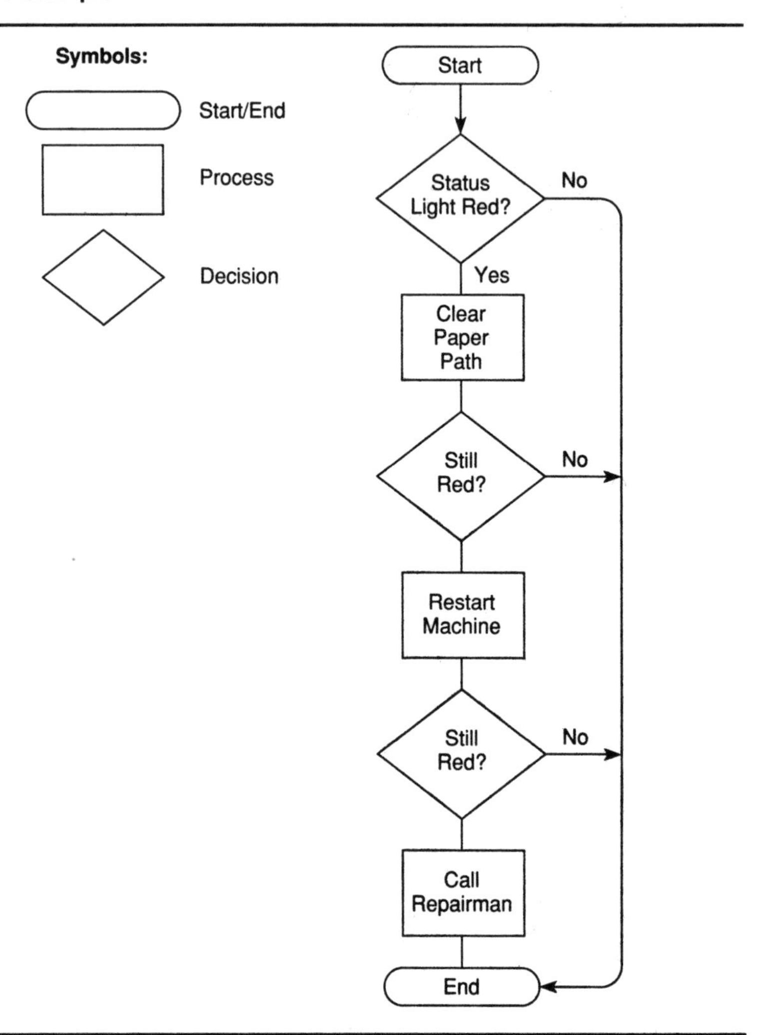

Process

*1. **Compose a context diagram.*** Context diagrams are a top-level input-output diagram, defining the scope, or context, of the work to be undertaken. It defines the major inputs and outputs. It consists solely of one process bubble or one input-output analysis process square.

2. Create a data flow diagram. The data transformation and information flow are next modeled using data flow diagrams. A top-down approach is usually taken. That is, executive level data is considered first, then what departments are involved in creating that information, and then what branches or individual functions are involved. A simple to more complex approach such as this avoids letting small details cloud major issues.

The team should walk through the process after making an initial model. Details are bound to be forgotten. Be sure to first model the process as it is now, even if change is the goal. Only after understanding how the process currently works can it be transfigured. When working in a group context, put the chart on a large piece of paper such as a flip chart—chalkboards become crowded fast.

3. Make the data dictionary. The data dictionary should be made at the same time with the context diagram and the data flow diagramming process. Data dictionaries ensure that everyone involved in the project has a consistent way of denoting data. Changes and updates to the data dictionary should be done through a control librarian.

4. Model the processes. The process bubbles can now be modeled by descriptive narrative, flowcharts, or by some other means.

5. Revise as necessary. Present the initial data flow diagrams to peers and those directly involved with the process. Revise the diagrams to reflect their input.

Dos and Don'ts

- Model the "normal" sequence of events first, and then model the exceptions. But do not forget to model the exceptions. Handling exceptions is what distinguishes a mediocre service company from a great one.
- Clarity of the chart is essential. Don't worry if a systems design analyst would laugh. Their charts are incomprehensible. Anyone should be able to decipher your chart.
- Don't worry if you're mixing data and control. You don't have to make two charts, although it may be useful in determining process improvements.
- Allow more time than you think will be necessary—it will be used. Several sessions may be required.

WORK FLOW ANALYSIS

Aliases

Work improvement study.

Brief Description and Purpose

Work flow analysis examines the work process for potential improvements in performance and quality of work life.

Discussion

Work flow analysis is really a special case of flowcharting and related process examinations. The goal is to systematically overcome the "not invented here" and "we've always done it that way" and other excuses for not changing work habits on the part of the employee as well as on management's part. Before changing, it is important to model the current work structure.

Work flow analysis should be distinguished from work measurement systems. Work measurement systems have often been imposed on employees, who then promptly figure out ways to "beat the system." Work flow analysis is done in an employee/management partnership, with the twin goal of improving productivity and quality of work life.

Relation to Other Concepts and Tools

Work flow analysis requires as input, information that may be gathered by surveys, auditing, benchmarking, brainstorming, flowcharting, and past data on the work process.

Process

1. Gather data concerning the present operation. Do this by observation and inquiry, *not* by hauling out a copy of a "Standard Operations Plan" (SOP) or other such unrealistic documentation. This SOP may be compared to the present work flow, but it should not form the basis for the way work is done. Flowchart the process.
2. Gather ideas on how to improve the process. Collect these ideas from any source you can, even the old SOP.
3. Identify the desired performance versus the actual performance, and identify any gaps (or areas that are exceeded). Identify the causes of

the gaps and propose a change or changes. Do not be afraid to eliminate unnecessary or duplicative tasks.

4. Analyze and review these changes with a cross-functional team.
5. Prototype these changes on a small basis. Work out any bugs in the new operations. Try a lead-the-fleet person to implement changes in their work, or try a certain area or shift.
6. Implement the changes on a wide scale basis. Flowchart the new process, and revise the SOP to reflect the new reality.

Example

Widgco Inc. decided to conduct a work flow analysis. They selected a self-contained shop area that made highly specialized components on a moderate volume basis. This area was a self-managed team, and the entire six-person shop was on the work flow analysis team, as well as a facilitator (hired from outside the firm), an accounting supervisor, and a procurement specialist.

The team first modeled their work using a flowchart which started at the raw material supplier and ended at the dock of the manufacturer who incorporated the part made by Widgco. The flowchart of the process included the approximate time it took for each process and subprocess. This flowchart was then examined, and it was determined that certain operations could be rerouted to create a more even flow of work.

A Pareto analysis was done on the time for each process, and a brainstorm session was then held to determine if the areas which took the most time could be reduced. It was determined that the exchanging of dies, which required many minutes of work, could be reduced—someone had read about how Toyota had used SMED (single minute exchange of dies) and had reduced the time to change dies from several hours to one minute (this was accomplished and chronicled by Shigeo Shingo). A subteam was created to investigate and prepare an analysis of how such a system could be used at Widgco.

Additionally some quality of work life issues related to work flow were identified, such as improved ventilation, placement of an ear plug and safety glasses caddy at the entrance, restocking of the first-aid cabinet on a regular basis, and elimination of reporting minor cuts and scrapes not requiring medical attention.

Dos and Don'ts

- Identify choke points or bottlenecks in the process, whether they are procedural or people oriented. Work to eliminate the bottleneck, even if it means eliminating a position.

- Involve the people whose work is being analyzed. Any analysis that leaves them out is inaccurate.
- Ask people to analyze the work flow who don't know why "it can't be done that way."

BIBLIOGRAPHY

Organizing

Ishikawa, Kaoru. *Guide to Quality Control.* Tokyo: Asian Productivity Organization, 1986.

Sarazen, J. Stephen. "The Tools of Quality—Part II: Cause-and-Effect Diagrams." *Quality Progress* 23 no. 7 (July 1990), pp. 59–62.

Wadsworth, Harrison; Kenneth S. Stephens; and A. Blanton Godfrey. *Modern Methods for Quality Control and Improvement.* New York: John Wiley & Sons, 1986.

Check Sheets and Data Collection

Gitlow, Howard; Shelly Gitlow; Alan Oppenheim; and Rosa Oppenheim. *Tools and Methods for the Improvement of Quality.* Homewood Ill.: Richard D. Irwin, 1989.

Ishikawa, Kaoru. *Guide to Quality Control.* Tokyo: Asian Productivity Organization; 1986.

Juran, J. M. *Quality Control Handbook.* New York: McGraw-Hill, 1988.

Juran Institute. "The Tools of Quality, Part V: Check Sheets." *Quality Progress* 23, no. 10 (October 1990), pp. 51–56.

Data Presentation

Almost any introductory text on statistics will cover graphical ways to display data.

Flowcharts and Input-Output Analysis

Burr, John T. "The Tools of Quality, Part I: Going with the Flow(chart)." *Quality Progress.* 23, no. 6 (June 1990), pp. 64–67.

Mizuno, Shigeru. *Management for Quality Improvement: The Seven New QC Tools.* Cambridge, Mass.: Productivity Press, 1988. (Translation of Kanrishi to *sutaffu no shin-qc-nanatsu-dogu*, 1979.)

CHAPTER 17

PLANNING

If you don't know where you're going you can't get lost.

Zippy the Pinhead (Bill Griffith)

Planning tools aid defining what problem we should be working on. This may seem obvious, but with shifting customer requirements, this is not the case. Planning tools also identify where networking or integration with other people or departments is required.

Planning maps out requirements so that tactical and logistical considerations can flow smoothly from customer needs. With a map we have a clearer idea of the task's magnitude and direction. It is doubtful that any business could long survive if it followed Zippy's observation. Yet businesses often follow this Zen-like dictum (it does have a certain Bohemian charm, however). They don't know why they sold, what they did, or for whom, or to what purpose. Studies have shown that the best customer is the most recent customer—repeat business is what keeps you in business. Disneyland has about a 70 to 80 percent return, for example. To keep them coming back, you had better know the who-what-why-where-when-how of your customers.

Planning tools such as goal setting and the Plan-Do-Check-Act cycle help establish these customer needs and desires into workaday procedures and objectives. Quality function deployment is one of the most important tools in the manager's and planners toolkit. This umbrella tool can quite literally translate customer requirements, needs, and desires into production specifications. Input to these tools is no small task, and will require a prodigious effort in data collection, organization, and self-examination.

DEMING CYCLE

Aliases

Plan-Do-Check-Act (PDCA) cycle, Deming wheel, and Shewhart cycle.

Brief Description and Purpose

The Deming cycle provides a systematic way to view continuous improvement. Primarily used in the development phase as a planning method, it is nonetheless useful throughout the life cycle of a product or service.

Discussion

The elements of the Deming cycle are:

Plan—a tentative plan is developed.
Do—a trial is done.
Check—monitor the results.
Act—modify the process, return to plan.

and the "wheel" is illustrated in Figure 17–1. Not so much a tool as a way to view evolution of change processes, the Deming cycle is applicable to many situations.

FIGURE 17–1
Deming Wheel / PDCA Cycle

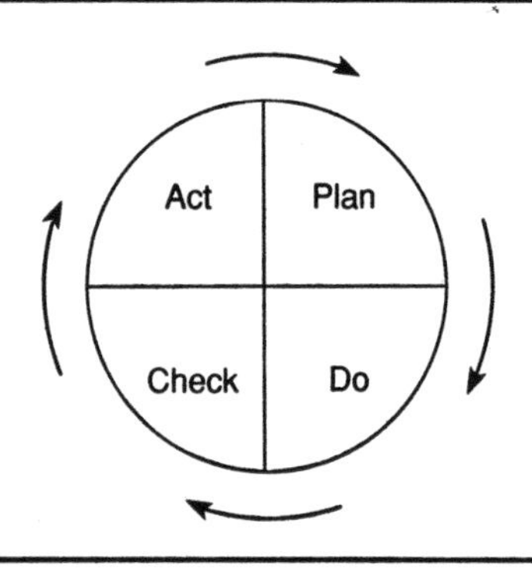

Relation to Other Concepts and Tools

The Deming wheel provides a feedback mechanism, and could well be an umbrella mechanism, under which many other tools are coordinated.

Process

- *Plan* a change, design, or test. Data must be collected on which to base this plan. The focus should be on overcoming the conformances, and not on vague numerical improvements (i.e., things that should be achieved, not 4 percent improvements versus 5 percent).
- *Do* the plan. Preferably on a prototype, test market, or sample. This is a trial plan, and should be done on a small scale prior to exporting the plan companywide. Might there be distortion when implementing the thing companywide? Quite probably, but if the characteristics of how it should work have been carried out on a small scale, then when exporting the plan only the scale factors will prove (newly) troublesome, not the operational aspects.
- *Check* or observe how the change worked. Did it reduce the difference between customer needs and performance of the process? Did the employees understand how and why the plan was supposed to work?
- *Act* on what was observed. Modify the plan—which brings us back to the *Plan* aspect.

Dos and Don'ts

- Try it on a small scale. If there is a failure, its effects can be minimized quickly. If it is successful, it is important to understand why it was effective, prior to implementing the process companywide.
- Exercise the use of the Plan-Do-Check-Act sequence whenever possible. The feedback critique is powerful in gaining competitive advantage, yet underused.
- Use personal computers that allow quick updates. The plan should be a "living document" that changes frequently.

FORCE FIELD ANALYSIS

Aliases

Push-pull analysis, positive-negative force analysis.

Brief Description and Purpose

Force field analysis is an accounting of positive and negative forces and their magnitude that have a bearing on a proposed solution or change in procedure. Analysis of the opposing forces generally occurs after a brainstorming and a cause-effect diagramming session. Force field analysis is also useful in determining which proposed solution, among several, will be least likely to meet with opposition.

Discussion

Force field analysis categorizes the forces affecting a given situation as positive or negative, and assigns a value to the degree that the force is negative or positive. The values can then be added up, if so desired. The closer the total is to zero, the more the status quo is likely to be retained as the desirable solution. The more positive the number, the more likely the proposed solution is the right one. A blank force field analysis worksheet is shown in Figure 17–2. Force field analysis aids in identifying those forces which must be reduced in order to proceed with the proposed solution.

Force field analysis is most useful on projects of modest scale. For dozens or hundreds of forces, a more sophisticated approach, such as the analytical hierarchy process (based upon linear programming techniques), may need to be used. Such techniques can be found in recent texts on operations research.

Relation to Other Concepts and Tools

Force field analysis requires a proposal and inputs. These may come from a data collection process, group techniques, and statistical tools. Assigning magnitudes and polarity to the forces may also require these same tools. Force field analysis may also feed into other tools, such as goal setting and quality function deployment.

Process

1. Identify the proposed solution.
2. Determine all forces that might impact the implementation of the solution. Do not assign polarities (positive or negative) or magnitudes at this stage.
3. Assign each force to be either a negative or positive, and determine the magnitude of the force. Use objective measures if possible to

FIGURE 17–2
Force Field Example

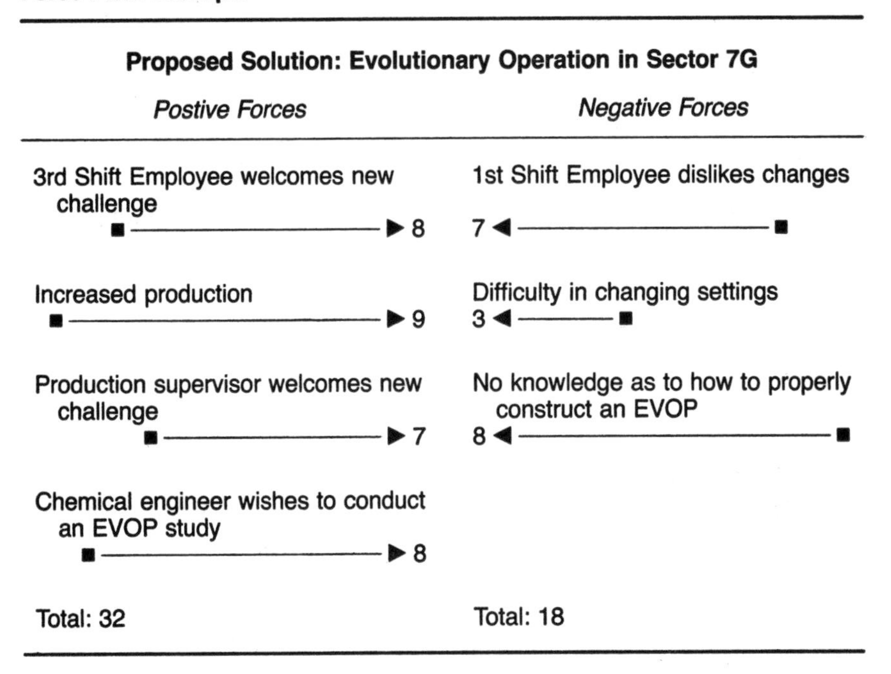

determine the magnitude of the force. If this is not possible, then use a technique such as Delphi to weight each factor as a group. The important thing is to reach group consensus, and to ensure that the same group assigns all magnitudes. Presumably, any bias may be equitable, assuming the composition of the group is an objective one.

4. Develop a strategy to lessen the negative forces and enhance the positive. This may require goal setting, or establishing a matrix, such as quality function deployment, to determine what factors are acting in concert. Root cause analysis may also expose the source problem (or benefit).

Example

Widgco desired to implement an evolutionary operations process (EVOP) in a production line creating a commercial fertilizer and herbicide mix. The process variables were temperature of the chemical process, time spent at each stage of mixing, and the amounts of three chemicals which could be varied slightly.

A brainstorming session with a chemical engineer, production line supervisor, and several production line workers identified various forces that might impact an EVOP implementation. The team then determined polarity and assigned weight factors to each.

A strategy was developed to defuse the negative forces:

- Start EVOP with the third shift. After production changes have been made, then implement them in the second and first shift.
- Send the chemical engineer to an evolutionary operation seminar, or hire a consultant, based upon a total life-cycle cost analysis.

Dos and Don'ts

- Assignment of the magnitudes is arbitrary in many situations, and a team effort is needed to determine the relative values of the forces.
- Force field analysis shows what bumps may lie ahead. Use it in advance of implementing changes, and for identifying underlying assumptions that everything will fall into place.
- Goal setting may benefit from a force field analysis which determines impediments (and aids) in implementing the goals.

GOAL SETTING

Aliases

Management by objectives (MBO).

Brief Description and Purpose

The setting of goals and objectives is at the heart of any planning effort. An *objective* is a quantitative statement of future expectations with an indication of when it should be achieved. By comparison, a *goal* is a nonquantitative statement of general intent, aim or desire; the end point or condition toward which management directs its efforts and the efforts of available resources. Goals guide the organization for years to come, objectives guide for 4 to 12 months.

Discussion

Goals and objectives reveal the purpose, ambitions, aspirations, intent, and design of the organization. Poorly set or unrealistic goals frustrate organizations as they do individuals. Goals must be translated into quantifiable and known objectives of the organization. Objectives that cannot be measured (or compared) in some way are useless, as are goals and objectives that are hidden from the work force. Goals are necessary as road maps and statements of purpose in organizations to ensure that everyone is moving in the same direction. It is important that goals be set and understood by everyone in the organization. Goals must flow down the Quality Management pyramid (Figure 17–3) from strategic goals to managerial objectives to operational strategies, and a feedback mechanism must ensure the communication of the disposition of these goals.

In 1954, Peter Drucker first espoused the idea of management by objectives (MBO). MBO is a process where the management of an organization identifies its goals and defines each individual's area of responsibility in terms of expected results. These objectives are used as guides for operations and assessment. MBO is premised on the beliefs that the clearer the idea of what you want to accomplish, the greater the chances of accomplishing it, and real progress can only be measured in relation to what one is trying to achieve. In other words, if you know where you are going, while you may get lost, you should know how far you have to go, and in which direction.

Goals should either compose the vision statement of the organization or be derived directly from it. While top management is responsible for this, in order for the goals to be realistic, input must be solicited from all levels of workers. Goals are shaped by one's culture, and goals incompatible with that culture must not be developed.

Objectives are dynamic, and must be reviewed and updated on a regularly scheduled basis, no less than once a year. Should achievement of an objective be easier than expected, identify the reason for that success. Apply successful techniques to other areas. Of course, if other objectives were missed, they must be thoroughly analyzed. Were the expectations too high? Was it unachievable at the present time? Was inadequate training to blame?

The primary advantage to be realized from MBO is the development of a systematic planning, monitoring, and improving approach to managing your organization. It forces you to regularly look at what your organization is doing and how it can be changed to better accomplish your mission. It gives workers clear purpose and targets. Managers who do not have systematic programs to plan and evaluate can only react to change. They then become managed by instead of managing the events impacting their organization. Obviously, even the best managers cannot foresee the future and prepare in advance for all changes. Thus flexibility needs to be built into the system. Leave 10, 20, or 25 percent of the time and goals to be determined during the period. Unforeseen and sudden changes will be devastating if some slack has not been built into the system.

An objectives-based management program is an excellent way to track operations. As your deviations become apparent between what is needed and what is being accomplished proactive measures can be taken to assure success.

Relation to Other Concepts and Tools

Inputs may be provided by brainstorming, cause-effect diagrams, input-output analysis, or a variety of other techniques.

Process

For each goal set, there will be one or more objectives. These objectives should have a corresponding strategy. Strategies are approaches that must be taken to accomplish the goals. Long-range strategies become the bases for mid-range goals, and mid-range strategies become short-term goals. In addition to goals and strategies established by management and/or a team, the individual must determine appropriate tactics and logistics. This is illustrated in Figure 17–3. The tactics and logistics portion are generally not addressed in most discussions on MBO, as they are in the domain of the individual worker. However, without proper resourcing and logistical considerations (such as secretarial support, work tables or rooms, and special material), MBO is sure to fail.

Goals must be realistic and attainable. Unrealistic expectations eventually severely damage an organization, just as they do individuals. Objectives must also be measurable, but not necessarily expressed as a "percent improvement." If goals are not understood and agreed upon by the individual, most likely they will find ways around *the system*. For example in a northeastern state social workers' *productivity* was measured by the number of cases per unit time. Workers seeking a high rating found clever ways to open and close cases quickly, ignoring the problems of those who might take more time than usual. This simple measure was inadequate. In a real system, there will be cases which require a lot of time, and cases which are simple. Data collection and measuring the standard deviation would have aided in determining who was "cheating" and most likely giving clients little or no attention. Unfortunately, such measurement systems require a knowledge of statistics, which is a rare asset in the United States today.

FIGURE 17–3
From Goal to Implementation

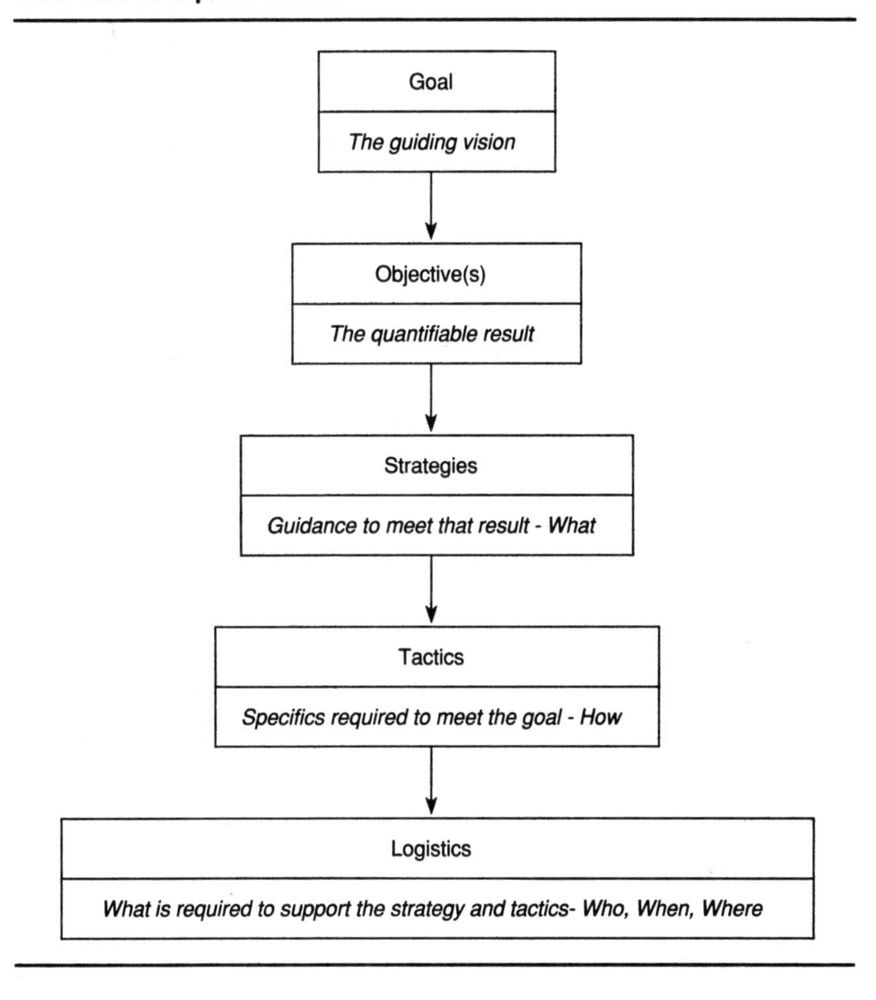

The following provides a step-by-step process to formulating clear and attainable goals and objectives:

Goal statements. Define the current situation and what it should be upon successful completion of the goals and objectives. Be as specific as possible. This step forms the basis for everything to follow. The broadest of goals should be simple enough to fit in one short sentence—"We're the cheapest and fastest small parcel delivery service," or "Absolutely, positively overnight.'' These are so simple they might almost be corny—if they were not backed up by hundreds of employees believing in fulfilling their commitment.

In preparing a vision, the organization can perform a situational analysis which provides an appraisal conducted in four areas: strengths, weaknesses, opportunities, and threats. Strengths and weaknesses are considered to be internal to the organization, while opportunities and threats are external. A summary of these factors is summarized in Figure 17–4.

FIGURE 17–4
Situation Analysis Factors

I. Internal factors
   (strengths and weaknesses)
   A. Management
      1. Leadership
      2. Planning
      3. Development of personnel
      4. Delegation
   B. Functional areas
      1. R&D
      2. Production
      3. Inventory management/production control
      4. Finance/accounting
   C. Resources
      1. Financial
      2. Facilities
      3. Employees
      4. Information

II. External factors
   (opportunities and threats)
   A. Macroenvironment
      1. Economic conditions
      2. Political conditions
      3. Regulations and policies
      4. Social influences
      5. Demographic data
   B. Microenvironment
      1. Services or products—competition
      2. Technology changes
      3. Labor market
      4. Other organizations

Objectives. The objectives should flow from the goals. Well-developed objectives should have the following characteristics cited in the checklist of Figure 17–5.

Juran in his *Quality Control Handbook* has stated that:

> Only when men know their responsibilities can they devote their energies to carrying out their duties. Lacking this knowledge, it is inevitable that some or much of their energy is devoted to discovering just what is their responsibility.

In order for objectives to work, they must be clearly communicated to the individual involved. The individual must "buy in" to the objective. This requires two-way communication, and a good many questions from the individual.

The format of an objective can be developed according to Figure 17–6. For example:

Start off with an action phrase: "The XYZ division will . . ."

FIGURE 17–5
Checklist for a Well-Stated Objective

- ☐ Do any of the objectives conflict?
- ☐ Does it state what the objective hopes to achieve?
- ☐ Does it state who has responsibility for doing it and who is responsible for supporting it, if appropriate?
- ☐ Does it give a completion date?
- ☐ Is there a justification for the objective?
- ☐ Does it support the organization's vision?
- ☐ How will it be known that the goal has been achieved?
- ☐ Is effort required to achieve the goal beyond "business as usual?"
- ☐ Are the goals clearly stated, realistic, attainable, and easy to remember?
- ☐ Are the goals limited in number? (Seven or fewer is ideal.)
- ☐ Is the relative importance of the goals clear? Is it stated which is first in importance, and so on?
- ☐ Did the goal result from participation by those responsible for carrying them out?
- ☐ Is the goal controllable by the individual responsible for its accomplishment?
- ☐ Is authority equal to responsibility?
- ☐ Is there a way the goal can be revised?
- ☐ Are the strategy, tactics and logistics left to the individual? If not, is it clear who has such authority and responsibility?
- ☐ Does it clearly identify what is to be done?

FIGURE 17–6
Format for a Well-Stated Objective

1. Identify the responsible person(s).
2. Action verb, such as develop, manufacture, publish, implement and test.
3. Specific, measurable end result. The person responsible must be able to determine when they have accomplished the objective.
4. Specific time period and/or date of completion.
5. Resources to be committed (i.e., dollars, manpower, time, space, computer time).
6. Be sure that the who-what-when and how much are covered, but not the method.

Identify a single-key result: ". . . produce a working prototype of the ABC project."

Identify time frame: ". . . within six months of completed schematics."

Identify calendar time: ". . . (by December 1, 199x)."

Identify costs (i.e., dollars, time, materials, and equipment): "This will require 60 percent of the manpower within the division, and require a capital investment of $210,000, and $40,000 of on-hand inventory. It will require 20 percent of the machine shop labor for two months."

State the "All-done" criteria: "The prototype will meet or exceed all operating and reliability characteristics set forth in the Statement of Work (SOW) in government contract DAA 0000. The critical factor in the SOW is the operation of the prototype for 1,000 hours continuously with no failures."

Objectives may be categorized. After developing the goals, the organization may define the few critical success factors. These critical success factors are those which are absolutely required to achieve/sustain a goal. Other objectives may be classified as directly or indirectly supporting these critical objectives.

Service objectives. Peter Drucker in *The New Realities*, 1989, stated:

> It is commonly argued that public service institutions aim at intangible results, which defy measurement. That would simply mean that public service institutions are incapable of producing results. Unless results can be appraised objectively, there will be no results. There will only be activity, that is costs. To produce results it is necessary to know what results are desirable and be able to determine whether the desired results are actually being achieved.

The problem with determining quantifiable goals in white-collar and service areas is that you may have to think. The old factory floor piece rate goal translates poorly to service work ("Answer 30 phone calls per hour—But be polite to everyone") and, in fact, doesn't work very well for the factory. Data collection and applying statistics may be in order. Alternatives may also be required, such as user satisfaction surveys. Care must be taken in this instance as surveys are usually light on serious criticism. There may be no better cross check on MBO than MBWA (management by walking around) in determining how well objectives are being attained.

Strategy. Strategies are the means of achieving the objective, and may be as diverse as creativity allows. Strategy should be realized by a team or partnership. Strategies begin to determine the "How" of the objective. Further refinement may be needed to define specific tactics and/or logistics needed to get the job done. The goal and objective may need to be revisited and updated when the logistics have been fully profiled.

Many of the tools described in this section are conducive to formulating strategy.

Systematic diagrams. The systematic diagram provides a graphical look at goals and objectives, and can be plotted throughout a product life cyle. Figure 17–7 depicts a systematic diagram for a manufacturing process improvement scheme at Widgco, Inc. Note that systematic diagrams primarily show the context of the objectives and strategies, rather than establishing individual work goals.

Example

An example of a goal, one of its subsequent objectives, and resultant strategies, is depicted:

Goal

- All employees doing continuous process improvement.

Objective

- Division XYZ will initiate four improvement projects within the next quarter, costs not to exceed $20,000 total.

Strategies

1. Evaluate continuous improvement (CI) training.
2. Evaluate techniques conducive to CI.

FIGURE 17–7
Systematic Diagram for Widgco

Purpose/Vision	*Goal*	*Objective*	*Strategy*
To Develop a High-Quality Microwave Oven	Improve performance	Improve seal	Use piano hinge
			Use plaster lining material
		Reduce door warranty repairs	Use push instead of pull door handle
			Replace humidity sensor with T43
	Improve productivity	Reduce waste	Kanban course
			SCC course
			Automed variance analysis
		Produce 20 more units/ week	Work flow study
			Facility layout study
	Improve user friendliness	Increase ease of programming	Use food names on settings
			Increase number of single-key operations
		Increase evenness of cooker	Smoother liner
			Provide tray with unit rotating rotator

3. Train all employees in CI techniques.
4. Establish teams to determine potential CI projects.
5. Examine, evaluate, and select at least four projects.
6. Initiate at least four projects.
7. Establish feedback mechanism to determine efficacy of CI techniques.
8. Modify and repeat strategies 1–5 as needed.

Some additional examples of objectives:

• The head of the maintenance department and the chief of computer operations will act as a team to be responsible for developing and implementing a computerized program for building maintenance by October 31 at a cost of no more than $2,000 and 40 work-hours.

- The executive team working with the personnel director, shall take actions to reduce the organizational absentee rate from 9 percent to 5 percent by September 1, at a cost not to exceed 35 work-hours and with no increase in the existing budget.
- The training director will develop a communication skills program for mid-level managers, costing no more than 40 work-hours to develop and no more than $500 per manager to operate (including the training department's budget as well as the cost of the manager's time, with a maximum allowable time change of $300 per manager). The program will begin by October 1, 1988 and be 80 percent completed by February 15, 1989.

Dos and Don'ts

- Seek input from those implementing the goals. In this way a bilateral contract is established, and mutual understanding has a hope of being achieved.
- Goals and objectives make us grow. They are not simply a "laundry list" of things to do in a quarter or during the year. Goals which do not stretch our capabilities are not worthy goals. As in sports, stretch gently. Overreaching goals creates negative self-impressions, which can reduce creativity and confidence.
- Be creative. The rather bland "all-done criteria" in the above example could have been stated that: "The customer will be so delighted with our product that they will showcase it at every opportunity, and will favor us with continued business." This may sound more like a vision statement than an attainable goal, but try different ways of expressing the desired outcome.
- Integrate the systematic diagram into goal setting and project planning.
- Don't be afraid to post the systematic diagram on the wall. It's a useful tool for management overview and allows workers to see how their work interconnects with others.

PROCESS DECISION PROGRAM CHART

Aliases

PDPC, contingency plan.

Brief Description and Purpose

The process decision program chart (PDPC) presents in a graphical way, much like a flowchart, contingencies and countermeasures that may be engaged in a process. This tool provides for anticipating problems, and provides an avenue of approach should the problem occur. It may also lead to design changes in the process.

Discussion

The PDPC formalizes what might often be done in planning a process—planning for the contingencies, flukes, and surprises that may arise. By being presented in a graphical format, it appeals to both cognitive and affective domains of the mind.

It is quite similar in concept to failure modes, effects, and criticality analysis (FMECA), in that both seek to enumerate the ways in which something could go wrong. FMECA has preventive maintenance as its end goal, and is not particularly concerned with changing the product or process, whereas PDPC seeks to improve the product or the process making the product. Further, a PDPC can be readily done during the early stages of design—a FMECA is generally difficult to conduct until the product has been completely designed. Hence PDPC is considered a planning tool for our purposes rather than a self-examination tool.

Relation to Other Concepts and Tools

The items that create a PDPC would come from group techniques and data collection. A flow-chart of the process may also be highly useful in creating the PDPC. An audit or review of the contingencies may need to be made for feedback.

Process

1. Brainstorm regarding the process at hand, focusing on the way things can go wrong.
2. Create a cause-effect diagram and/or a flow diagram relating the stages of the process.
3. Create the PDPC.
4. Determine ways that the process or product could be redesigned to avoid encountering traps.
5. Follow through to see if other unforeseen problems arose.

Example

Figure 17–8 is adapted from the PDPC example given by Shigeru Mizuno in *Management for Quality Improvement*. In this case, Widgco wishes to deliver a bulky fragile item to a newly industrializing country. Problems have been encountered with the item being crushed in transit. A cross-functional team

FIGURE 17–8
Widgco PDPC for Fragile Item Delivery

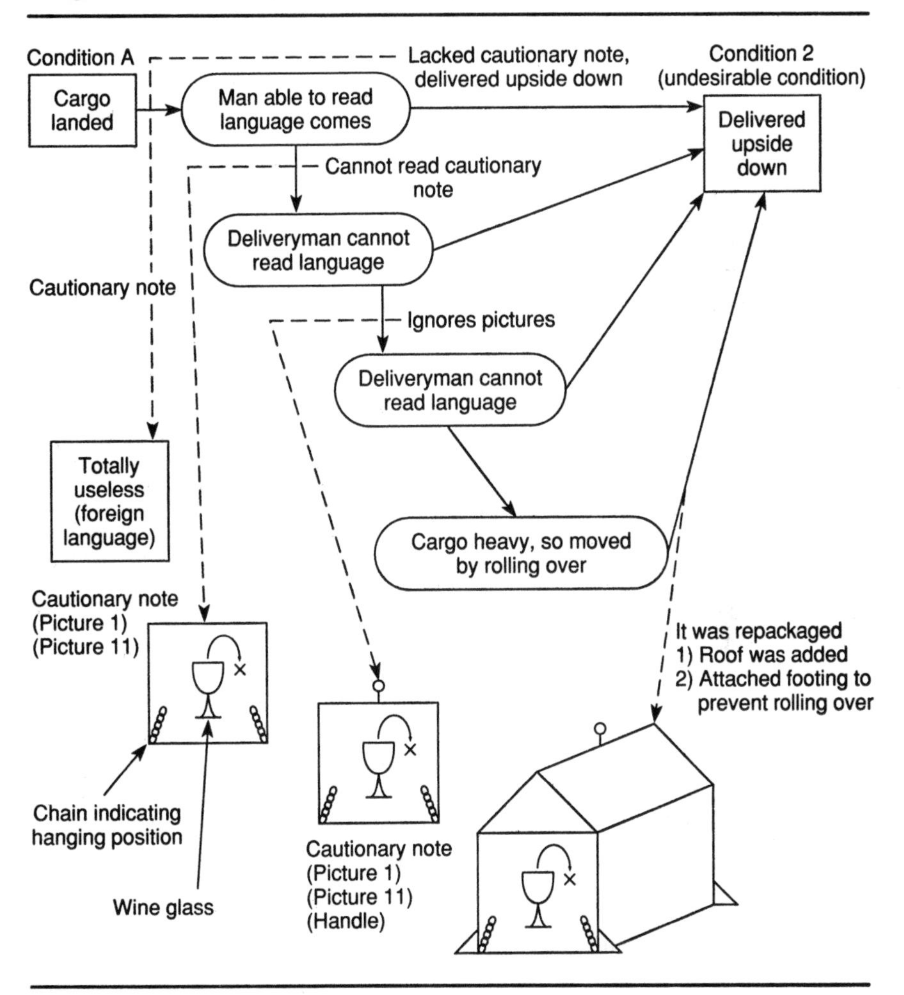

Source: Adapted from Shigeru Mizuno, ed., *Management for Quality Improvement: The Seven New QC Tools* (Cambridge, Mass.: Productivity Press, 1988).

from production, shipping, engineering, and procurement has developed the PDPC. The end result was a redesigned shipping container.

Dos and Don'ts

- Audit or foolproof the diagram by using it and deliberately try to misinterpret it.
- During the creative process, do not cost justify or be concerned with technical details.
- Test the end result on workers directly involved.

QUALITY FUNCTION DEPLOYMENT

Aliases

House of quality, matrix diagram, quality function evolution.

Brief Description and Purpose

Quality function deployment (QFD) is a matrixlike planning tool capable of integrating customer requirements into design features, which in turn cascade into production requirements. QFD may (and should) be used by every function in the producing organization in every stage of the product or service development. QFD begins with basic research and continues through sales and after marketservice.

Discussion

QFD was first used in an industrial application by Mitsubishi Kobe in 1972. QFD integrates the requirements of the customer because that is where the QFD starts. Side benefits of QFD are reducing engineering changes (experience indicates as much as 30 to 50 percent), reducing the design cycle (30 to 50 percent), and reducing start-up costs (20 to 60 percent). QFD is not something done out of the quality department. It is a companywide tool requiring cross-functional teams. Other benefits of QFD are:

- Reduced number of changes after the product is on-line.
- Hears the voice of the customer.

- Proactive instead of reactive.
- Prevents things from "falling through cracks."
- Fast and economical.
- Easy to learn (but not necessarily easy to do).

Figure 17–9 illustrates the process of translating the customer requirements into production requirement. Each *house of quality* (Figure 17–10) is a matrix of How and Why elements. The *roof* indicates the relationship between the elements (correlation matrix)—whether they support, counteract, or are neutral. A *chimney* might be added to prioritize the What segment. A sidebar may be added for comparing your product to the competitor's, and additional sidebars may be added for target values, competitive assessment of the What and How elements, and an importance weighting of the How elements.

American Supplier Institute gives the following example of translating a customer requirement into a production requirement:

1. Customer requirement:
   Paint job that gives years of durability.
2. Design requirements:
   No visible exterior rust for three years.
3. Paint characteristics:
   Paint weight: 2-2.5 gm/m2.
   Crystal size: 3 maximum.
4. Manufacturing operations:
   Dip tank, three coats.
5. Production requirements:
   Time: 2.0 minutes minimum.
   Acidity: 15–20.
   Temperature °C: 48–55.

FIGURE 17–9
Translating Customer Attributes into Production Requirements

	Phase 1	*Phase 2*	*Phase 3*	*Phase 4*
What	Customer attributes	Engineering characteristics	Parts characteristics	Key process operations
How	Engineering characteristics	Parts characteristics	Key process operations	Production requirements

FIGURE 17–10
House of Quality Template

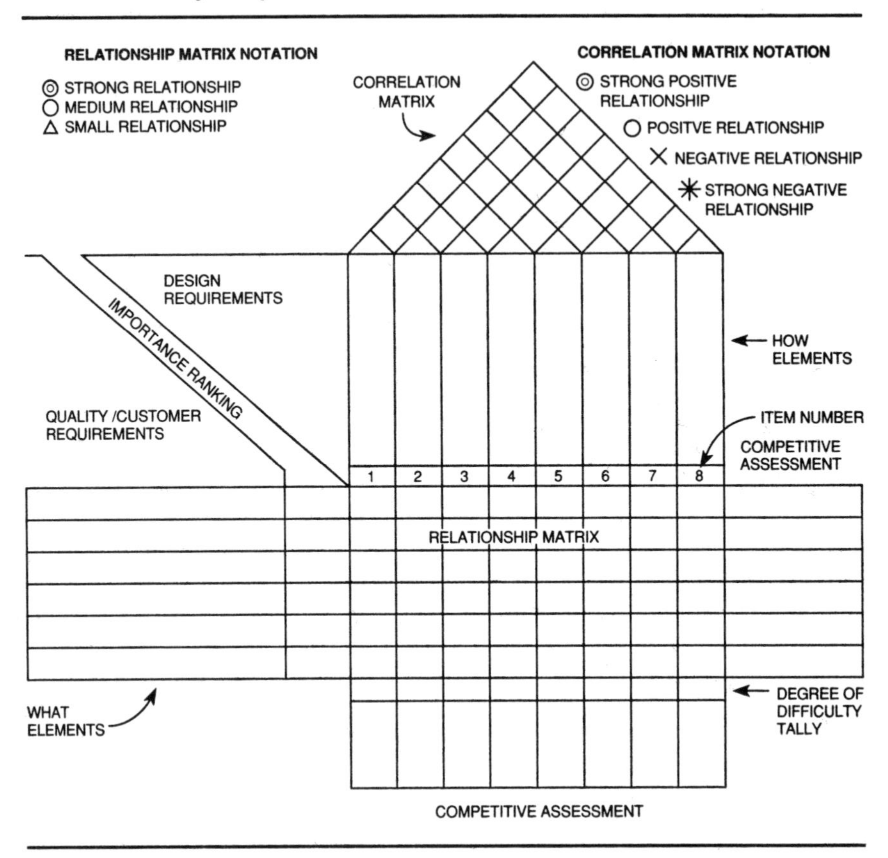

Relation to Other Concepts and Tools

Input to the QFD will require team sessions engaged in brainstorming, variance analysis, Pareto analysis, cause-effect analysis, and other techniques that may aid in gleaning information about the customer requirements and the corresponding design and manufacturing requirements.

Process

1. Determine customer requirements and prioritize. Start with a list of what the customer wants. This is the What, and will be placed in the rows of the house. A customer importance weighting should be associated with each item.

2. List technical design requirements and correlate them. The How columns are the technical points to the product under consideration. Each requirement has either a target value, or is to be maximized or minimized. This is summarized in the bar below the *roof,* which assesses how well (or poorly) the technical aspects correlate. Target values can be placed in the bar below the main body of the *house*. The design requirements are compared in the *roof,* and can correlate strongly positive, positively, negatively, or strongly negatively. Strongly negative relationships indicate areas that may need to be researched—changing a strongly negative correlation to a positive one could be a decisive competitive advantage.

3. Determine What and How relationships. The main body of the matrix can be filled out by determining the nature of the relationship of the What element to the How element. The relationships are weighted. In the example, a strong relationship is weighted at nine, medium is three, and small is weighted at one. The relationship weightings may be revised (e.g., using 5,3,1) as necessary.

4. Tally the "scores" of the relationships. The tally for each design requirement (column) can be made by multiplying the type or relationship weighting factor by the customer importance (thus it is essential to have higher customer importance be a higher number). Doing this prioritizes the technical requirements, and could be considered the *end result* of a single house of quality, although there are many quality factors that can be explored from the benchmarking assessment.

5. Perform a competitive benchmark assessment. This competitive assessment can be done on both the technical design requirements and the customer requirements. Note too that customer complaints are logged. Competitive advantages can be readily seen, and areas of improvement can also be identified.

6. Repeat until customer requirements are translated into production requirements. Now use the design requirements as the What, and part characteristics or manufacturing characteristics as the How, and so on, until specific production requirements are identified.

Example

A coke foundry example is presented in Figure 17–11. Note that the requirements of the customer and the design are categorized. This aids in determining niche market appeal. A competitive benchmarking has been

FIGURE 17-11
Coke Foundry Example House of Quality

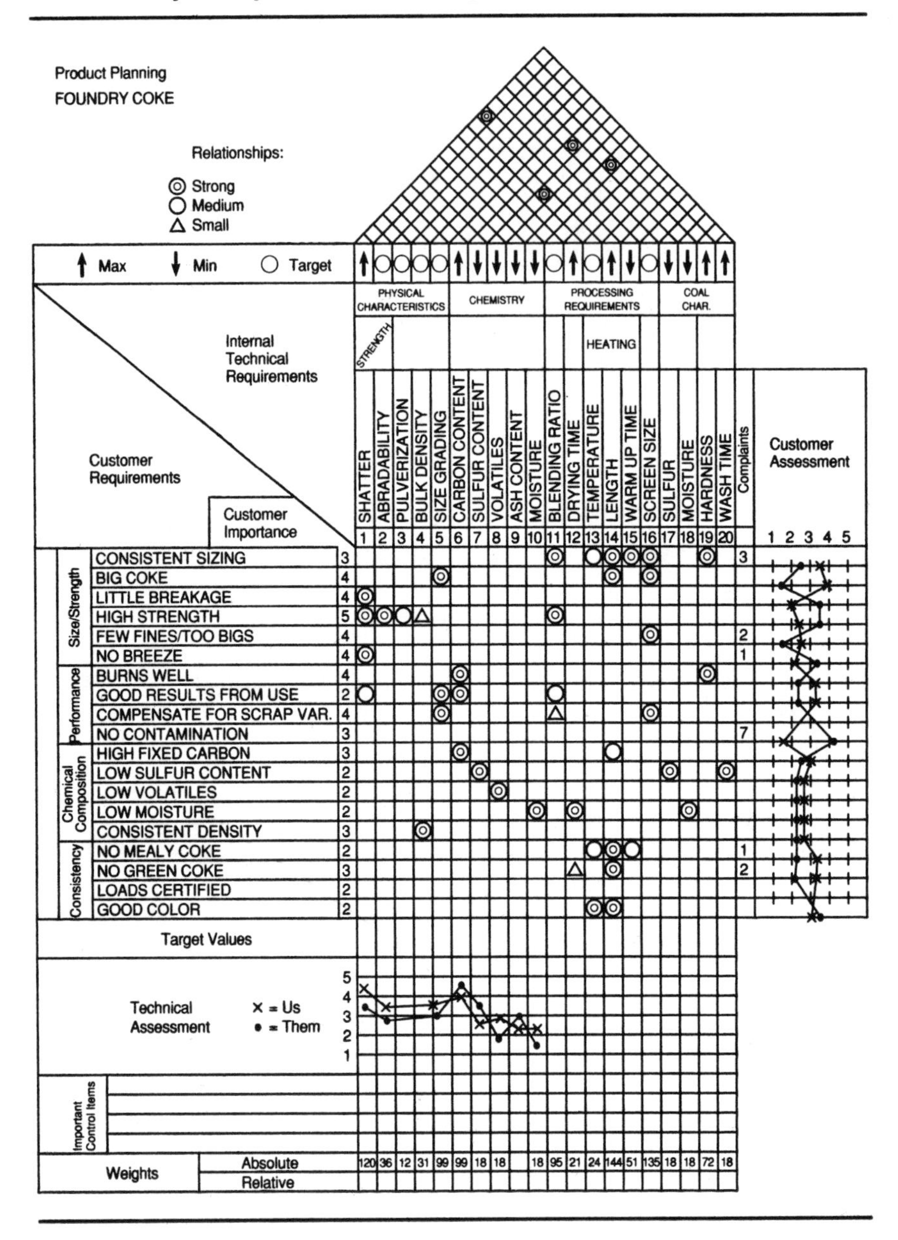

made on technical requirements as well as the customer requirements. In benchmarking the customer requirements, the customer's assessment was used. Perception is a key factor here. A second customer requirement assessment from the manufacturer's point of view might reveal where discrepancies between perceived and actual quality lie. It may be that advertising must get the word out about quality levels that actually exceed the perceived levels.

Dos and Don'ts

• Developing a QFD, especially a multiphased one, is difficult and requires a lengthy time commitment.

• Take courses specifically on QFD (sources may found in the annual *Quality Progress* service directory in the August issue). The more examples of QFD you are exposed to, the better your own QFD will be. There are many other sidebars and additional bits of information that can go into a house of quality.

• Start off with a "toy" project before using QFD on a "real" project. Use an example from a book, or use a made up example pertaining to the organization. Assess problems encountered.

BIBLIOGRAPHY

Deming Cycle

Any work on Deming's philosophy discusses the Plan-Do-Check-Act cycle.

Force Field Analysis

GOAL/QPC. *Memory Jogger*. Lawrence, Mass.: GOAL/QPC, 1988.

Juran, J. M. *Quality Control Handbook*. New York: McGraw-Hill, 1988.

Paton, Scott M. "Force Field Analysis Proves Money Isn't Everything." *Quality Digest*, April 1989, pp. 21–25.

Goal Setting

Drucker, Peter. *The New Realties*. New York: Harper & Row, 1989.

Drucker, Peter. *People and Performance: The Best of Peter Drucker*. New York: Harper & Row, 1977.

Latham, G. P., and Locke, E. A. "Goal Setting—A Motivational Technique that Works." *Organizational Dynamics*, 8 (Autumn 1979), pp. 68–80.

Marlow, Edward, and Richard Schilhavy. "Expectation Issues in Management by Objectives Programs." *Industrial Management*, January/February 1991, pp. 29–32.

Sherwin, D. S. "Management of Objectives." *Harvard Business Review* 54 (May/June 1976), pp. 149–60.

Process Decision Program Chart

Mizuno, Shigeru, ed. *Management for Quality Improvement: The Seven New QC Tools*. Cambridge, Mass.: Productivity Press, 1988.

Quality Function Deployment

Akao, Y. *Quality Deployment: A Series of Articles April 1986–March 1987*. Lawrence, Mass.: Growth Opportunity Alliance of Greater Lawrence, 1987.

American Supplier Institute. *Quality Function Deployment*, 1989.

Cohen, Louis. "Quality Function Deployment: An Application Perspective from Digital Equipment Corp." *National Productivity Review*, Summer 1988.

King, Bob. *Better Designs in Half the Time: Implementing QFD Quality Function Deployment in America*. Lawrence, Mass.: GOAL/QPC, 1987.

Sullivan, Laurence P. "Quality Function Deployment." *Quality Progress*, June 1986, pp. 39–50.

Sullivan, Lawrence P. "Policy Management through Quality Function Deployment." *Quality Progress* 21, no.6 (June 1988), pp. 18–20.

Vasilash, Gary S. "Hearing the Voice of the Customer." *Production*, February 1989.

Chapter 18

SELF-EXAMINATION

I keep six honest serving-men
(They taught me all I knew);
Their names are What and Why and When
And How and Where and Who.

Rudyard Kipling

Self-examination or diagnostic aids are a primary facet of any feedback mechanism. Organizations need to understand where they fit in compared with other organizations in their field. This benchmarking exercise has been used as a quality driver for some companies, such as AT&T and Alcoa. By comparing and contrasting one organization to another, strengths and weaknesses are exposed. Once laid bare, strategic advantage can be taken of such knowledge.

Auditing, benchmarking, and quality costs can be companywide in scope. Failure modes and effects analysis (FMEA), and foolproofing are generally applied to a single product. FMEA is generally applied to complex products, with multiple layers of assemblies. Foolproofing works well for one assembly or relatively simple item or unit. Who-what-when-where-why-how is a technique useful in almost any situation at any time.

AUDITING

Aliases

Reviews, walkthroughs, and internal audit.

Brief Description and Purpose

Auditing is a feedback mechanism that can be applied to departments, companies, and vendors. Audits are examinations or reviews of records, products, or processes, and can be a self-assessment tool or as a supplier certification process, or in a number of other ways.

Discussion

For some, auditing evokes images of solemn green eye shaded accountants. Auditing as a Quality Management tool need not be. It is an independent look at practices, processes, and procedures of an individual, department, division, company, or supplier. It can be used as an internal self-assessment tool, performed either by oneself on his or her own work, or by a team of peers, or by management. This sort of audit is generally termed a *competitive benchmark*, and is popular and important enough to merit its own section (see "Benchmarking"). The goal in an audit is not only to determine deficiencies or noncompliance, but opportunities to improve individual and organizational performance.

A properly conducted audit is not adversarial—they are not "witch hunts" seeking blame. The audit team makes observations (and perhaps recommendations) to management, but the audit team should have no direct authority of the area being audited. Both the auditors and the audited area should be aware of any standards to which they will be audited, and a preaudit review may be necessary by both parties prior to an audit, although audits must be done in the standard working environment. No extraordinary effort should be expended because an audit will be conducted.

Process

Figure 18–1 illustrates a sample audit form. It is relatively uncluttered, and uses positive language. Although techniques can vary dramatically among the different types of audits and even within differing operational areas, the following eight-step process for conducting an audit remains the same:

1. Determine objectives and scope (who-what-when-where-why- how) of the audit.
2. Establish and train an auditor/auditing team. The team may be internal, external, or mixed.
3. The audit team establishes any baselines, manuals, or comparison tests. A preaudit review of documentation (checklists, standards, and so on) may be appropriate.

FIGURE 18–1
Sample Audit Form

Area:	Location:	Audit No.	
Assessment category		Subject:	
Key Positive Observations			
Key Opportunity Observations			
Recommendations			
Auditors	Date	Area Representative	
Response / Action Plan			
Area Management (signature)	Date	Approval	Date

4. Conduct the audit.
5. Review findings with the area supervisor and individuals involved.
6. Agreement (between management and audited area) achieved upon actions to be taken.
7. Provide feedback from audited area to management regarding the audit team.
8. Follow up with another audit or other evaluation at an appropriate time.

Auditors should be trained in the area they are auditing, and they should be trained in human behavior in organizations. An exception to this would be any customers on the auditing team. It is beneficial to have their *raw* response, without any organizational assumptions about "What the customer wants."

It is not unusual to be able to find published checklists regarding the evaluation of a supplier for a given field, or to be able to obtain a copy of a vendor evaluation program from a major manufacturer. Unfortunately, there are few industry standards in this area. Certainly checklists and manuals are no panacea (and they can be daunting in size and scope), they do provide a baseline for tailoring a checklist to individual situations.

Some examples of different types of audits are sketched below. This is by no means exhaustive, but gives the flavor of some possible types of audits. Types of audits are only limited to the imagination of the organization.

Supplier Audit

Purpose. Determination of vendor quality and capability. Many manufacturers now have highly structured vendor quality audits and ratings, and are willing to pay more (or purchase more) for improved quality.

Who. Manufacturing or buying organization.

What. Audit of items under consideration, but also a review of the plant personnel and environment.

When. Prior to signing a contract.

Where. Vendor's site.

How. Enlist cooperation; don't bully. Use statistical sampling, checklists, past data, and/or customer surveys. Share your findings with the vendor. Their quality may not be adequate now, but watch their development.

Physical Configuration Audit or Product/Service Audit

Purpose. To determine if the product conforms to the internal documentation and the documentation provided to the customer.

Who. Operational staff, and operational staff from upstream and downstream in the organization. In other words the designers, requirements analysts, and sales engineers would participate as well as the manufacturing engineers.

What. The product and its documentation.

When. At the prototype stage, or ideally, throughout the product life cycle.

Where. Internal, and at the customer's site.

How. Tedious comparison of the documentation to the reality of the product. Do all of the start-up, shutdown, and operational procedures described for compliance.

Area Self-Assessment Audit or Location/Function Audit

Purpose. To examine an area for improvement and compliance to organizational procedures. (If compliance is difficult to achieve, examine the need for the policy.)

Who. The area under assessment, internal customers of that area, and possibly external customers.

What. A scope or subject should be selected, such as safety, operations, quality control, accounting, and personnel procedures.

When. Anytime.

Where. On-site.

How. The team should establish guidelines (which may be quite dynamic). Observation and communication are the real keys to success in this type of audit.

Program Audit

Purpose. To determine if a companywide or customer-required policy, such as the ISO 9000 series regarding quality, is implemented properly.

Who. Top management should be the auditors. Possibly an external auditor can be added, but be careful to avoid feelings of resentment among staff.

What. A single, identifiable policy, such as "Safety Policy 065" or "Executive Order 234." Be sure that the areas under survey are informed of this policy well in advance of the audit. Attempting to *catch someone napping* reflects poorly on everyone. Save this attack energy for tackling the competition.

When. Anytime.

Where. On-site.

How. The policy itself should provide guidance. In addition, barriers and roadblocks should be identified and removed.

Example

A hospital wished to audit the operations of Ward 3B. The purpose was to ensure the highest level of patient care. The focus was to be on the nursing staff. An audit team was established with members from Ward 3B, an intern, and two hospital administrators, one of whom would act as a facilitator (not leader) of the group. The team determined that the goal had three components:

Structure—Facilities, equipment, floor layout, and so on.

Processes—Nursing procedures, housekeeping, and so on.

Result—Patient health and education.

The structure could be examined and compared to layouts in other hospitals. Data collection of processes and results was required. The standards selected were ones that were mandated by hospital and regulatory policy. It was determined that an analysis of data records would be made in a three-month period. Sampling would be taken of the data records over an 18-month period to determine any seasonal effects. Sampling of current procedures would be taken and correlated with the records. All patients within the three-month period would be asked to fill out surveys and questionnaires regarding their management and understanding of their health problems. Nurses would be asked to describe the procedures they used to ensure compliance with hospital policy.

The results were:

Key positive observations:

1. The procedures either followed recommended hospital policy or deviated due to an improvement in technique not yet reflected in any written procedures. It was noted which items were in compliance, which deviated (with any justification given), and which were not applicable, thus giving a statement of record for the audit.
2. No significant deviation was found between the sampling and the records survey.
3. The layout of the ward was conducive to rapid action.

Key opportunity observations:

1. Clutter in the nurses area could be reduced.
2. Patients were not well informed on their disease.
3. Patients said they felt like they were in an assembly line.
4. The new second shift supervisor and employees were experiencing some friction.

Recommendations for improvement:

1. Provide organizers and more closet space for nursing staff.
2. Provide videotapes to patients regarding their illness.
3. Stress the importance of managing their illness once they are out of the hospital.
4. Practice and review these illness management techniques.
5. Provide training to new supervisor on supervision techniques.
6. Provide training to all workers on team management.
7. Provide a counselor on staff for patients.

Management response:

1. Organizers and closet space provided.
2. Combination VCR/TV provided in rooms. Medical videotapes provided free of charge. Other video rentals available from hospital volunteer organization.
3. Available supervisors sent to Management Development Seminar, "Leadership in Nursing" (7/12–14), and all supervisors will be scheduled for course in the fall.
4. "Teambuilding and quality circles" course taught on-site.
5. Increased contacts with clergy and volunteers to provide counseling. Funding for a full-time staff position requires further justification.

Dos and Don'ts

• Don't rely too heavily on paperwork. Certainly don't rely on paperwork over good judgment. Walk about the office, plant, or supplier under audit. Does the impression left from the audit match your personal findings?

• Throw out any notions of "We've always done it that way" or "That's the other company's way." Steal any and all ideas you can. Change the status quo if it may result in increased quality. Change back if doesn't work out.

• Don't overlook "irrelevant" items such as housekeeping, employee motivation, employee situations outside of work, work atmosphere or culture, and office equipment.

BENCHMARKING

Aliases

Self-assessment and competitive benchmarking.

Brief Description and Purpose

Benchmarking analyzes how the firm is doing against competitors, excellent companies in noncompetitive fields, and third-party assessment measures, such as the Baldrige Award and the Deming Prize.

Discussion

Self-reflection on the state of the firm can result in dramatic renewal. The means of self-assessment can be internal or external, comparing the company to competitors or noncompetitors. Put in the vernacular, benchmarking answers the question "How am I doing?"

Each time a colleague of ours went to the doctor with severe flu symptoms the doctor would prescribe a broad-base antibiotic. The doctor talked to the patient for some four to five minutes at the most. After having bronchitis four times in one season, the patient went to an allergist who patiently listened to the history of symptoms for 45 minutes. The allergist then conducted several allergy tests, and was able to conclude that nonallergic rhinitis was the root cause of the bronchitis. The doctor suggested techniques to manage the illness which greatly reduced severe bouts of bronchitis.

Like the allergist, auditors must take time to look at the whole system, and understand why noncompliances occur: Is the defect rate high because the lighting is poor? Is the slow production rate and high reject rate due to an aging machine for which the worker is compensating? Auditors must also look for positive ideas to reinforce or to export elsewhere. The management team at Wal-Mart visits competitors' stores with much less market share than they, figuring that if the store is still in business, it must be doing something right.

Relation to Other Concepts and Tools

Benchmarking will require many of the analytical tools in this section. The most important tools are those involving teams. The CEO *cannot* unbiasedly assess the state of the organization against a competitor or benchmark.

Process

• Form a team to investigate the state of the firm. It must be broadly cross-functional. The effort may require that a central facilitating steering board be set up with a number of departmental boards reporting to it. The goal of this central committee is to share information, not only to garner it.

• Select a criteria. Examples of criteria are in Figure 18–2. The Malcolm Baldrige National Quality Award uses the following categories: leadership, information and analysis, strategic quality planning, human resource utilization, quality assurance of products and services, quality results, and customer satisfaction.

• Collect data on your firm and on the other firm/guideline. The Baldrige Award has become so popular as a self-assessment tool that there are even classes on using the guidelines in this manner.

• Analyze the data, prepare an action plan, and implement it. Try to implement as many new procedures as possible. Give the changes ''sunset'' times if necessary to squelch the grousing among those opposed to change, but implement them quickly. Be sure that a feedback/change assessment stage is implemented. Measure the effect of all changes.

FIGURE 18–2
Benchmarking Criteria

Type of Benchmark	*Example*	*Comments*
Internal	Aggressive collecting of suggestions and ideas.	Results in a deluge of suggestions which should be acted upon quickly, or a lack of confidence will result. Blind spots may still occur. Suggestions may suffer from the "way it works around here" syndrome.
External	Company XYZ public records.	Your benchmark is only as good as the competitor you select. Copious data may be hard to come by, even if they are public and have an active PR firm. Don't overlook data in trade magazines or data from trade associations.
Third Party	Baldrige Award Guidelines.	Helps prevent blind spots. May give areas to consider doing—that is, "nobody in our business does this." Questions you answer not applicable may be gold mines of information

Johnson Edosomwan proposes a four-stage assessment model shown in Figure 18–3. Note that it integrates the self-assessment effort into strategic quality planning and continuous improvement.

Example: The Baldrige Award

The U.S. Department of Commerce has issued the Malcolm Baldrige National Quality Award (MBNQA), Public Law 100–107, named after a former secretary of the department. The award itself is a gold medallion encased in a 14-inch high piece of crystal. The prestige of the award has grown phenomenally since its inception in 1988. Figure 18–4 lists the winners since the first year. Although tens of thousands request the guidelines (over 180,000 for the 1991 prize), only about 100 (97 in 1991) complete the forms and file. This indicates a tremendous interest, if not the stomach (undergoing the self-examination process will probably require a full-time, dedicated staff) to

FIGURE 18–3
Four-Stage Assessment Model

Stage 1: Pay attention to the voice of the customer.
- *a.* Identify needs, wants, desires.
- *b.* Obtain marketing information on customer perception for existing products and services.
- *c.* Translate the needs, wants, and desires into products and service requirements.

Stage 2: Evaluate effectiveness of mechanisms in place to satisfy customer requirements.
- *a.* Focus on product, service, and process characteristics.
- *b.* Specify requirements by operational needs.

Stage 3: Analyze customer requirements and market information and synthesize data for relevance.
- *a.* Develop interrelationships between product/service characteristics.
- *b.* Select the requirements that depict customer input.

Stage 4: Develop product and service characteristics and technical measurements that meet the customer requirements.
- *a.* Review final specifications with development teams.
- *b.* Implement the recommended product and service specifications.

Source: Adapted from Johnson Edosomwan, "The Baldrige Award: Focus on Total Customer Satisfaction," *Industrial Engineering,* July 1991, pp. 24ff.

FIGURE 18-4
Baldrige Award Winners

Year	*Manufacturer*	*Small Business*	*Service Business*
1988	Motorola Westinghouse Nuclear Fuel Division	Globe Metallurgical	None
1989	Xerox Business Products	Milliken & Company	None
1990	General Motors—Cadillac IBM—Rochester	Wallace Co.	Federal Express
1991	Solectron Zytec	Marlo Industries	None

submit a 50–75 page report and bare one's corporate soul to Baldrige examiner's on-site visits. A copy of the guidelines is included in Appendix B.

Even if one doesn't compete, examining the guidelines and conducting an internal self-assessment can prove highly revealing. Hire consultants, or if possible, trade your internal *consultants* for a friendly corporation's consultants. Many request the guidelines and self-assess for survival: Motorola (the 1988 winner) has controversially required 3,600 of its largest suppliers to follow the guidelines—200 who failed to comply were abandoned. Winning is no free ride. Winners must share their information on quality with others: Motorola, for example, made over 300 speeches, answered over 1,000 queries and held monthly briefings. *Losers* who capitalize on criticisms may find themselves in better competition, or even winners the next year.

Curt W. Reimann, director of the Baldrige Award, points out eight essentials success factors:

- A plan to keep improving all operations continuously.
- A system for measuring these improvements accurately.
- A strategic plan based on benchmarks that compare the company's performance with the world's best.
- A close partnership with suppliers and customers that feeds improvements back into the operation.
- A deep understanding of the customers so that their wants can be translated into products.
- A long lasting relationship with customers so that their wants can be translated into products.
- A focus on preventing mistakes rather than merely correcting them.

- A commitment to improving quality that runs from the top of the organization to the bottom.

An additional list of essential features discovered by the U.S. General Accounting Office's study of 20 top entries to the MBNQA is listed in Chapter 2.

Dos and Don'ts

- Don't enter the MBNQA for publicity purposes, or because you think your competitor is doing it, or because you think it will look good to your customer.
- Do enter the MBNQA (or at least undertake the rigourous self-assessment as if you were going to enter) in order to benchmark or baseline how your company stands on Quality Management.

FAILURE MODE AND EFFECTS ANALYSIS

Aliases

Failure mode, effects, and criticality analysis (FMECA).

Brief Description and Purpose

Failure mode and effects analysis (FMEA) formally collects information regarding possible piece, component, assembly, and system failure modes, their effect, probabilities, and corrective and preventative action.

Discussion

FMEA can and should be done early in the design stage. It should be done iteratively, when revisions from new components and field data can be input. While identifying potential failure modes may be a useful exercise in itself, the real power of FMEAs lies in their ability to establish a preventive maintenance schedule during the design or prototype stage, if reliabilities of the components are known. FMEA formats should be uniform and computerized within a company, if possible, to promote their ease of use. Defense contractors may be bound to use the automated formats specified in the military standard on *Failure Mode Effects and Criticality Analysis* (FMECA), MIL-STD-1629. Nondefense contractors may also find the format suitable for adaptation. The form given in the example is similar to the military standard.

Relation to Other Concepts and Tools

Inputs to FMEAs come from brainstorming, nominal group technique, and cause-effect diagrams. If the system has been in production, additional inputs may come from analysis of statistical process control results.

Process

1. Identify failure modes. The ways in which a system, assembly, component, or piece part can fail must be specified. The lowest repairable unit should be considered the end point of the failure mode process. That is, if a circuit board is replaced completely, then there is no point to considering piece part failures on the board, except in generating ways the board itself can fail. It may be best to work from the bottom up—analyzing piece part failure modes first, then assemblies, and so on until the system levels. Identification of failure modes should be a cross-functional team effort, with participants from design engineering, reliability, maintenance, sales, and procurement. Some examples of generic failure modes are listed in Figure 18–5.

2. Collect data. Examine historical records of maintenance actions if the product being considered is similar. If not, supplier's data may need to be collected.

3. Assign causes, effects, and codes. Determine and assign causes, effects, safety, probability of occurrence, criticality level, and recommended action. The safety level is often categorized into four levels:

Level I: Catastrophic—Failure possibly resulting in death or total system loss.

FIGURE 18–5
Generic Failure Modes

Premature operation
Failure to operate on time
Intermittent operation
Failure to cease operation on time
Failure during operation
Degraded output
Out-of-cycle failure
Dimensional problem

Level II: Critical—Failure that may result in severe injury, and/or major property damage.

Level III: Marginal—Failure that may cause minor injury, minor property damage, or otherwise delay or degrade the system.

Level IV: Minor—Failure may result in unscheduled maintenance or repair.

The probability of occurrence should be handled in bands, as exact numbers are unlikely to be available during the design stage. One scheme is presented in Figure 18–6. The criticality factor can be determined by multiplying the safety hazard factor by the probability of occurrence. Analysis of how to correct the problem may lead to design changes.

4. Revise. FMEAs are necessarily living documents. As new information becomes available from the field, the FMEA should be revised accordingly.

Example

A sample form is shown in Figure 18–7. Note that these forms should be automated in either a data base, spreadsheet, or proprietary form especially designed for FMEA or FMECA. In this example, a criticality analysis is used.

Dos and Don'ts

• Cross-functional teams are critical to the success of a FMEA. If possible, involve the end users.

• Apply FMEA to office work as well. There are causes, effects, and corrective actions that can be taken in administrative and service work as well as complex manufacturing work.

FIGURE 18–6
Probability of Occurrence Codes

Level	*Probability %*
A	>20
B	>10
C	>1
D	<1

FIGURE 18–7
FMECA Example

Block No.	*Function*	*Failure Mode*	*Effects of Failure*	*Cause of Failure*	*Safety Code*	*Probability of Occurrence*	*Criticality Code*
Diagram A3	Containment collar	1. Wrong dimensions	1. Loss of pressure 2. Loss of fluid	Improper manu-facture	3	D	3D
		2. Metal fatigue	1. Loss of pressure 2. Loss of fluid 3. Sudden loss may damage other compo-nents	Poor quality Contamination Failure to replace at maintenance interval	3	C	3C
		3. Improper installment	1. Loss of pres-sure 2. Loss of fluid	Operator error	3	C	3C

FOOLPROOFING

Aliases

Idiot proofing and *Poka-yoke*.

Brief Description and Purpose

Foolproofing attempts to ensure that "anyone" can use a product with as little instruction as possible. The user may be maintenance personnel as well as the consumer. It also tries to prevent a user from harm due to misuse of the product or from destroying the product.

Discussion

The driving force behind foolproofing is not that we have fools who work for us or buy our products, but that by foolproofing we have greatly reduced the ability to make an error, and have accelerated production, and added convenience. Foolproofing involves the producer as well as the customer. Foolproofing can be prevention/safety oriented, such as providing a cover for an emergency switch, so that it cannot be activated accidently, or it can be value-added oriented, such as the new style of microwave dinners that only require to be placed in the microwave—no venting is even needed. It can also be used to advantage during production. Figure 18–8 shows two angle irons to be attached. Note that in Figure 18–8, A, the drill holes are *not* symmet-

FIGURE 18–8
Foolproofing an Angle Iron

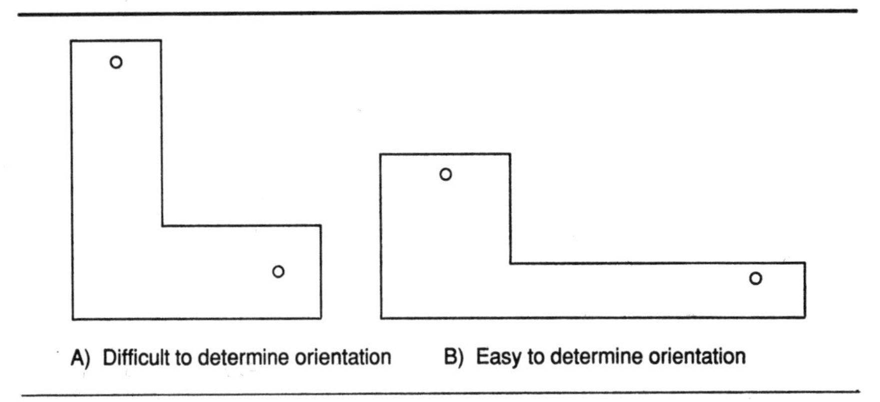

ric, even though the arms are almost symmetric. Figure 18–8, B, shows a foolproofed version. It should be quite clear to the installer how to orient the bracket now.

Foolproofing can be applied to virtually any system—even those with no traditional product. A report or instruction manual can and should be foolproofed, for example.

One system in desperate need of foolproofing is the videocassette recorder (VCR) delayed recording system. Only 20 percent of Americans can program their VCR to tape at a given later time. This is a lot of people, considering some 70 to 80 percent of U.S. households have a VCR. The VCR manufacturers tend to blame customers who don't care to read instructions. Who is right? Just because an industry has adopted an apparently indecipherable standard is no reason to follow along. The manufacturer who breaks away from the pack on this issue will be certain to see a market share increase—at least until the other manufacturers follow suit. Whether or not the customer should read instructions to an incredibly complex piece of machinery is beside the point. VCRs are made for the mass consumer market, and it is therefore unreasonable to assume that they will read *anything* regarding its operation.

Relation to Other Concepts and Tools

Foolproofing could be considered while using almost any other tool or task. It is closely related to process decision program charts (see section "Process Decision Program Chart," Chapter 17). Foolproofing is generally the goal, rather than a tool per se. The example gives some pointers as to what tools can be used in foolproofing.

Process

- Form a brainstorming group to create ideas on how a process or product could be misinterpreted. Relate these ideas via a cause-effect diagram. Determine if other tools, such as evolutionary operations, fault tree analysis, flowcharting, or auditing are needed for collection or to analyze more information.
- Determine which changes are necessary now, and which should be delayed or added later. Implement these changes.
- Be sure the changes have been realized properly, and that further modifications are/are not necessary.
- Incorporate a foolproofing session into the formal design process.

Example

The NewWave Software Co. has had customer complaints regarding its software. This has created overwhelming incoming calls to its (toll free) help line, and the company has begun to see its market share decline, even as total unit sales have improved. A quality circle tiger team was formed to examine this issue, and was given time and resources to accomplish its tasking within 30 calendar days. They were to identify and suggest corrections to the problems, but not implement changes until management and peers had reviewed their findings.

The team developed a multipronged attack:

Phase I: Data collection:

1. Conduct a physical audit of the documentation supplied to the customer.
2. Prepare and send out a closed- and open-ended survey to 400 users.
3. Collect data on the types and frequency of questions posed to the answer line over a three-week span.
4. Internal users created a "hands-on" diary while making extensive use of the software.

Phase II: Data analysis:

1. Analyze physical audit results, creating a cause-effect diagram.
2. Create a Pareto diagram of the problems encountered in the answer line, and a separate chart for the customer survey.
3. Create a failure modes and effects chart from all available data.

The physical audit revealed a number of discrepancies between the documentation and the way the software actually worked. In fact, this accounted for some 70 percent of the incoming calls. The irony here is that the documentation was incorrect, so the users who "winged it" (i.e., did not read the book) were happier users than those who carefully read the book.

The Pareto analysis verified that the results of the physical audit identified the major items of dissatisfaction. The major remaining problems were identified from user surveys and the internal walkthrough of the software. These problems were caused by the competing of memory space and "hotkeys" from "terminate-and-stay-resident" program that many users have installed. This interference could be avoided by programming around these spaces and common hotkeys in the future. The cause-effect diagram, while quite pretty, did not seem to illuminate any new problem areas. The tiger team recommended that a foolproofing session be added to the next design upgrade, and on subsequent redesigns.

Dos and Don'ts

- Don't feel that you're patronizing by foolproofing. You're adding value.
- Be creative, and *play* with the product or process. Approach it as if it were the first time you had seen or used the product or service.
- Services can be foolproofed as well. Consider streamlining operations, reducing training overhead by simplifying procedures and forms.
- Ask the people who directly use the product or service what confuses them, or could be improved. Observe them using the product, preferably with the user unaware of the observer.
- If you feel the need to call common sense by a new name, use Poka-yoke, the Japanese term for foolproofing. It may sound less patronizing as well.

QUALITY COSTS

Aliases

Cost of quality.

Brief Description and Purpose

Quality costs provide a way of tracking the expense and benefit of a quality program. It can be considered a figure of merit, and one that can be compared at least within similar types of companies, if not for all companies.

Discussion

Scrap, rework, recalls, and other such things all contribute to the high cost of quality. But quality costs do not have to be high, especially if greater effort is expended than the historical norm at the beginning of a project. Japanese cost of quality is often quoted at somewhere around 5 to 15 percent for their automotive industry. The cost of quality of U.S. automakers is around 35 to 50 percent. There are obviously tremendous advantages to be gained by reducing the cost of quality—which is to increase the quality of the product or service. Clearly tracking cost of quality alone will not reduce it. But it may point out areas where a greater return on investment could be made.

Feigenbaum would seem to be the originator of the concept of the cost of quality. Chapter 7 of his work *Total Quality Control* is dedicated to this topic, and the reader wishing to explore the ideas in this chapter further are encouraged to start with Feigenbaum.

For those who are driven more by dollars than humanistic management or tools, or even by product quality, the cost of quality can be a useful umbrella philosophy to bring about a quality transformation.

Relation to Other Concepts and Tools

Determining quality costs may entail data collection and auditing. Analysis of the costs may require a competitive benchmarking.

Process

1. Identify the quality costs. Quality costs can be divided into four major segments: (1) prevention costs, (2) appraisal costs, (3) internal failure costs, and (4) external failure costs. The first two are considered to be the cost of control, the last two the cost of control and failure. Internal failures result in scrap and rework; external failures include recalls, customer complaints, warranty work, and anything that occurs once the item leaves the factory or office. Some examples of the above types of costs are provided in Figure 18–9.

While prevention and appraisal costs are high, failure costs can be catastrophic.

FIGURE 18–9
Cost of Quality: Example Factors

Prevention:
- Quality engineering.
- Quality planning.
- Design review and verification.
- Training.
- Quality improvement projects.
- Statistical process control activities.

Appraisal:
- In-process inspection.
- Set-up for testing.
- Testing.
- Audits.
- Administrative costs.

Internal Failure:
- Scrap.
- Rework.
- Reinspection.
- Incoming material inspection.
- Downtime caused by defects.
- Investigation of failure.

External Failure:
- Warranty work.
- Repairs.
- Customer service.
- Returned goods.
- Investigation of defects.
- Product liability lawsuits.

2. ***Develop and implement a method of collecting the data on a regular basis.*** Watch for intangible and indirect costs, such as lost customers due to bad publicity, vendor quality, and equipment quality.

3. ***Identify areas of the most significant costs, then identify their cause and cure.*** Compare the costs on several different bases, such as direct labor, contributed value, net sales, or whatever types of base unit make sense. Partition the analysis by different areas, shifts, and the like. Be guarded against seasonal or daily variations that may be natural to your business.

4. ***Implement the solutions.*** Be sure to establish a feedback mechanism. Fixing one problem may reveal others, or the solution may not work.

5. ***Continuously and regularly monitor the cost of quality, and identify ways to reduce it.***

Dos and Don'ts

- Ask for the help of accounting professionals, but be alert for entrenched thinking not conducive to measuring quality costs. The information being sought is very different than that required for taxes or regulators, and requires creative thinking.
- Seek new ways to measure quality costs. Develop several bases for comparison and track them. Does one method have more information than the others? Is one too sensitive?
- Reducing the cost of quality may paradoxically require additional spending on training, equipment, or other preventative or appraisal methods.

WHO-WHAT-WHEN-WHERE-WHY-HOW

Aliases

Five Ws, One H; Six Ws (counting the w in How).

Brief Description and Purpose

These six terms are drummed into every journalism major in their first course, if not into every elementary school child learning about report writing. Because they have been mentioned so frequently throughout this book and others on TQM, it seemed appropriate to single them out as a technique.

Discussion

The following gives a sampling of questions in the Who, What, When, Where, Why, and How tradition.

Who	*When*
Who should do it? Who should *not* do it? Who else could do it? Who else should participate?	When is it due? When should it be due? Why should it be due then? When will time be "available?"
What	*Why*
What has been done? What should we do? What will happen if we don't? What more can be done? What *less* can we do and still have the same thing? What can be done now, this week, this year?	Why is this our job? Why *isn't* this our job? Why do it this way? Why do it here? or there? Why do it now?
Where	*How*
Where to do it? Where not to do it? Where should it be done, ideally? Do it here, or contract out?	How to do it? How often? How can we improve? When? How can we do it differently?

Root cause analysis is benefited by a technique called the *Five Whys*, which asks the question why five (or more) times in order to ferret out the root cause. For example:

Observation: Our revenues were down 12 percent this quarter.

Why?: Because we sold fewer products, and the price has stayed the same.

Why?: Our advertising presence was down 25 percent.

Why?: The ad budget request wasn't received in time.

Why?: There was no advertising manager.

Why?: The position wasn't posted as open for two months after the initial opening.

At this point, it would seem to point to a problem in personnel. Additional Whys? might reveal that the manager of the advertising area was delinquent in sending the opening to personnel, or some other reason may be revealed.

BIBLIOGRAPHY

Auditing

Aquino, Michael A. "Improvement versus Compliance: A New Look at Auditing." *Quality Progress*, October 1990, pp. 47–49.

Arter, Dennis R. *Quality Audits for Improved Performance*. Milwaukee, Wisc.: ASQC Press, 1989.

ASQC Quality Audit Technical Committee. *How to Plan an Audit*. ed. Charles B. Robinson. Milwaukee, Wis.: ASQC Press, 1987.

Danzer, H. H. "Audit—A Center of Future Quality Assurance." *EOQC Quality* 32, no.1 (March 1988), pp. 13–17.

Farrow, John. "Quality Audits: An Invitation to Managers." *Quality Progress*, January 1987.

Marquardt, Donald W. "Youden Address: Quality Audits in Relation to International Business Strategy—What Is Our National Posture?" *Statistics Division Newsletter*. Winter 1989, pp. 10–13.

Mills, Charles A. *The Quality Audit: A Management Evaluation Tool*. Milwaukee, Wis.: ASQC Press, 1989.

Reynolds, Edward A. "The Science (Art?) of Quality Audit and Evaluation." *Quality Progress* 23, no. 7 (July 1990), pp. 55–56.

Robinson, Charles B. "Auditing a Quality System, Part 2: Audit Policy and Protocol." *Quality Progress* 23, no. 2 (February 1990), pp. 54–58.

Sayle, Allan J. *Management Audits: The Assessment of Quality Management Systems*. 2nd ed. Milwaukee, Wis.: ASQC Press, 1988.

Shimoyamada, Kaoru. "The President's Audit: QC Audit at Komatsu." *Quality Progress*, January 1987.

Willborn, Walter. *Quality Management System: A Planning and Auditing Guide*. Milwaukee, Wis.: ASQC Press, 1989.

Benchmarking

The MBNQA guidelines may be obtained from: Baldrige Award Consortium, POB 443, Milwaukee WI 53201, Phone: (414) 272-8575, Fax: (414) 272-1734.

Bemowski, Karen. "The Benchmarking Bandwagon." *Quality Progress* 24, no. 1 (January 1991), pp. 19–24.

Bhote, Keki R. "Motorola's Long March to the Malcolm Baldrige National Quality Award." *National Productivity Review* 8, no. 4 (Autumn 1989), pp. 365–76.

Camp, Robert. "Benchmarking: The Search for the Best Practices that Lead to Superior Performance. Parts 1–5." *Quality Progress*, January–May 1989.

Edosomwan, Johnson, and Wanda Savage-Moore. "Assess Your Organization's TQM Posture and Readiness to Successfully Compete for the Malcolm Baldrige Award." *Industrial Engineering*, February 1991, pp. 22–24.

Edosomwan, Johnson. "The Baldrige Award: Focus on Total Customer Satisfaction." *Industrial Engineering*, July 1991, pp. 24ff.

Leibowitz, Michael R. "Baldrige Winners Start Ahead of the Pack." *Electronic Business*, October 16, 1989, pp. 80–82.

Main, Jeremy. "How to Win the Baldrige Award." *Fortune*, April 23, 1990, pp. 101–16.

Pryor, Lawrence S. "Benchmarking: A Self-Improvement Strategy." *The Journal of Business Strategy*, November/December 1989, pp. 28–32.

U.S. Department of Commerce. *Application Guidelines for the Malcom Baldrige National Quality Award*. Milwaukee, Wis.: Baldrige Award Consortium (Annual).

Xerox Corporation. "Baldrige Quality Award Winner." *National Productivity Report*, 18, no. 24 (December 31, 1989), pp. 1–4.

Camp, Robert C. *Benchmarking: The Search for Industry Best Practices that Lead to Superior Performance*. Milwaukee, Wis.: Quality Press, 1989.

Failure Mode and Effects Analysis

Procedures for Performing A Failure Mode Effects and Criticality Analysis, MIL-STD-1629. Washington, D.C.: Government Printing Office, 1980.

Foolproofing

Shingo, Shigeo. *Poka-Yoke: Improving Product Quality by Preventing Defects*. Cambridge, Mass.: Productivity Press, 1989.

Shingo, Shigeo. *Zero Quality Control: Source Inspection and the Poka-Yoke System*. Cambridge, Mass.: Productivity Press, 1986.

Quality Costs

ASQC Quality Costs Committee. *The Management of Quality: Preparing for a Competitive Future*. Ed. John T. Hagan. Milwaukee, Wis.: ASQC Press, 1984.

ASQC Quality Costs Committee. *Guide for Reducing Quality Costs*. Milwaukee, Wis.: ASQC Press, 1987.

ASQC Quality Costs Committee. *Quality Costs: Ideas and Applications*. vol. 1. 2nd. ed. Ed. Andrew F. Grimm. Milwaukee, Wis.: ASQC Press, 1987.

ASQC Quality Costs Committee. *Quality Costs: Ideas and Applicatios*. vol. 2. Ed. Jack Campanella. Milwaukee, Wis.: ASQC Press, 1989.

ASQC Quality Costs Committee. *Principles of Quality Costs, Implementation, and Use.* 2nd ed. Ed. Jack Campanella. Milwaukee, Wis.: ASQC Press, 1990.

Feigenbaum, Armand V. *Total Quality Control.* New York: McGraw-Hill, 1983.

Greenwood, Frank. "How to Survive by Raising Quality While Dropping Costs." *Journal of Systems Management* 39 (September 1988), pp. 36–43.

Schrader, Lawrence J. "An Engineering Organization's Cost of Quality Program." *Quality Progress,* January 1986, pp. 29–33.

Sullivan, Edward. "OPTIM: Linking Cost, Time and Quality." *Quality Progress,* April 1986.

Who-What-When-Where-Why-How

Imai, Masaaki. *Kaizen: The Key to Japan's Competitive Success*. New York: Random House, 1986.

CHAPTER 19

GROUP TECHNIQUES

It's difficult to work in a group when you're omnipotent.

"Q" from "Star Trek: The Next Generation"

All of the group techniques in this chapter are predicated upon the principles found in Chapter 14, "Team Building" section. The tools all rely on group interaction to formulate positive and productive results. Unfortunately, American businesses have long operated in the mode of the alien being "Q"—the Boss was omnipotent, thereby making group decision making irrelevant. Competitive businesses are realizing that they can no longer afford to ignore the opportunities that may lie dormant within their own staff.

Groups work best when there are enough people to bring several perspectives to bear on a problem, but not so many that some members feel comfortable in not participating. This optimum number lies somewhere between four and eight. Participation in groups should be voluntary, with clear ground rules on when tenure is terminated. Facilitators or group counselors are valuable assets in properly implementing a group process. Facilitators should be well trained in small-group behavior in order to steer groups to success. Appointing someone from the quality assurance department as a companywide facilitator may be an optimal approach.

BRAINSTORMING

Aliases

Storyboarding.

Brief Description and Purpose

Brainstorming is a group process wherein individuals *storm* or generate ideas in an unfettered way, free of criticism and second thoughts. The purpose is to create and detail ideas on a focused problem. It formulates a group consensus on strategy, planning, directions, and problem solving. As all group members are equals, it is a useful tool in arriving at compromise solutions in potentially antagonistic situations, such as union negotiations. Brainstorming does not determine a solution, but proposes many. Other techniques, such as nominal group technique, may be used to select a solution.

Discussion

Brainstorming started with Walt Disney (he called it *storyboarding*) in creating the 1928 animated film "Steamboat Willie" (the first appearance of Mickey Mouse). The Disney Company still uses the technique, and has used it in planning Walt Disney World and Epcot Center.

Brainstorming has been used informally for many years in a business context, and is being used more and more frequently. It is easy to learn—a one-day workshop, including several trial storyboards should be sufficient. Done properly, it can coax bashful yet creative people into yielding wonderful ideas. Brainstorming is a way to rapidly generate ideas for further consideration by the other tools.

For important brainstorming sessions, a facilitator is required. Facilitators act as helpers in the storyboard process, and should be highly knowledgeable about brainstorming and human behavior, but should have little or no stake in the outcome of the storyboard. A good facilitator will ensure equal participation among members, and cajole the maximum number of ideas from a group.

Relation to Other Concepts and Tools

Brainstorming is closely related to the other group techniques. It is useful as an input technique to planning and organizing tools.

Process

The Setting

A facilitator ensures the proper conduct of the storyboard, but does not provide input of any sort. The storyboard belongs totally to the group. The participants are equal players, regardless of their working relationships when not on the storyboard. The facilitator must ensure that no criticism is made during the creative phase, and that the tone of criticism is kept professional and noninflammatory during the criticism phase. Therefore, the facilitator should have no vested interest in the storyboard.

The supplies needed are:

Large surface to affix cards.

Pushpins or tape (pushpins are faster).

Cards for ideas (3″ × 5″).

Cards for categories (4″ × 6″).

Phase I: The creative phase. Pin the topic card (the reason for the brainstorm session, or the title of the project) on the board. The group then generates ideas, writing them with a marker on a card, one idea per card. The idea creator says the idea aloud so everyone hears, and hands the card to the facilitator to attach to the storyboard (no particular order necessary). No criticism or comments are allowed. The facilitator may encourage modifying or combining other participants' ideas, thus "piggybacking" on other ideas to obtain fresh ones. The facilitator could review the ideas out loud when the ideas begin to dwindle. It is important to pin the cards up instantly so that the group can read them and ruminate upon their content to come up with other ideas.

Phase II: The criticism phase. After pausing to examine the cards (no criticism yet), the group develops categories the same way it previously generated ideas. Then each card is examined and placed under a category. A card may be a duplicate of another, and should be combined (or discarded). Do not assume that the ideas are the same because they use the same or nearly the same words. Ideas may be discarded that the group disagrees with. The ideas belong to the group, not to any individual. Whole categories may be discarded as needed. The discarded cards may be placed on the floor, or wherever is convenient. When the group has arrived at a final form, the entire storyboard may be kept by taping each column together. A sketch of a categorized, abstract, storyboard session is shown in Figure 19–1.

Example

Customer service has a problem. They spend a lot of time answering technical questions on the phone (they also take orders over the phone), and these conversations seem to take a long time, and the caller may be routed several times. What can be done about this situation?

FIGURE 19–1
Storyboard Set-up—After Ideas Have Been Categorized

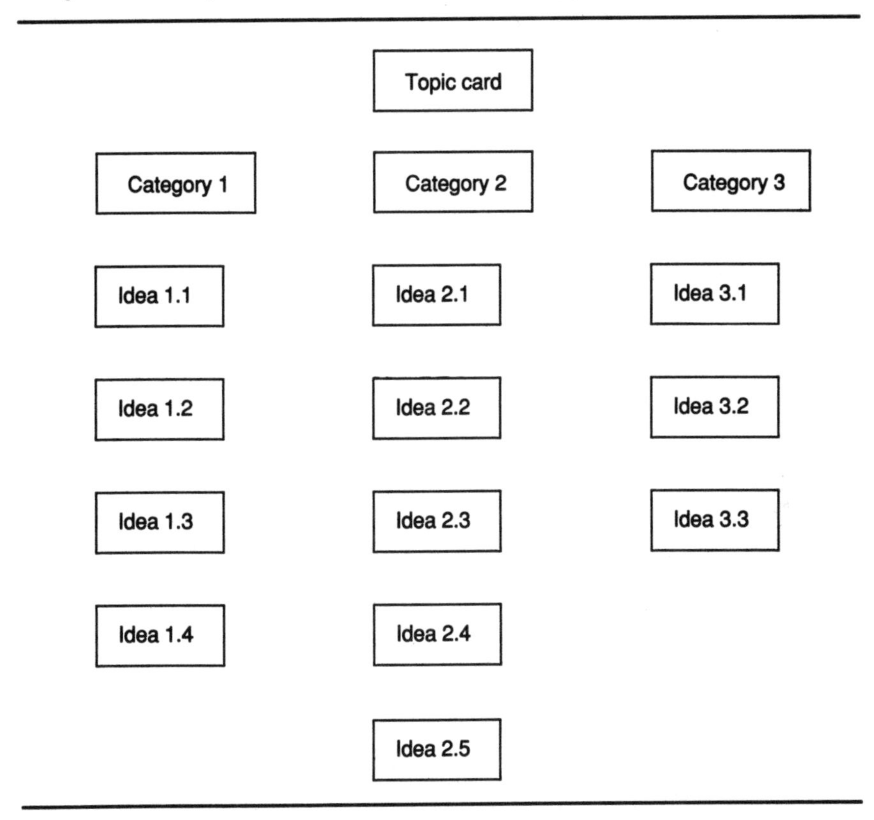

Our brainstorming session has come up with the following, in no particular order:

- Dedicate a phone answer line.
- Provide better instructions on how to use the products.
- Make different numbers for orders and for technical questions.
- Make our product easier to use.
- Make the caller read the instructions first.
- Make different numbers for different categories of technical questions.
- Get rid of the phones.

- Provide technical training to phone-answering staff.
- Find better educated customers.
- Provide an "answer book" to phone answerers.
- Make a list of the top 10 problems.
- Have a guru available in the background.
- Hire more people.
- Don't worry about the time spent.
- Live with the situation.
- Give us more vacation time.

Upon refining and categorizing, the group has now come up with the following categories and entries:

Phone mechanics:

- Dedicate a phone answer line.
- Make different numbers for orders and for technical questions.
- Make different numbers for different categories of technical questions.

Training:

- Provide better instructions on how to use the products.
- Provide technical training to phone-answering staff.
- Provide an "answer book" to phone answerers.
- Have a guru available in the background.

Product design related:

- Make our product easier to use.
- Make the caller read the instructions first.
- Find better educated customers.

Human resources:

- Don't worry about the time we spend.
- Give us more vacation time.
- Hire more people.

Other:

- Make a list of the top 10 problems.
- Live with the situation.
- Get rid of the phones.

Next, a study must be done to examine the costs (both human and material) and roadblocks for each idea. Don't assume which ones will be the "easiest" or "cheapest." Don't ignore the "Other" suggestions. A Pareto analysis and/or a cause-effect diagram logically would be the next step. For example "Get rid of the phones" would seem to be an outcry from someone needing a vacation. But perhaps they have the germ of an innovation: Perhaps installing a fax machine or voice mail could be used. These devices are becoming inexpensive and indispensable.

Dos and Don'ts

During the creative phase:

- The facilitator must not have vested interest, and must pace the group.
- No criticism is allowed by anyone.
- Takeoff on other ideas. Modify or combine other ideas.
- When the ideas start to slow down, the facilitator should review, aloud, the list already created.
- Generate as many ideas as you can. Quantity, not quality, counts.
- Generate wild ideas. The wilder the better. No idea is bad.
- Do not pause to evaluate your own idea or that of others.
- Try warm-up session on some nonwork-related, trivial issue.

During the criticism phase:

- Review each idea.
- Clarify and revise ideas, rewriting cards as necessary.
- The group assigns the meaning to ideas. It means what the group says.
- The group validates ideas. If the group consensus is to get rid of an idea, then get rid of it.
- Don't use "killer" phrases that stifle creativity, such as: "We've done that before"; "It won't work"; "The boss won't go for it"; and "It's against company policy."

DELPHI TECHNIQUE

Aliases

Small group consensus or expert opinion consensus.

Brief Description and Purpose

The Delphi technique is an iterative approach to arrive at a consensus of a group of experts. Getting experts to agree is a difficult task, especially when they are in the same room. The Delphi technique is a relatively simple way to arrive at a group consensus. Notice that it is a consensus technique—not a technique to combine expert opinion. The analytical combining of opinions is a relatively new field of great interest to artificial intelligence developers as they are often faced with the task of sorting out conflicts in expert opinion.

Discussion

The experts receive whatever it is they are to review and comment upon. Without discussion with one another, they send in their response to the facilitator. All comments are annotated to the work, and redistributed to the group. This continues until a consensus is reached, and may require three, four, or more trials.

Relation to Other Concepts and Tools

This technique is similar to other group consensus techniques, but is particularly useful in eliminating personality clashes in technical areas. It is also useful whenever powerful personalities are likely to dominate the discussion.

Process

1. Define the task. Clearly define the task you want the experts to comment or work on. Establish a monitor group to determine objectives, design questionnaires, tabulate results, and so on. This monitor group should be well grounded in team techniques, and tools such as brainstorming and Pareto analysis.

2. Establish selection criteria. Establish criteria to select between legitimate multiple alternatives. For example, is the cheapest solution always the best? This criteria should be made known to the experts.

3. Identify the experts. Identify the experts. Conventional wisdom places experts at least 50 miles away from where you are. This is silly, of course. Don't overlook experts in your own backyard. Experts can be found by examining trade and society publications, classified advertisements, and word of mouth. Ask local professors or others who the experts might be (don't overlook the professor as an expert either). Then ask yourself (and the monitor team) if you really need the best expert in a field, or just an expert in the field. For example, if we want to guess the rate of inflation for the next 12 to 60 months, can we probably cull information from magazines, journals, and newsletters and combine this information (being careful to note which sources are reporting from other sources, and which sources are originating their forecasts). At the other extreme, do we require an expert in electronic system environmental stress screening? There are probably only a dozen or so recognized experts in this field. Be skeptical of experts with vested interests (i.e., never ask a siding salesman if your house could stand a facelift).

4. Round one of opinions. Send out the tasking for round one of opinions. Consolidate the opinions after receiving all opinions. This must be done in a way that in the next review process, the opinions are anonymous.

5. Repeat until a consensus is reached. Send out the task again, this time with all the comments. Examine subsequent rounds for consensus, and repeat until a consensus or at least clear alternatives are reached; there can be legitimate alternative views. Another round may be appropriate to determine how to select among the alternatives.

Example

A small monitor team of four people is established internal to a company that produces parts to vehicle manufacturers. This team is to "crystal ball" the near term (18 to 24 months away) on external impacts to the company. This team will oversee the Delphi team of several experts. A questionnaire is designed that has both open and closed form questions. After each round the experts are given feedback on the survey, including how many (but not who) responded to each answer on the closed form questions. Comments and open form questions are disguised as much as possible. After three rounds, the panel of experts has arrived at the following:

1. Raw material prices will increase 5–7 percent.
2. Inflation will be 5–6 percent.
3. Major buyers will establish quality rating schemes. No central scheme, therefore each buyer will have its own scheme, leading to increased burden in administration.

4. Labor rates will rise 10–15 percent.
5. Foreign competition will increase its market share 3–9 percent.
6. Market share must be increased to 12 percent to maintain status quo.

Notice that the last point is a conclusion drawn from the first five observations. Also note that point 3 is rather open ended, but may be something that the company can influence, either by pressing for industrywide standards, or by organizing the manufacturers they deal with, or by establishing a regional standard.

Also note that these are the final conclusions. It is reasonable to expect the experts to reference their work, so that a highly popular article (which may later prove inaccurate or otherwise distorted) does not skew the results. The experts also offered no solution to increasing market share. When it was becoming apparent that this alternative was going to be chosen, the monitoring team could have expanded the "How" questions and pared down the "What" questions.

Dos and Don'ts

- The facilitator should not have expertise in this area. A small amount of knowledge may lead to assumptions regarding similarity of opinion. It is best to recraft all of the responses (if they are handwritten) into an anonymous format, completely unedited.
- Be wary of multiple alternatives. Perhaps the objective was not clear.
- Don't shortcut by using a "majority rules" on the first round. A minority opinion may emerge as the subtle winner over conventional wisdom.

NOMINAL GROUP TECHNIQUE

Aliases

Small group technique.

Brief Description and Purpose

Nominal group technique (NGT) is similar to brainstorming and the Delphi technique, but is more structured, decision oriented, and suitable to specialized problems requiring a degree of expertise. A small to moderate size (10 to 15) group is needed.

Discussion

NGT could be categorized as a structured, silent brainstorming session with a decision analysis process. The structure opens lines of communication and ensures (and requires) participation of every member. During the decision process, a consensus is formed, rather than a majority rule. Many of the points made in the "Brainstorming" section are relevant in this section as well, and the reader is encouraged to read that section in conjunction with this one.

Relation to Other Concepts and Tools

NGT has fewer applications than brainstorming, but allows all members to have a say. It is similar to the Delphi technique in that regard, but the Delphi technique is used when nonresident expertise is needed, or strong personalities make forming a small group interaction consensus difficult.

Process

1. Idea generation. A facilitator presents the problem or issue, and provides instructions. The team then quietly generates ideas for 5 to 15 minutes. No discussion is allowed. No one should be allowed to leave the room (putting pressure on those remaining to finish) or otherwise distract those who need more time.

2. Idea presentation. The facilitator gathers the ideas one at a time, round-robin fashion and posts them on a flip chart. Tape the used charts to the wall. No discussion at this time.

3. Idea discussion. Eliminate duplicate ideas. Clarify ideas, limiting discussion to brief explanations of the reasoning behind an item. After it is clear what the ideas are, discussion regarding the pros and cons of each alternative may be made. Facilitators should guard against strong personalities dominating the discussion. If necessary, cut off discussion, and proceed to the voting.

4. Establish priorities—make a decision. Establishing priorities may consist of simple rankings, or they may consist of formulas: If we do items 3 and 4, this requires item 8, which must be done *first*. Usually though, a priority list, and not a formula, is what is desired as the outcome of the NGT session. It is unlikely the group will immediately reach a consensus, so a ranking of the alternatives needs to be established. In a simple ranking, each member "votes" on all of the alternatives, and gives each alternative a score. For example, if there are 10 choices, the member has 10 votes. The votes are weighted however, and range from a weight of 1 to a weight of 10. For each alternative, the total weighted votes may be added, and the winner determined. In the case of a tie, the process can be repeated, using only the two alternatives. This process is illustrated in the example, Figure 19–2.

Another way to rank choices is by the method of pairwise ranking. In this scheme, each alternative is compared with each other, with the winner recorded in the cell. The alternative with the most number of instances wins. In the event of a tie, the box is examined where the two alternatives were compared. This technique is illustrated in Figure 19–2.

Depending upon the complexity, a weighted ranking scheme that can maneuver a large number of factors, such as the analytical hierarchy process may need to be used. This process, developed by T. Saaty, has been used for such things as the *Places Rated Almanac,* which rates cities by a large number of factors.

Example

A maintenance problem is occurring in some mechanical parts in a subassembly. The subassembly is often stored for long periods of time, and then experiences a large amount of stress. A group of five and a facilitator have arrived at the following ideas in Figure 19–3, column one. After discussion of the pros and cons of each proposed solution, the rank votes were taken (using a secret ballot), as shown in column two, the lower number being the better. These votes are tabulated in column three, which gives us the overall ranking of the ideas. In this case, the choice would be idea A, using the nylon washer.

Dos and Don'ts

• Instructions should be given to the group regarding body language. Even though the thought process is silent in NGT, body language can convey a great deal. Loud yawning, stretching, finger drumming, and the like should be avoided.

• Rank the outcomes in more than one way. In one round, assign no weight factors. In another round, allow weight factors to be assigned. Allow each member to assign a weight. The weights can be open ended, or closed. In an open weighting, a scale is first devised, then the weights assigned. There is no penalty for favoring all choices equally, or strongly favoring a few. In a

FIGURE 19–2
Pairwise Ranking

	1	2	3
2			
3			
4			

The pairwise matrix

	1	2	3
2	2		
3	3	2	
4			

Compare choice 1 to 2 (2 was better), choice 1 to 3 (3 was better), choice 3 to 2 (two was better), and so on, until the matrix is complete.

	1	2	3
2	2		
3	3	2	
4	1	2	3

In this example, choice 2 is the winner.

FIGURE 19–3
Results of a Nominal Group Technique Session

Alternative	*Votes*	*Totals*
A. Use nylon washer	1,1,2,3	7
B. Use felt gasket	4,2,3,2	11
C. Remove the feature	3,4,1,4	12
D. Encapsulate feature—maintenance free	2,3,4,1	10

closed weighting each member has only so many weight "coins" that they can spend in total. This prevents a group member from "liking" or "hating" every idea.

QUALITY CIRCLES

Aliases

QC circles, team building, TQC teams.

Brief Description and Purpose

Quality circles are a small group or team that is composed of employees who meet regularly to identify quality problems that have to do with their own work, and to generate possible solutions to these problems.

Discussion

All of the techniques and caveats of the section "Team Building" in Chapter 14, Part 3, "Management Dynamics" apply to quality circles. Quality circles are a specialized type of team, and are closely associated with Quality Management, although many other types of teams have equal value.

Since the first QC circle was registered with the Japanese Union of Scientists and Engineers (JUSE) in 1962, there have been over a million registered quality circles in Japan. Japanese managers highly praise the system for the vast number of highly useful, practical solutions they produce. In the United States, quality circles have had a spotty track record. This is probably due to a society that traditionally values heroic individual efforts to the exclusion of teams (there is no reason why both cannot be valued). The only teams introduced in the school systems are generally in the physical education classes. Here teams are often selected by children who are "team captains." This can result in demeaning those "picked last." The resultant emotional scarring and lack of having seen a team function properly would seem to carry over into the adult work life. The tremendous benefits that can come from team building will have to wait for either almost constant training of adults already employed, or a Herculean change in the way American schools perceive teamwork over individual behavior.

Relation to Other Concepts and Tools

Quality circles require knowledge of the various tools discussed in this section. Members should also be well grounded in team building techniques.

Process

1. Present training. Present QC circle training to all employees in the area to implement the QC circle.

2. Select participants. Ask for volunteers, but limit participation to 5–15 members. The ideal group size for this purpose is probably 6–9. Volunteers may resign at any time. A manager should not be a member.

3. Select a facilitator. The facilitator should be selected from the group, and should perhaps rotate. Ideally, there would be a dedicated facilitator who oversees several circles.

4. Start small. No more than a few QC circles at first. Work out the bugs in the system, and change corporate culture, if necessary. Empower and trust the QC circle. Meetings should last about an hour and occur weekly or every other week.

5. Act on the results. Listen to the team's results. If not highly supported by management the team will fade. The team must be empowered by management and supervisors. Go slow on rewards. It is tempting to reward initial successes, but after all, the workers are doing the job they are paid to do. It may be highly rewarding to employees that they have a voice that management listens to. Adding monetary rewards may brand the team as fawning and lead to disintegration of the area's morale. When workers who are slow to volunteer realize that the QC circles can fulfill their promise (if implemented properly), they too will seek a voice on the QC circle.

6. Ongoing analysis and feedback. After operating for a year, reexamine the structure of the circle. Should terms on board be time limited? Should the size be increased or decreased? Should minor purchasing authority be given to the circle? Consider establishing a steering committee to coordinate cross-departmental or companywide problems (ever see a suggestion rejected because it affected the whole company and "not our job"?). Be careful that this committee does not add a layer of bureaucracy.

Dos and Don'ts

- Avoid the word *meeting* when referring to a quality circle session. This word has been so abused, and so many shoddy meetings have been conducted that it leaves a pall over the task of the group.
- Don't let quality circles degenerate into a "Circle Jerk" where little is accomplished except blowing off steam. Poll individuals separately about the progress. Ask specific questions. Be sure to act promptly on any decisions the group has.
- Let the QC circle become the fulcrum for establishing self-managing teams. It would be disastrous to go into work tomorrow and fire the vice president of production and tell 12 managers that they were now self-managing. It must be demonstrated that the company is ready to implement team concepts in a meaningful way.
- Quality circles have often failed when implemented in the United States and other countries that do not have a strong group context in their early educational experiences. Figure 19–4 lists a number of points to do and to avoid to successfully establish quality circles.

SERVICE QUALITY

Aliases

Customer-oriented culture.

Brief Description and Purpose

Service quality can be considered a tool as well as a foundational issue (Chapter 3) in Quality Management. It can even be the driving force in a

FIGURE 19–4
Quality Circle Targets

1. Membership is strictly voluntary.
2. Involve all members as a team, stressing leadership building skills.
3. Provide training for circle members. Members must be trained to listen and to communicate.
4. Focus on work problems.
5. Utilize the creative capability inherent in everyone.
6. Do not focus on the trivial as being "safe"—that is, the quality circle has as its primary recommendation that the entryway plant be repotted for enhanced quality of work life.
7. Power and authority must be granted to the group; management should support the groups' actions.
8. Non-QC members may be disappointed at their noninvolvement, and must be given a chance to serve.
9. The QC must have a common understanding of the purpose and genuine participation by the members.
10. A facilitator is necessary to augment group cohesion and diminish conflicts.

Quality Management system; Robin Lawton has developed what he terms *Total Performance Management®*[1], which is a TQM system driven by customer service, whether the customer be internal or external to the company. Kaoru Ishikawa defines a customer as the next person in the process. Thus one's supervisor could be considered a customer, or another department, or whoever receives your work.

Discussion

Each individual in the organization has a customer and is a customer. Internal customers are as important as external from a service quality viewpoint. Supervisors and employees are customers of one another—when the supervisor tasks someone, there is a service rendered—communicating the task. When the employee completes the task then there is also a service (and product) rendered, whether it be a financial report or a breadboard prototype. If the artificial distinction between service and products is dropped, it becomes easy to apply principles of product quality to traditional service quality.

---

[1]Total Performance Management is a registered trademark of Innovative Management technologies.

Relation to Other Concepts and Tools

Service quality is infused throughout all tools and concepts.

Process

Robin Lawton in his work on creating a customer-centered culture for service quality outlines six steps:

1. Define the product. Products can be as diverse as a contract or an electronic assembly. The customers must also be identified. They tend to belong to one of three categories:

Users—Actual end users of the item.

Agents—Promotes the product to the user.

Fixers—People who "make things right."

In the case of a toy train, the consumer is the user, perhaps a hobby store acted as agent, and either a repair shop, customer service answer line, or even a trade magazine or hobby group would act as a fixer.

2. Identify customer requirements. Requirements may be identified using brainstorming, surveys, quality function deployment and other requirements tools. Yes, identify what your boss wants prior to starting on that report. Do they want it in color graphics, tables, and 35mm slides? How long should it be? Should it have summaries?

3. Compare the product with requirements. A simple and easy step to overlook. This process aids in targeting future improvements. Features which were ***not*** requested by the customer should be examined. They may add unneeded complexity.

4. Describe the process. By describing the process extraneous steps to the development process may be revealed. It also provides additional opportunity to identify future refinements.

5. Measure. Measure productivity, quality, and profitability. How do they correlate? For the long and short term? Are the right things being measured? How do your numbers compare with the rest of the industry for the same type of product?

6. Include customers in product development. This is perhaps the same as the first step. Feedback is the key to continuously improving service quality.

Dos and Don'ts

• Tom Peters is constantly exhorting us to get "close to the customer." Send them surveys. Take them out to lunch. Give them awards for suggestions. Observe their behavior. Conduct market research. Do whatever it takes to do that.

• Considering "peers" and "managers" as customers can be a new, refreshing way of looking at our old colleagues. It will also help focus service quality initiatives into everyone's work.

• Use audits and benchmarking to establish current baselines of customer satisfaction, whether the customers are internal or external.

• It is not facetious to think of our peers as internal customers. The term *customer* connotes a life-giving force to a business. Unless we take our peers as serious customers or consumers just as we take external customers seriously, work quality to internal customers will be dismal.

• Doggedly pursue customers' complaints until the customers are satisfied. Dissatisfied customers can undo thousands of dollars worth of advertising by conducting their own word-of-mouth negative ad campaigns.

BIBLIOGRAPHY

Brainstorming

Ackoff, Russell. "Creativity in Problem Solving and Planning: A Review." *European Journal of Operational Research* 7, no. 1, pp. 1–13.

Rickards, Tudor; Simon Aldridge; and Kevin Gaston. "Factors Affecting Brainstorming: Towards the Development of Diagnostic Tools for Assessment of Creative Performance." *R&D Management* 18, no. 4 (October 1988), pp. 309–20.

The Storyboard Process, GE Co. Engineering and Manufacturing, 1285 Boston Ave., Bridgeport CT, 06602.

Storyboard Facilitator Training Guide, Saginaw Division—GMC, Saginaw, Mich.

Delphi Technique

Dalkey, N. C., "The Delphi Method: An Experimental Study of Group Opinion." *Research Paper RM-5888-PR*, June 1969. The RAND Corporation, Santa Monica, California.

Khorramshahgol, Reza; A. Ason Okoruwa; and Hossein Azani. "Capital Budgeting in a Multiple Objective Environment." *Journal of Information and Optimization Sciences* 10, no. 3 (1989), pp. 567–77.

Nominal Group Technique

Davis, J. H. *Group Performance*. Reading, Mass.: Addison-Wesley Publishing, 1969.

Delbecq, Andre L.; Andrew H. Van de Von; and David H. Gustafson. *Group Techniques for Program Planning: A Guide to Nominal Group and Delphi Processes*. Glenview, Ill.: Scott Foresman Co, 1975.

Fisher, B. A. *Small Group Decision Making: Communication and the Group Process*. New York: McGraw-Hill, 1974.

Gregerman, Ira. *Knowledge Worker Productivity*. New York: American Management Association, 1981.

Lewin, Kurt. "Group Decision and Social Change." *Readings in Social Psychology*. New York: Holt, Rinehart, & Winston, 1958, pp. 197–211.

Shaw, M.E. *Group Dynamics: The Psychology of Small Group Behavior*. New York: McGraw-Hill, 1976.

Quality Circles

Adam, Everett E., Jr. "Quality Circle Performance." *Journal of Management* 17, no.1 (1991), pp. 25–39.

Baird, John E., Jr. *Quality Circles Leader's Manual*. Milwaukee, Wisc.: ASQC Press, 1982.

Baird, John E., Jr. *Quality Circles Participant's Manual*. Milwaukee, Wisc.: ASQC Press, 1982.

Baird, John E., Jr., and David J. Rittof. *Quality Circles Facilitator's Manual*. Milwaukee, Wis.: ASQC Press, 1983.

Barry, Thomas J. *Quality Circles: Proceed with Caution*. Milwaukee, Wis.: ASQC Press, 1988.

Berger, Roger W.; David L. Shores; and Mary Thompson, eds. *Quality Circles*. Milwaukee, Wisc.: ASQC Press, 1986.

De Vries, J., and H. van de Water. "Quality Circles and Quality of Working Life: Results of a Study in Seven Large Organizations." *EOQ Quality* 2, no. 2 (July 1990), pp. 4–10.

Gibson, P. *Quality Circles: An Approach to Productivity Improvement*. Elmsford, N.Y.: Pergamon Press, 1983.

Gryna, F. *Quality Circles: A Team Approach to Problem Solving*. New York: American Management Association (AMACOM), 1982.

Ingle, S. *Quality Circle Master Guide*. Englewood Cliffs, N.J.: Prentice Hall, 1981.

International Association of Quality Circles. *QC Sources, Selected Writing on Quality Circles*, IAQC, 1984.

Ishikawa, Kaoru. *Quality Control Circles at Work*. Tokyo: JUSE, 1984.

Japanese Union of Scientists and Engineers. *QC Circle Koryu, General Principles of the QC Circle*. Tokyo: JUSE.

Li-Ping Tang; Peggy Smith Tollison Thomas; and Harold D. Whiteside. "Quality Circle Productivity as Related to Upper-Management Attendance, Circle Initiation, and Collar Color." *Journal of Management* 15, no. 1 (March 1989), pp. 101–113.

Mohr, W., and H. Mohr. *Quality Circles: Changing Images of People at Work*. Reading, Mass.: Addison-Wesley Publishing, 1983.

QC Circle Headquarters. *How to Operate QC Circle Activities*. Tokyo: JUSE, 1985.

QC Headquarters. "QC Circle Koryo—General Principles of the QC Circle." Tokyo: JUSE, 1980.

Thompson, P. *Quality Circles: How to Make Them Work in America*. New York: American Management Association (AMACOM), 1983.

Trepo, Georges X. "Introduction and Diffusion of Management Tools: The Example of Quality Circles and Total Quality Control." *European Management Journal* 5, no. 4 (Winter 1987), pp. 287–93.

Service Quality

"Customer Satisfaction Powers Everything Motorola Does." *National Productivity Report* 19, no.6 (March 31, 1990). pp. 1–4.

Davidow W., and B. Uttal. *Total Customer Service—The Ultimate Weapon*. New York: Harper & Row, 1989.

Desatnick, Robert L. *Managing to Keep the Customer*. Milwaukee, Wisc.: ASQC Press, 1987.

Eastman Kodak Company. *Keeping the Customer Satisfied—A Guide to Field Service*. Milwaukee, Wisc.: ASQC Press, 1989.

Finkelman, Daniel. "If the Customer Has an Itch, Scratch It." *The New York Times*, May 14, 1989.

Hensel, James S. "Service Quality Improvement and Control: A Customer-based Approach." *Journal of Business Research* 20, no. 1 (January 1990), pp. 43–54.

Lash, Linda M. *Complete Guide to Customer Service*. Milwaukee, Wisc.: ASQC Press, 1989.

Lawton, Robin L. "Creating a Customer-Centered Culture for Service Quality." *Quality Progress* 22, no. 5 (May 1989), pp. 34–36.

Liswood, Laura A. *Serving Them Right: Innovative and Powerful Customer Retention Strategies*. Milwaukee, Wisc.: ASQC Press, 1990.

Sellers, Patricia. "Getting Customers to Love You." *Fortune*, March 13, 1989.

Zeithaml, Valarie; A. Parasuraman; and Leonard L. Berry. *Delivering Quality Service —Balancing Customer Perceptions and Expectations*. Milwaukee, Wisc.: ASQC Press, 1990.

CHAPTER 20

STATISTICAL TOOLS

"Lies, Damned Lies, and Statistics."

Mark Twain

Unfortunately most Americans have taken Mark Twain's attitude toward statistics to heart, and consider anything having to do with statistics to be disingenious at best, most likely boring, and impossible to understand. This ignorance as a society has cost us dearly in attempting to catch up to world-class standards of quality. At a Japanese-owned plant in the Southeast a quality position was filled by a person with a masters degree in statistics. In Japan, the equivalent position was filled by a high school graduate.

The purpose to this chapter is to acquaint the reader who is not familiar with statistical applications to see where and if these tools could be used in their operation. Statistical tools have a place not just in manufacturing, but in services and administrative functions as well. In fact sampling and Pareto analysis may have greater applications in the nonmanufacturing sector than it does in modern manufacturing plants. Evolutionary operation, control charts, and the design of experiments are used more often in a manufacturing environment than in the office.

Whatever your setting, peruse the tools here and see how they can be integrated into your organization. The tools are required to collect data and make sense of it; without data collection other tools are essentially worthless or hollow. Once you have selected a tool, take a course in its use, and read more about it.

STATISTICAL MEASURES AND SAMPLING

Aliases

Descriptive statistics.

Brief Description and Purpose

Statistics can provide a simple and quick way to generalize about a process without resorting to highly complex models. Unfortunately, statistical models are easy to abuse, misinterpret, or be made misleading. Used properly, and with an eye to underlying assumptions, statistics need not "lie" or otherwise misrepresent a situation. Perhaps the best-known example of carefully used statistics in the United States are the exit polls conducted by the television networks during major national elections. Based upon a small sample size relative to the population voting, highly accurate predictions are made.

Discussion

In the theoretical world, things can behave *nicely*—according to simple algebraic formulas. In the real world, things never seem to work this way. There are always minor deviations, even if the behavior of the object we are studying is supposed to behave in a pleasant, theoretically predictable way, such as ballistic calculations. Things such as wind speed, heat, humidity, and other environmental factors can affect the predicted accuracy. In an application that is theoretically chaotic, our neat formulas are of little, if any, value. Many have tried to predict the stock market, but to no avail. It is affected by complex, dynamic, and unknown factors. Not even statistics can predict the market. But it can tell us much about what has happened, perhaps aiding our intuition. Hopefully your organization's statistical applications will be better behaved than the volatile stock market.

Statistics are based upon accumulated data, or measurements. In general, the more the better (providing the quality of the data is high), but one does encounter a point of diminishing returns—1 more or even 100 more measurements will add little to our knowledge. This helps us to determine an appropriate sample size—the number of data elements we should collect to arrive at a given measure of confidence that our statistics will be within a certain range of the reality. The power of sampling allows some 1,500 "Nielsen families" determine what millions of Americans are watching. (Note that a census is an enumeration of all data points or the entire population, as opposed to a sample of the data points, or part of the population.) This sampling aspect is one of the most powerful aspects of statistics. It enables us to predict the behavior of a process, and when combined with other tools such as control charts, can actually enable us to control the process as well. Statistical measures can be used even when the "underlying distributions" (i.e., the governing physical laws) are unknown or uncertain. These measures may allow us to determine this underlying theoretical structure.

Descriptive Statistics

To understand what the data is telling us, it is useful to arrange our data points in order, and to plot them on a graph. This plot could have the values along the horizontal axis, and the number of occurrences along the vertical axis. Another way to plot the data is cumulatively—plotting the values along the horizontal axis and the cumulative number of occurrences along the vertical axis. This is very useful in quickly determining the percentage or number of occurrences above and below a certain value(s), such as upper and lower tolerance limits. This cumulative frequency curve is often encountered in statistical quality control. This curve is also referred to as the ogive curve due to its shape looking like the architect and woodworker's ogive curve. An example of this type of curve is given in Figure 20–5. In this example, 40 percent of the measurements are at 307 or below; 10 percent are below 303, and so on. This sort of information is not obvious from the frequency plot.

While having a shape of the distribution is highly useful, it is cumbersome to use, and it would be of value to be able to describe the shape using as few measures as possible. This is the utility of descriptive statistics. Instead of trying to match and compare shapes, we can compare descriptive measures. One measure is the central tendency, which indicates where most of the observations occur, and another measure is how spread out the data is, which is a measure of the dispersion.

Measures of Central Tendency

Measures of central tendency attempt to assess just where the "middle" of the observations lie. One such measure is the arithmetic mean or average:

Average = Sum of data measures divided by the number of items

or:

$\Sigma_i = 1^n x_i / n$, where x_i is data element number i,
n is the total number of items.

The problem with using only this measure of central tendency is that it is affected by extremes. For example, if five incomes were: $25,000; $30,000; $32,000; $23,000; $250,000; the average would be: $72,000. This measure is biased by the single, high salary. So perhaps some other measure should be used in addition to the average.

There are two other common ways to find out this central point: the median and the mode. The median is the midpoint value. There will be an equal number of measurements above and below this value. In the salary example, the median would be: $30,000. (The next time you see typical salary figures given, pay attention to whether they use the median or the mean—generally they will use the median value.) This value probably makes more "common sense" than the average value did for this instance. The mode is the most frequently occurring value. There can be more than one mode, or there may be no mode (as in our list of salaries). Multiple modes can make for a strange looking distribution; the modes could be far away from the center. But if they "balance" each other, the wooden model would spin about the central point, even though there may be no values there. Some properties of these measures of central tendency can be summed up as follows:

FIGURE 20–1
Properties of Central Tendency

	Mean	*Median*	*Mode*
Easy to calculate	Yes	Yes	Yes
Must exist	Yes	Yes	No
Unique	Yes	Yes	No
Affected by extremes	Yes	Usually no	Not applicable

While the average is a good general measure, note that it is affected by extremes, and so should be used with caution when extremes are possible and likely, such as income surveys. In judging ice skating and in other judged sports, the high and low scores are often cast aside. This is an attempt not to have the average be influenced by judges who perhaps unfairly give high or low marks.

Measures of Dispersion

Dispersion measures the variability or spread of values in our data set. Hence, dispersion is the enemy of quality control. One simple measure of dispersion is the range of values, which is the spread between the two extremes of high and low values. The range of the salaries listed before would be $225,000. A different set of salary data could have given us the same average, but a much narrower range—the set $78,000; $66,000, $71,000, and $73,000 yields an average of $72,000 as before, but a range of only $12,000. Ranges are clearly susceptible to extremes. The range could also be expressed as a low and high point such as $66,000–$78,000.

A measure less susceptible to extremes is the *standard deviation,* a term that would seem to be a contradiction in terms at first blush. This measure is the square root of the variance, the variance given by: $\Sigma (x - x')^2 / n$, or, in English, the sum of the squares of each data element from the average (X'), divided by the number of observations, n. To make the standard deviation slightly more conservative (larger) when dealing with samples, $n-1$ is used instead of n. A useful computational form of the standard deviation (denoted σ) is:

$$\sigma = \text{Square root } \{(n\Sigma\, x^2 - (\Sigma x)^2)/n(n - 1)\}$$

In the 'normal' curve (a name for a bell shaped curve in which the mean, median and mode are the same) this number can be used to determine how likely an observation will fall with a standard deviation range. If we accept all products within a $\pm 3\sigma$ range, this would mean that we are rejecting 0.027 percent of the material. The key to improving quality here is how to reduce the standard deviation itself—changing acceptance from $\pm 3\sigma$ to something else does nothing to improve the quality, it will simply change how much we accept or reject. This number is normalized, and has no unit of measurement associated with it.

Dispersion measures are crucial in understanding the variance from the average. One can have an acceptable average, but with data that is too disperse. A case of meeting specifications (the average) but providing poor quality (too much dispersion) is shown in Figure 20–2. In this example color density was the measured quality factor for television sets produced by the same company (hence the same specifications), but manufactured on both sides of the Pacific rim. The color density of TV sets produced in Japan follow a typical normal distribution with a large number of sets close to the target value with a few sets beyond the upper and lower control limit. On the other hand, the inverted U-shaped distribution of color density produced in

FIGURE 20–2
Wide versus Narrow Dispersion

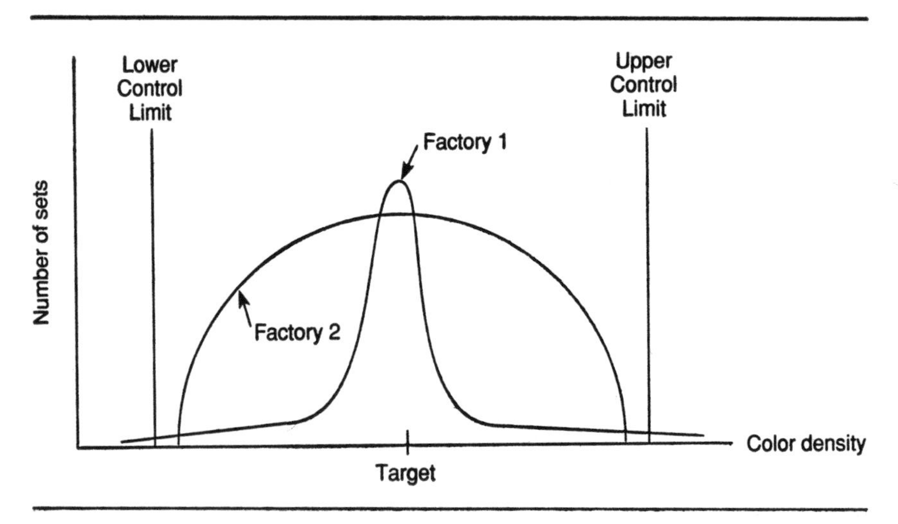

the U.S. sets produces zero defects, but the inverted U-shape results in poor perceived quality; there is a wide variation in the color density, resulting in the set not "looking right" to the consumer. This dissatisfaction increases as we move away from the center of the distribution or the target value. Therefore, there are more dissatisfied consumers with the American sets than the Japanese sets. This illustration demonstrates the inadequacy of simply meeting specifications or providing zero defects.

Sampling

Sampling predicts the nature of the whole from the examination of a few members of the whole; it can require little work, yet yield information concerning a very large amount of items. Not surprisingly, it is easy to misuse and abuse. The easiest way to misuse sampling is to not take a random sample. Always grabbing a sample of the material from the top of the bin, for example, will lead to nonrandom sampling.

There are two aspects about sampling that are of interest to Quality Management. One is acceptance sampling, where a plan can be concocted to accept or reject incoming material. Typically a quality engineer selects an appropriate sampling plan from a published table (or creates a sampling plan). The inspector can then be given straightforward directions in whether to accept or reject a lot—for instance, if one or fewer items out of a sample of five from a lot are bad, we still accept the lot. Acceptance sampling requires a considerable amount of time, and for high-quality (1 bad per 100,000), high-speed production, it is an archaic tool.

Operational. Because of poor training or instructions the forms are filled out incorrectly.

Nonresponse. The problem of missing data is a hot research topic in statistics, indicating that this is far from a simple problem. Do not assume that the missing data resembles the rest of the data.

Estimation. Perhaps the method rather than the collector has influenced the results. A classic case of miscollection of data occurred during the 1936 presidential election. A national magazine (Literary Digest) telephoned and asked for the voter's opinion. The predicted winner of the election was wrong, by a very wide margin. This was because few telephones were installed (they were relatively expensive) in private homes. The "rich person" candidate naturally won the phone vote. However, the "poor person" candidate won handily in a long-depressed economy. It is left to the reader to determine which four (possibly five) biases this faux pas entailed.

Example

A single dimension was measured on 200 items. A tally table was used, and the results are shown in Figure 20–3. Multiplying the measurement (in thousandths of an inch) by the number of times it occurred, adding these numbers together, and dividing by 200 (the number in the sample) yields an average of: 307.665. Note that the mode is 308, as is the median. It would appear that the normal distribution is being followed. This is made even more obvious by plotting the data (Figure 20–4). A cumulative or ogive plot appears in Figure 20–5. The standard deviation is 1.2.

Dos and Don'ts

- If your problem is intricate, seek the advice of a professional statistician. Statistics can become very complicated in real-life situations.

FIGURE 20–3
Sample Measurements

Measurement	*Count*	*Cumulative Count*
290	1	200
291	1	199
292	1	198
297	1	197
300	1	196
301	3	195
302	6	192
303	5	186
304	12	181
305	12	169
306	19	157
307	23	138
308	34	115
309	24	81
310	22	57
311	12	35
312	8	23
313	4	15
314	7	11
315	2	4
316	1	2
318	1	1

FIGURE 20-4
Frequency Plot of Sample Data

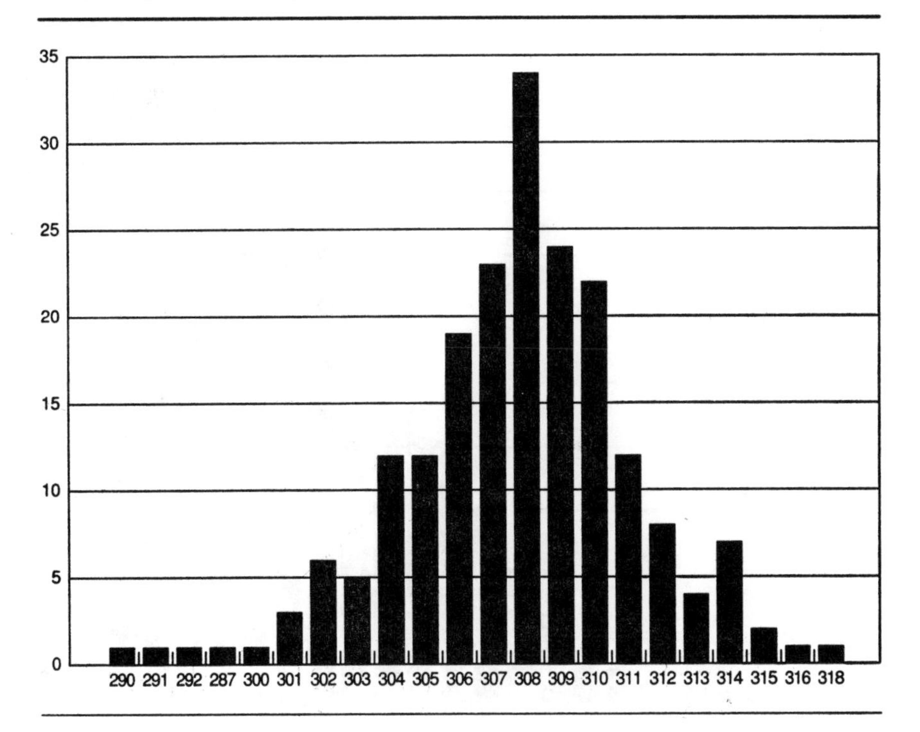

- Contemplate the physical meaning of all measures. Do they make sense? For example, the standard deviation should not exceed the range; the average also should lie within the range.
- Use screening rather than sampling if possible. Many manufacturing innovations have taken place in continuous inspection and expert system diagnostics.

CONTROL CHARTS

Aliases

Shewhart charts, attributes charts, and variables charts.

FIGURE 20–5
Cumulative Frequency Plot of Sample Data

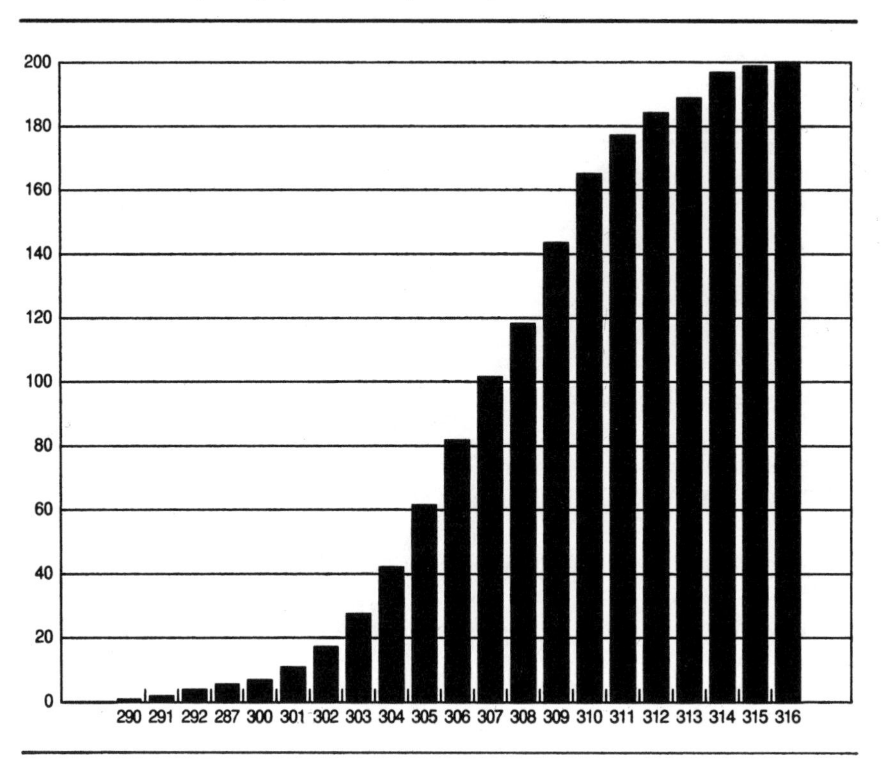

Brief Description and Purpose

Control charts provide a graphical means of determining whether or not a specific process is performing within acceptable parameters.

Discussion

Statistical quality control (or process control) traditionally consists of control charts and acceptance sampling. Other statistical techniques, such as evolutionary operation, analysis of variance, and design of experiments can also be considered an integral part of statistical quality control. Control charts aid in monitoring the quality and stability of an established process. They are readily constructed, and can be used without extensive knowledge of statistics.

Control charts provide a means of assessing quality without regard to the underlying cause. This cause can be determined at some later time. The control chart can give us real-time feedback as to how well the process is behaving; that is, if it is staying within predefined process limits. These limits may be adjusted as the process matures.

There are two types of control charts in common use: attributes charts and variables charts. Attributes are based upon characteristics that are not readily measurable: it fits or does not, go/no-go, smooth, dented, or scratched. These attributes charts are sometimes referred to as *p* charts because they track the fraction defective, denoted by *p*. Variables are measurements made on the process or product under study. Two statistics are tracked concurrently: *X*-bar, the sample average, and *R*, the sample range.

Note that the chart tells us only if the process is in control or not. It is silent regarding the underlying causes. To determine the causes, other techniques such as brainstorming and cause-effect analysis may be necessary. Often an operator who has a chart in their area can determine the cause because of the way a chart looks (i.e., certain causes lead to an out-of-control process in characteristic ways that an operator will soon be able to recognize).

There are two sources of variability in any process: chance causes and assignable causes. If a system is stable and predictable, the causes of variation influencing a process produce a usually constant pattern. Such a pattern can be used to predict future quality. The causes of variation which do not belong to a stable system are referred to as assignable causes of variation. Such causes can be identified and eliminated. If we eliminate these assignable causes, it leaves only the chance causes and, consequently, a constant pattern of quality variability. Once a constant pattern of variability is attained, it must be maintained by prompt detection and correction of assignable causes if and when they appear.

Variation comes in two flavors: chance and assignable. Chance causes of variation are due to random variations in the process. Chance causes are:

- Always present.
- Inherent in the nature of the process.
- Neither identifiable nor removable.
- Impossible to control.

So if chance alone is causing the variations in quality, the product cannot be improved without some fundamental change in the process. Chance causes are predictable in a statistical way.

Assignable causes of variation are due to nonchance variations. Assignable causes are:

FIGURE 20–6
Generic Control Chart

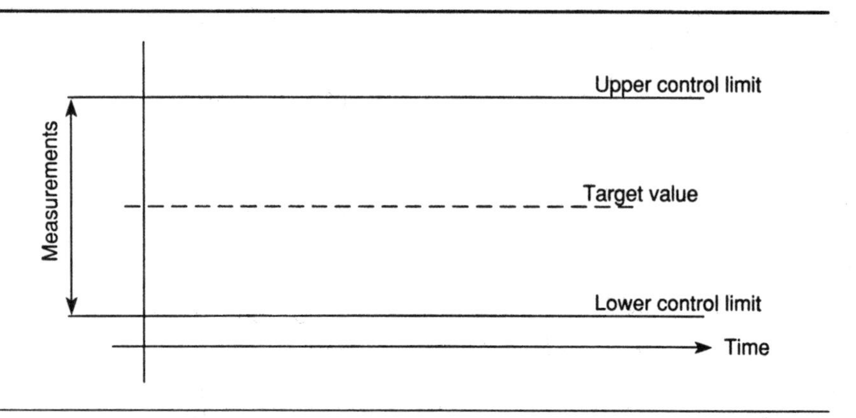

- Potentially identifiable and removable.
- Capable of being controlled.
- Require immediate action.

Assignable causes are unpredictable; however the product can be improved by detecting, identifying, and eliminating these causes.

Figure 20–6 illustrates the concept of a control chart. The normal curve has been superimposed on the chart to show that the center line is the target value, and that as you approach the control limits, fewer and fewer points can be expected. Figure 20–7 shows sample control charts that merit investigation. Rules may be formed as to when an investigation is necessary.

Relation to Other Concepts and Tools

Control charts rely on a knowledge of statistical measures such as range and dispersion, and distributions.

Process

Attributes Charts

To construct an attributes chart, the average fraction defective, p, must be known or estimated. This value may be expressed as:

$$p = \frac{\text{Total number of nonconforming (or defective) units from all samples}}{\text{Number of samples} \times \text{Sample size (total number of items inspected)}}$$

FIGURE 20–7
Out-of-Control Examples

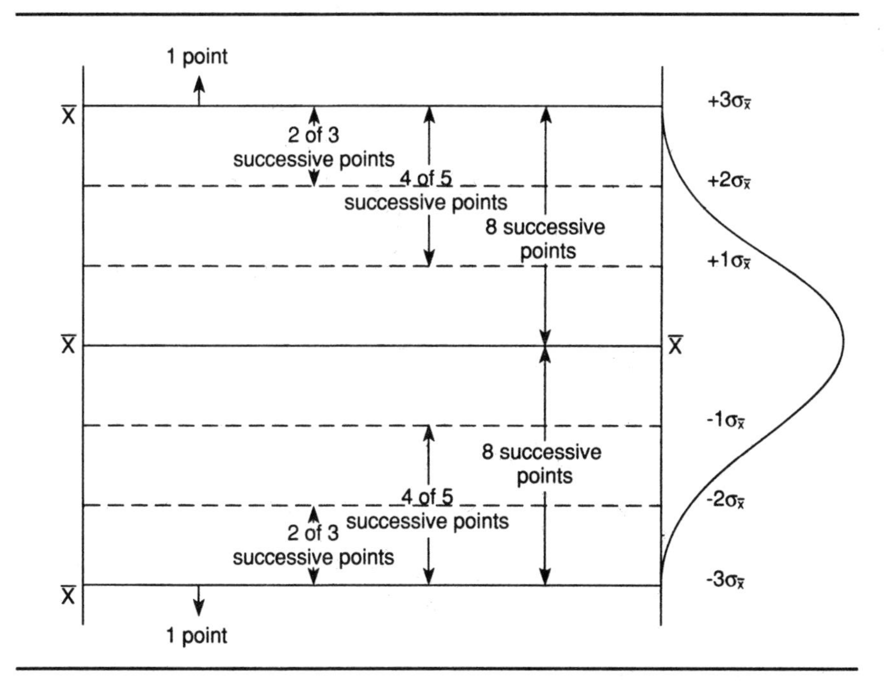

The standard deviation of p, denoted by s, is:

$$s = \sqrt{p(1-p)/(n-1)}$$

where n is the sample size. The upper and lower control limits are then expressed in terms of the fraction defective, its standard deviation, and the parameter z, which is the number of standard deviations required for a specific confidence. For example, 99 percent confidence requires a z value of 2.58. Selected z values and their corresponding confidences may be found in Figure 20–8.

This standard deviation value is used to quickly determine whether a process is in control or not. A sample of out-of-control rules of thumb are presented in Figure 20–7. Note the intervals are expressed in terms of a multiple of the standard deviation. For example, if the target value was 4.55 cm, and the standard deviation was 0.01 cm, four out of five successive points at greater than 4.56 cm or less 4.54 cm would indicate that the process was out of control. Note that with eight successive point at 4.55 cm or greater would be cause for concern, even though the measurements may be very

FIGURE 20–8
***z* Values**

	Percent value
1	68%
2	95
3	99.7
5	99.99994

close to the target value. This is of concern because in a random environment, we would expect data points to occur on both sides of the target value (but not in a perfect zigzag, either — 14 or more points alternating back and forth is a cause for concern).

There are several other types of attributes charts such as:

Number of defectives, *np* chart.

Number of defects per sample, *c* chart.

Number of defects per unit, *u* chart.

The latter two are especially useful when more than one defect per unit is possible. Their uses and equations for defining control lines is given in Figure 20–9.

Variables Charts

Variables charts require the consideration of four items:

- Sample size: Keep it small. The process may change while a sample is being taken. Common practice is to keep the size to about four or five.
- Number of samples: At least 25 samples should be taken to establish the control limits for the charts (this is true of *p* charts, as well).
- Sampling frequency: A trade-off between the cost of sampling and its utility must be studied. This will also vary with the variance of the product.
- Control limits: The standard is to set the control limits at three standard deviations (three sigmas - the sigma sign denotes standard deviation) above and below the mean (six sigmas total). This means that 99.7 percent of the sample means should lie between the control limits.

FIGURE 20–9
Attributes Charts Types—Control Charts for Attributes

If You are Charting . . .	*And the Sample Size* Varies, *Use a*	*And the Sample Size* Is Constant, *Use a*
Defectives A defective is a unit containing one or more defects. For example, a memo with five misspelled words and incorrect margins could be considered defective. The customer determines the criteria that establish a unit as defective.	*p Chart* (to chart the fraction or percent defective)	*np Chart* (to chart the number of defectives in a subgroup)
Defects A defect is an individual failure to meet a single requirement. For example, a misspelled word in a	*u Chart* (to chart the number of defects per unit)	*c Chart* (to chart the number of defects in a subgroup)

***p* Chart**

$$p = \frac{\text{Number of defectives in a subgroup}}{\text{Size of subgroup } (n)} = \text{Fraction or percent defective}$$

$$p = \frac{\text{Total defective}}{\text{Total inspected}} = \text{Centerline} = \text{Average fraction defective}$$

$$\text{UCL} = \bar{p} + 3\sqrt{\frac{\bar{p}(1-\bar{p})}{n}} = \text{Upper control limit (varies by subgroup)}$$

$$\text{LCL} = \bar{p} - 3\sqrt{\frac{\bar{p}(1-\bar{p})}{n}} = \text{Lower control limit (varies by subgroup)}$$

You can use an average subgroup size to obtain a single set of control limits if:

- The largest subgroup size is less than twice the average subgroup size.
- The smallest subgroup size is more than half the average subgroup size.
- Plot p for each subgroup.

***np* Chart**

$\bar{p}$, n = Same as for p chart, except n must be constant

$n\bar{p}$ = Centerline = Average number of defectives

$\text{UCL} = n\bar{p} + 3\sqrt{n\bar{p}(1-\bar{p})}$

$\text{LCL} = n\bar{p} - 3\sqrt{n\bar{p}(1-\bar{p})}$

- Plot np for each subgroup.

FIGURE 20–9
(*concluded*)

***u* Chart**

$$u = \frac{\text{Number of defects per subgroup}}{\text{Number of units per subgroup}}$$

$$\bar{u} = \frac{\text{Total number of defects for all subgroups}}{\text{Total inspected}} = \text{Centerline}$$

$$\text{UCL} = \bar{u} + 3\sqrt{\frac{\bar{u}}{n}}$$

$$\text{LCL} = \bar{u} - 3\sqrt{\frac{\bar{u}}{n}}$$

- Plot *u* for each subgroup.

***c* Chart**

c = Number of defects per subgroup.

$$\bar{c} = \frac{\text{Total defects}}{\text{Total number of subgroups}} = \text{Centerline}$$

$$\text{UCL} = \bar{c} + 3\sqrt{\bar{c}}$$

$$\text{LCL} = \bar{c} - 3\sqrt{\bar{c}}$$

- Plot *c* for each subgroup.

Source: Electronics Systems Division, Hanscom AFB, Mass.: 1991.

The upper and lower control limits may be found by using:

For means: Upper limit, X-bar = $X'' + A_2R$-bar
Lower limit, X-bar = $X'' - A_2R$-bar
For ranges: Upper limit, $R = D_4R$-bar
Lower limit, $R = D_3R$-bar

R-bar is the average of the ranges (the sum of the ranges divided by the number of samples), X is the average of the averages. The constants A_2, D_3, and D_4 are based upon the normal distribution, and depend upon the number of observations in the subgroup, or n. A table of these values is supplied in Figure 20–10.

Process Capability

The process capability is a figure of merit for the process. For a normal distribution with six sigma control limits, the process capability index may be computed as:

Process capability = USL − LSL/(6R-bar/d_2).

FIGURE 20–10
Factor for Upper and Lower Control Limits

Number of Observations in Sample n	A_2	D_3	D_4	d_2	E_2
2	1.880	0	3.267	1.128	2.660
3	1.023	0	2.575	1.693	1.772
4	0.729	0	2.282	2.059	1.457
5	0.577	0	2.115	2.326	1.290
6	0.483	0	2.004	2.534	1.184
7	0.419	0.076	1.924	2.704	1.109
8	0.373	0.136	1.864	2.847	1.054
9	0.337	0.184	1.816	2.970	1.010
10	0.308	0.223	1.777	3.078	0.975

Where USL is the upper specification limit and LSL is the lower specification limit. The factor d_2 is also listed in Figure 20–8. The meaning of process capability number can be gleaned from Figure 20–11. When the process capability is less than one, the process is not capable. When it is one or (ideally) more, then the process is a capable one.

Example

Figures 20–12 and 20–13 give an example of the construction of an *X*-bar and *R* chart.

Dos and Don'ts

- Before implementation, take a course, or read Grant and Leavenworth's book. Their work is the classic work in the field. Other, newer works have recently come out that apply Quality Management tools with statistical quality control, such as those by Gitlow, et al., Ryan, or Pyzdek.
- Don't impose blanket SQC or control chart requirements on all suppliers. Some suppliers may be operating in a job shop or a one-item-at-time. Control charts require moderate to high volume to work well.

FIGURE 20–11
Process Capability

This is an *uncapable* process. The process variability is greater than the specification limits, so a large number of nonconformances will be made.

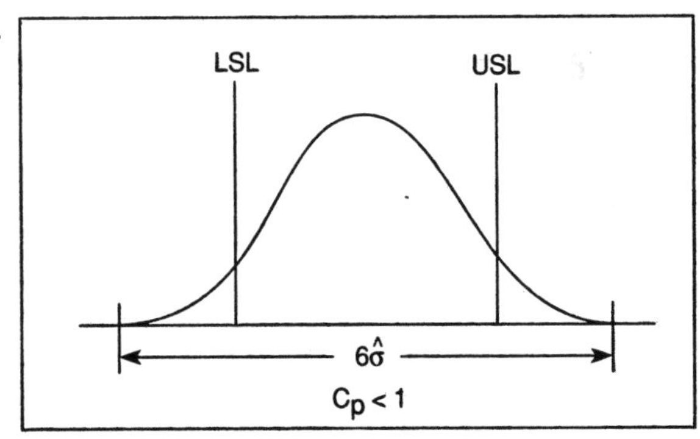

This is a *capable* process. The process variability is equal to the specification limits. If the process remains centered, specifications will be met more than 99 percent of the time.

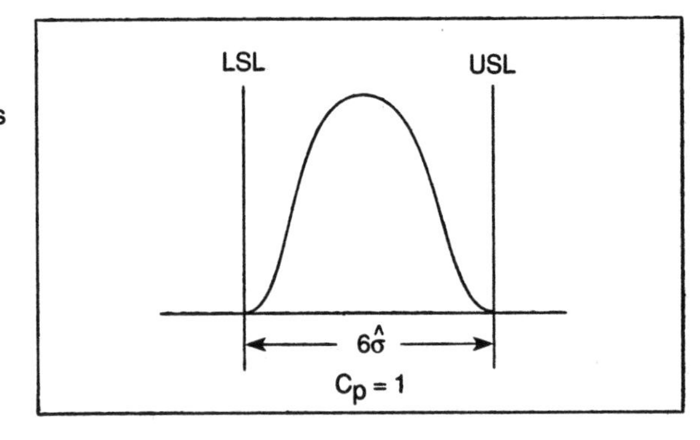

This is an *ideal* process. The process variability is much less than the specification limits. Even if this process shifts off-center, you can detect the shift and correct it without creating nonconformances.

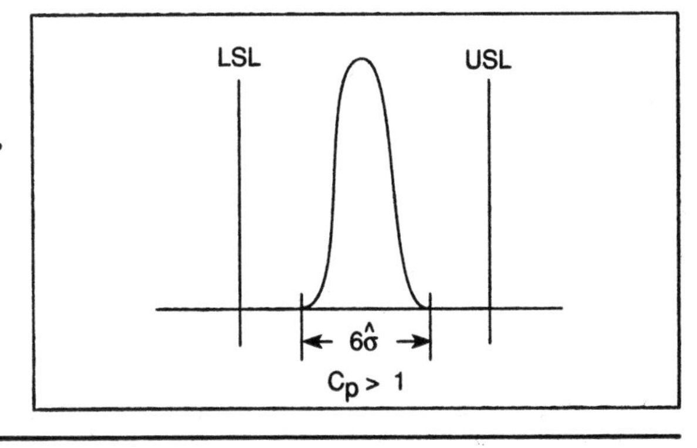

Source: Electronics Systems Division, Hanscom AFB, Mass.: 1991.

FIGURE 20–12 Control Chart Example—Establishing Control Limits

Example: X and R Chart

A manufactured component has a particular dimension with a tolerance of 50 ± 8 mm. It is the beginning of production and it is desired to determine if the process is in control. An X and R chart will be constructed using samples of four. These samples will be taken hourly. The results of the first 20 samples are as follows:

Sample No.	Sample Readings X_1	X_2	X_3	X_4
1	47	50	50	45
2	51	48	41	44
3	43	45	43	45
4	48	45	47	46
5	51	52	52	53
6	43	50	46	48
7	48	44	45	46
8	43	47	42	44
9	45	57	45	43
10	40	47	46	49
11	49	47	44	46
12	44	43	46	43
13	43	49	44	46
14	47	51	48	47
15	45	42	46	51
16	42	41	43	40
17	47	47	52	46
18	43	46	46	48
19	49	48	54	49
20	46	50	45	45

Solution: The $\overline{X}$ and R chart is constructed as follows:

a. The mean ($\overline{X}$) and range (R) are found for each sample.
b. The central lines and control limits are computed.

Central Lines

$$\overline{\overline{X}} = \frac{\Sigma \overline{X}}{\text{Number of samples}} = \frac{928}{20} = 46.40$$

$$\overline{R} = \frac{\Sigma R}{\text{Number of samples}} = \frac{113}{20} = 5.65$$

Control Limits

For $n = 4$, $A_2 = 0.729$, $D_3 = 0$, $D_4 = 2.282$

$$UCL_{\bar{x}} = \overline{\overline{X}} + A_2\overline{R} = 46.40 + (0.729)(5.65) = 50.52$$

$$LCL_{\bar{x}} = \overline{\overline{X}} - A_2\overline{R} = 46.40 - (0.729)(5.65) = 42.28$$

$$UCL_R = D_4\overline{R} = (2.282)(5.65) = 12.89$$

$$LCL_R = D_3\overline{R} = (0)(5.65) = 0$$

These values are indicated in the upper right-hand corner of the chart (Figure 20–13) and drawn on the respective charts. The central lines are shown as solid lines, and the control limits are shown as dotted lines.

Source: *Army Management Engineering College Statistical Quality Control Coursebook*, Rock Island, Ill.

FIGURE 20–13
Control Chart Example—Sample Chart

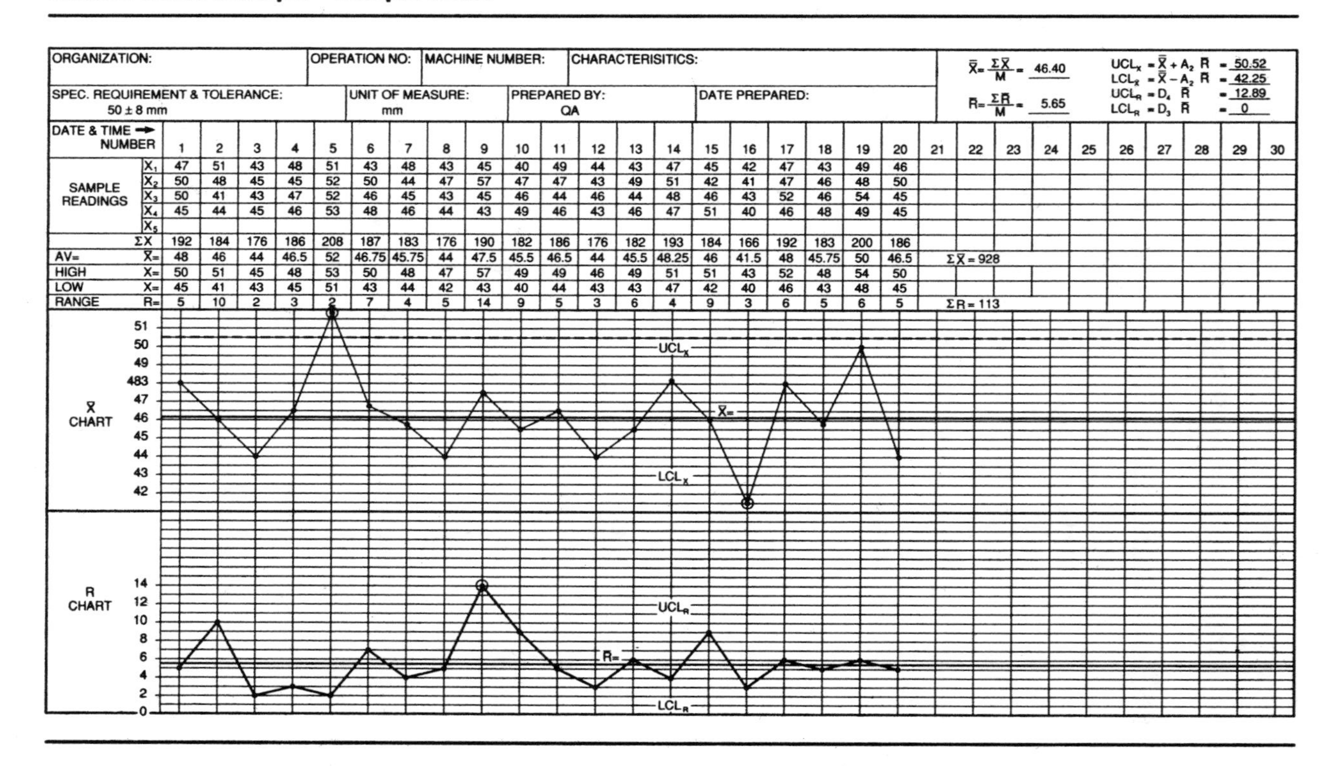

ORGANIZATION:	OPERATION NO:	MACHINE NUMBER:	CHARACTERISITICS:
SPEC. REQUIREMENT & TOLERANCE: 50 ± 8 mm	UNIT OF MEASURE: mm	PREPARED BY: QA	DATE PREPARED:

$\bar{\bar{X}} = \frac{\Sigma \bar{X}}{M} = 46.40$

$\bar{R} = \frac{\Sigma \bar{R}}{M} = 5.65$

$UCL_{\bar{x}} = \bar{\bar{X}} + A_2 \bar{R} = 50.52$

$LCL_{\bar{x}} = \bar{\bar{X}} - A_2 \bar{R} = 42.25$

$UCL_R = D_4 \bar{R} = 12.89$

$LCL_R = D_3 \bar{R} = 0$

DATE & TIME → NUMBER		1	2	3	4	5	6	7	8	9	10	11	12	13	14	15	16	17	18	19	20	21	22	23	24	25	26	27	28	29	30
SAMPLE READINGS	X_1	47	51	43	48	51	43	48	43	45	40	49	44	43	47	45	42	47	43	49	46										
	X_2	50	48	45	45	52	50	44	47	57	47	47	43	49	51	42	41	47	46	48	50										
	X_3	50	41	43	47	52	46	45	43	45	46	44	46	44	48	46	43	52	46	54	45										
	X_4	45	44	45	46	53	48	46	44	43	49	46	43	46	47	51	40	46	48	49	45										
	X_5																														
	ΣX	192	184	176	186	208	187	183	176	190	182	186	176	182	193	184	166	192	183	200	186										
AV=	$\bar{X}=$	48	46	44	46.5	52	46.75	45.75	44	47.5	45.5	46.5	44	45.5	48.25	46	41.5	48	45.75	50	46.5		$\Sigma \bar{X} = 928$								
HIGH	X=	50	51	45	48	53	50	48	47	57	49	49	46	49	51	51	43	52	48	54	50										
LOW	X=	45	41	43	45	51	43	44	42	43	40	44	43	43	47	42	40	46	43	48	45										
RANGE	R=	5	10	2	3	2	7	4	5	14	9	5	3	6	4	9	3	6	5	6	5		$\Sigma R = 113$								

DESIGN OF EXPERIMENTS

Aliases

Design and analysis of experiments.

Brief Description and Purpose

An experiment is a systematic approach for acquiring information, with a minimum amount of data collection (thus reducing cost), on the effect of one or more variables on a process or product. This information is generally obtained by sampling or experimental techniques. It is essential to clearly define the purpose and scope of the experiment at the outset. The purpose may be to optimize a process, to determine the reliability of a system, or to evaluate the effects of variability.

Discussion

The experiment must be planned so that the effect of changing a factor can be measured and distinguished from the effects that other factors have, and these effects can be measured. There are dependent variables and independent variables. Dependent variables are the responses of interest; for example, temperature (the dependent variable) varying over time (the independent variable). In this case the independent variable is time. Independent variable are the factors we want to study, and are usually associated with the horizontal scale in a graph. Dependent variables are usually displayed across the vertical axis. Experimental error must be known or estimated from the outset, in order to discern if the effects are real or are within tolerance or error limits. Sources of error include:

- Measurement precision.
- Number of replications.
- The design to be used.

Figure 20–14 lists some definitions of a few terms used in the design and analysis of experiments. The purpose of this chapter is to acquaint the reader with the design of experiments. Further study of the references, or a course in design of experiments would be necessary before conducting an experiment. Many closely associate design of experiments with "Taguchi Methods," as Taguchi has been a champion of design of experiments for quality control uses. Although there are many designs (e.g., randomized, full fac-

FIGURE 20–14
Some Design of Experiments Terms

Subject: The subject of a designed experiment is the dependent measurable characteristics.

Factor: A variable which may be brought into the experiment in order that relevant information can be obtained; an independent variable. A factor may be either of two types:

Qualitative factor: A factor whose levels cannot be arranged in order of magnitude.

Quantitative factor: A factor whose levels can be arranged in order of magnitude.

Levels: Those "values" of the factor which are under consideration in the experiment.

Treatment or treatment combination: The combination of the levels of the different factors in a given trial.

Response: The observation or numerical result of a given trial based on a given treatment combination.

Effect: A change in the response due to a change in the level of the factor.

torial, fractional factorial, block, incomplete block, Youden squares, Latin squares, and nested), there are three types that are commonly used in quality control:

The ***One-Factor-at-a-Time*** method varies only one of the factors at a time. Presumably then, the factor that was changed was the factor responsible for any effect. One experiment per factor must be conducted. The problem with this approach is that it can require many trials, and more importantly, it may be difficult to change only one factor without involving the others. This phenomenon is sometimes called the *tuning problem*: tweaking one control requires tweaking of another (i.e., simply returning the first control back to the original position is insufficient for the result to be the same). Another potential drawback is that this approach does not mimic reality, where all factors could be changing at once.

The *full factorial* experiment tests all possible combinations of factors at each factor level. This method can easily lead to a large number of test runs or trials. Say that there were eight factors, each having just two levels. Using the formula below, this would lead to 2^8 or 256 trials!

$$\text{Levels}^{\text{Factors}} = \text{Number of trials}$$

Note that two levels gives a very coarse measurement. Thus, refining the overall experiment to, say, five levels and eight factors yields 390,625 trials conducting one trial per hour, it would take 195 years to complete this experiment!

Fortunately, not all of the combinations are of interest, so this problem can be overcome by using the *fractional factorial* design, where only a specific subset is considered. Taguchi proposes that the design should be balanced—each factor is tested at each level an equal number of times.

Relation to Other Concepts and Tools

Design of experiments requires a thorough grounding in elementary statistics.

Process

1. Define the purpose and scope. In order to draw proper inferences about situations which are beyond the scope of the experiment, the scope should provide enough coverage to mimic a realistic operation. This is particularly true if operator error is likely to occur. Underlying assumptions need to be exposed. For instance, if a production run were made from a single lot of raw material, then the conclusions are applicable *only to that lot.*

2. Determine the variables. There are several types of variables, outlined below.

Primary variables are those whose effects are to be evaluated directly; these variables probably caused the need for experimentation. The number of levels to be selected depends upon the type of relation expected: a straight-line relationship requires two points or levels. Other relationships require three levels or more. Be aware that as the levels increase, the number of trials can expand tremendously.

Variables can interact with one another—one variable effects the response of another variable. The lower right quadrant of Figure 20–15 shows no interaction between factors A and B, the upper left quadrant shows a mild interaction, and the lower left quadrant shows the results of a strong interaction. Interactions can impact the reliability of the results. It may not be known until the experiment is conducted if factors interact.

Background variables are those which cannot be (or have chosen not to be) held constant. This might occur when different operators or instruments are used. Background should not be varied with the primary variables. If one used thermometer A on the first day, and thermometer B on the second, how much of the difference was due to the thermometers being different? In this case, it may be possible to calibrate the instruments. But what if different (human) operators were used?

FIGURE 20–15
Interacting Variables

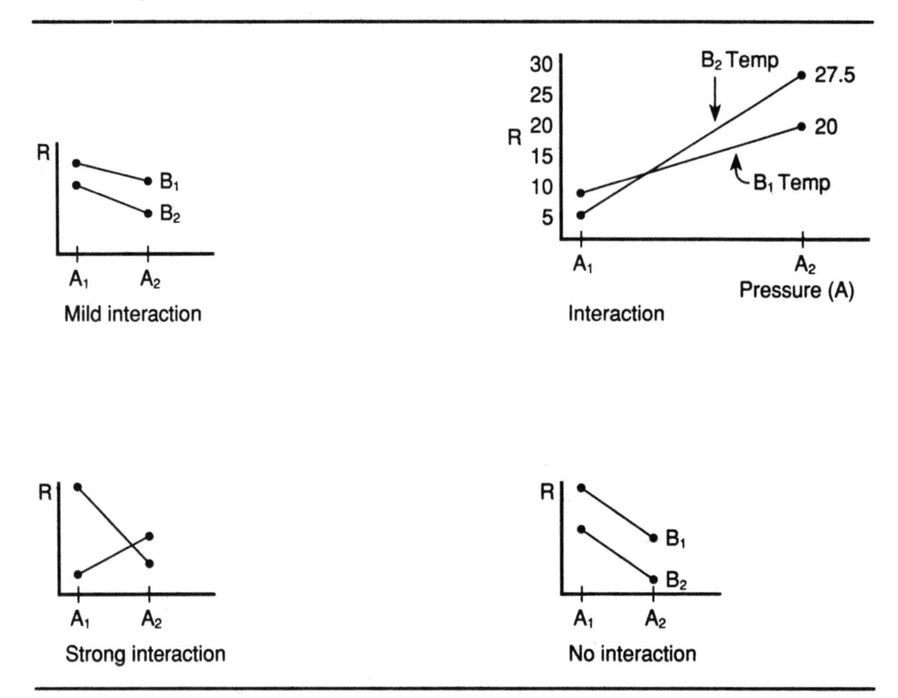

Properly mixing the background variables so that their effects (hopefully) cancel out is termed *blocking*.

Uncontrolled variables crop up as ambient conditions and cannot be controlled, although they may be measurable or identifiable. To overcome these conditions, a random environment must be created. Randomization is often easier to do than blocking.

Constant variables are those that are held constant; ambient temperature in a factory setting, for example. While this may be convenient, constant variables may mask actual use however. In the factory, temperatures may range from 50–120 degrees Fahrenheit. This wide range is bound to affect something or someone.

3. Select levels of each factor.

4. Select orthogonal array. Figure 20–16 shows the array for three factors, A, B, and C for two levels. Figure 20–17 has selected an orthogonal or balanced array from the nine possibilities. This array could be termed an $L_4(2)^3$, as it has three factors, two levels, and four trials.

FIGURE 20–16
Array for Three Factors, Two Levels

Trial	A	B	C	Response
1	1	1	1	R1
2	1	1	2	R2
3	1	2	1	R3
4	1	2	2	R4
5	2	1	1	R5
6	2	1	2	R6
7	2	2	1	R7
8	2	2	2	R8

FIGURE 20–17
Orthogonal Array

Trial	A	B	C	Response
1	1	1	1	R1
2	1	2	2	R2
3	2	1	2	R3
4	2	2	1	R4

5\. ***Conduct the trials and record responses.***

6\. ***Calculate main effects.*** A simple average could be taken, adding together responses R1 through R4 (A at level 1), and R5 through R4 (A at level 2) and dividing by 4. A more sophisticated analysis of variance could also be performed, and is advisable on large experiments.

7\. ***Select the parameters to optimize.*** After a few initial trials have been run, it may be beneficial to reexamine the variables for interactions, prior to optimizing parameters.

Dos and Don'ts

• Do get assistance from statisticians. If you don't have one available, most colleges have at least one professor who understands statistics. Even if you do have a statistician on board, two or more heads can sometimes see assumptions overlooked by one individual.

EVOLUTIONARY OPERATION

Aliases

The acronym, EVOP, is used interchangeably with the term, *evolutionary operation*.

Brief Description and Purpose

Evolutionary operation (EVOP) is based on the idea that the manufacturing process itself can yield information on how to improve the quality of a process. EVOP involves small changes in process variables. The effect of these changes is noted, and the process modified when quality improvements can be made. Acceptable material is produced during an evolutionary operation.

Discussion

Evolutionary operation was introduced in 1957 by Dr. George Box of the University of Wisconsin. The concept was developed from experimental design (see "Design of Experiments" section, in Chapter 20) methodologies, but differs from design of experiments (DOE) in that:

- EVOP is in a manufacturing, or ongoing process, rather than an experimental setting.
- One or a few variables at a time are involved in EVOP, there may be many in DOE.
- EVOP is done by production personnel, on a long-term basis, carried out under real manufacturing conditions.

EVOP differs from statistical process control (SPC) in that SPC is a monitoring program, and does not necessarily reveal ways to improve the process. EVOP and SPC are complementary tools and work well together. Like SPC, EVOP works well on a long-running, high-volume process; it is too complex for the craftsman or artisan approach of making a limited quantity.

While it may require time to administer (and perhaps even more scary is the idea that something may change as a result of EVOP), companies such as Dow Chemical Co., Monsanto, American Cynamid, Tennessee Eastman Co., and Imperial Chemical industries have all seen production efficiencies increase from 4 to 37 percent and have had savings of up to $250,000 per year using EVOP.

Relation to Other Concepts and Tools

EVOP requires an understanding of statistical concepts, as well as design of experiments.

Process

1. Select two or three variables likely to affect quality.
2. Change these variables in *small* increments, according to a prearranged experimental plan. Figure 20–18 shows five points in an EVOP plan. The point labeled 1 is the current operating or target value.
3. Implement the plan and observe the effects.
4. When one or more of the effects is meaningful, change the midpoint and adjust the incremental changes accordingly.
5. Repeat until the optimal conditions are found.

Example

A large-volume bakery wishes to use EVOP on its snack cake line to maximize its yield by varying the time and temperature to bake its cupcakes (if baked too long, the cakes stick to the pan and can become burnt—too short, and the cakes are too gummy to sell). The current operating conditions of the variables are: 18 minutes and 375° F. It is determined, after talking to the

FIGURE 20–18
EVOP Plan for a Bakery

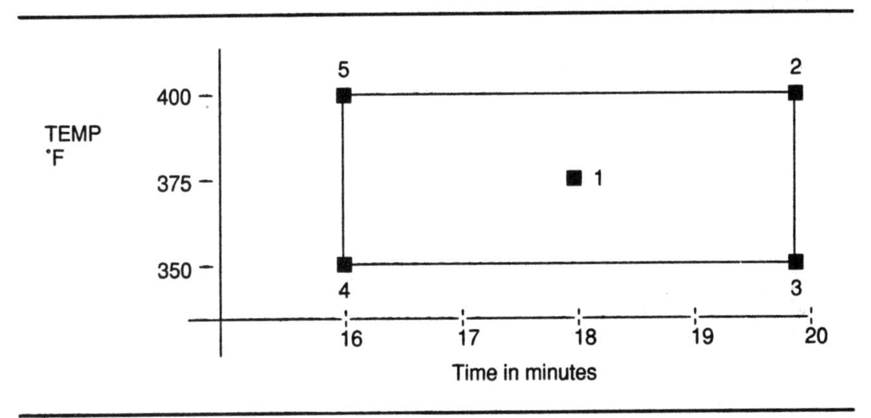

bakers, that the temperature can be varied by ± 25°F and the baking time by ± 2 minutes. The initial plan is illustrated in Figure 20–18. After the batches have been made, the following results were found:

Batch	*Temperature*	*Time*	*Usable Cupcakes*
1	375	18	73
2	400	20	72
3	350	20	76
4	350	16	71
5	400	16	81

It would appear that batch 4 is marginally better, and batch 5 is almost 10 percent better than the current point, number 1. But this is only one *sample*. Just as it would be foolish to predict a political winner from one exit interview from five different wards, so also is it foolish to conclude much of anything from one run. The next step in this experiment might be to use point 5 as the new current point, and choose closer settings, perhaps using only one minute variations and 10 degree changes. A more statistically oriented approach would be to take several samples, and determine the levels of significance.

Just as in control charts, a control limit can be established here as well. Instead of three standard errors, two standard errors from the average will be used. These limits will encompass 95 percent of the area under the normal curve. These limits can be expressed as:

$$\text{Significance limits} = \pm\ 0.667 \times A_2 \times R\text{-bar}$$

The A_2 factors are listed in Figure 20–19. The cumulative averages for six runs are tabulated in Figure 20–20. The average range of the yields was found to be 2.3 (the raw data is not shown here). The significance limits for this example are ± 0.667*0.4823*2.3, or ±0.739. Note that the difference in increasing the time was 0.7, slightly below the threshold for significance, but the temperature factor was −1.6, which is beyond the limit. So we could conclude that *decreasing* the temperature (the temperature factor was negative) would improve the process yield. The next step might be to move the operating point, centering it about a decreased temperature.

Bear in mind that EVOP is a slow process, and that the example here was fairly simple. Even so, six runs at each operating points, assuming that each run takes about half an hour (each run requires a setup and takedown), then this experiment would take 15 person hours, or two days of labor.

FIGURE 20–19
A_2 Factor for Significance Limits

Sample Size *n*	A_2
2	1.880
3	1.023
4	0.729
5	0.577
6	0.483
7	0.420
8	0.373

FIGURE 20–20
Yield Averages after Six Runs

Time	*16 min.*	*18 min.*	*20 min.*	*Average*
400°F	76.3		71.0	73.65
TEMP: 375° F		75.1		Change = -1.6
350° F	73.3		77.2	75.25
Average	74.8	Change = 0.7	74.1	

Expanding the experiment by one more variable, for a total of three variables, for example, would require an illustration using a cube, and would have at least nine points to test for each run.

Dos and Don'ts

- Start simple and slow. Become familiar with the process of charting or planning an EVOP.
- Be conservative on the significance limits. Take more measurements to be certain.
- Make several trials, over several days. Just as in design of experiments, be sure to account for uncontrollable variables that may affect results.
- Unlike sampling and design of experiments, EVOP is comparatively straightforward. Wait to call on a consultant until several experiments have been tried and verified.

PARETO ANALYSIS

Aliases

80-20 Rule, vital few, trivial many, and "hit parade."

Brief Description and Purpose

A Pareto analysis reveals which causes among the many are responsible for the greatest effect. Quite often some 20 percent of the causes produce 80 percent of the effect. The technique is based on the Pareto principle, which states that a few of the causes often account for most of the effect. The Pareto chart makes clear which *vital few* problems (causes) should be addressed first. It is one of the most effective tools to find these problems and to estimate the magnitude of benefits possible. The Pareto chart is a bar chart with bars arrayed in descending order, with the category with the greatest frequency of occurrence to the left. Each bar represents a cause. The chart displays the relative contribution of each cause to the total effect. A line representing the cumulative effect is often added to Pareto charts.

Discussion

Inventory analysis in World War II found that 10 to 20 percent of inventory items accounted for 80 to 90 percent of the dollar value. Subsequent observation has shown a widespread applicability of the same phenomenon. For example, 20 percent of the population holds 80 percent of the wealth; 20 percent of employees account for 80 percent of tardiness; 10 percent of accounts may have 80 percent of dollars received, or 10 percent of a firm's engineers have 90 percent of the patents; 90 percent of scientific articles are written by 10 percent of the scientists. It appears that the application of Pareto's law is universal.

While the existence of this relationship is recognized by managers, it frequently has no influence on management decisions. In terms of return on our investment, and specifically for TQM, the uniform level of management is not sound. Effective management requires the isolation of the vital factors of an operation from the insignificant, and the development of management/improvement systems that are justified for each group.

In the late 1940s, J. M. Juran identified the wider application of a discovery by the 19th-century economist, Alfredo Pareto. Pareto had studied the distribution of wealth in the latter part of the 19th century and the first part of the 20th century in Italy and discovered that an extremely large proportion of the Italian national income was going to approximately 10 percent of the population. Simply stated, it is the principle of "the vital few and the trivial many." The proportion of the vital few and the trivial many in a vast majority of instances has been found to be approximately 20 percent for "the vital few" and 80 percent for "the trivial many." This 20 percent is responsible for the major portion of the effect that takes place.

When using Pareto's principle, never use unit costs, time per unit, or production counts. They will not normally reflect the "vital few." Pareto's principle should use data that reflects total costs or total time expended.

Relation to Other Concepts and Tools

Pareto charts are useful after brainstorming and then constructing a cause-effect diagram to identify those items that could be responsible for the most impact. Data gathering is usually necessary to create an accurate Pareto diagram; guessing at the vital few may be clouded by temperament or by the last cause encountered.

Process

1. Decide what to analyze. Determine the purpose of the Pareto chart, and what data will best suit this purpose.

2. Collect information: Use the same unit of measurement for each element. The data may be on hand, or it may need to be collected (see Chapter 16, section "Check Sheets and Data Collection").

3. Group information into categories: Either types or ranges or whatever is appropriate. Some typical categories might be:

1. Department or work center.
2. Product type.
3. Machines.
4. Personnel.

4. Plot the data. Each cause or category should have a frequency or number of occurrences associated with it. If the data is inputted into a computer spreadsheet, then the data can be sorted easily, making a Pareto chart simple to construct and revise. The most frequently occurring cause is usually listed to the left, with the rest in descending order, except for the category "Other" which is often listed last. An example of a completed Pareto diagram is shown in Figure 20–22.

5. Interpret the diagram. Pareto analysis usually suggests what one, two, or three classes, which corrected, could make the greatest contribution to solving the problem. Note that this does not point out the most economical solutions. Correcting one of the trivial many may be expensive or inexpensive. Correcting the largest contributor may be most expensive, or it may not.

Also note that classes or categories may require further study. The following example reveals that packaging problems were the culprit, not manufacturing. Now that the field is narrowed, we may need to further analyze what happens in shipping. How do they become damaged? Improper packaging, rough handling, poor packing box or tape? We cannot be sure without further study. This followup study should also be a Pareto analysis.

Example

A publishing company produces a magazine, which is sent monthly to subscribers. Occasionally, a copy is lost in the mail (or is damaged or otherwise lost), and is "claimed" by the subscriber. The publisher then provides a replacement copy at no charge to the customer. The circulation manager believes that some countries have an unusually high claim rate.

The category used is country (the first column). The number of subscribers from that country is also listed in column two. Any honored claim (i.e., a single journal issue) over the past 12 months is counted (column three). The raw data is presented in Figure 20–21, and is illustrated in Figure 20–22. Note that only countries that had a claim are listed. About 20 other countries had no claim. The figure is in descending claim amount order. Using solely the number of claims would not be a true reflection of the situation. The number of claims was then translated into a percentage of claims, based on total subscriptions (the fourth and fifth columns are percentages of the total number of subscribers and the total number of claims). The "Delta" is the difference between the percent claims and percent subscriptions, giving in effect a ratio of how many were claimed versus how many could be expected to be claimed. This is graphically represented in Figure 20–23. Note that some are negative, indicating better service than might be expected from the average. Country 6 would appear to have exceptional service, while country 1 poses a problem. Interestingly, when the deltas of all of the countries are added together, they add up to zero, suggesting that the claims are not disproportionate, overall. Of course, this does not mean that we can't do better, and it was decided that country 1 would have an extra sealing tape on the envelope flaps. A follow-up Pareto analysis revealed that this did indeed reduce claims from country 1.

FIGURE 20–21
Data for Pareto Example

Country	*No. of Subs*	*No. of Claims*	*Percent Subs*	*Percent Claims*	*Delta*	*Cumulative*
1	23	19	5.2%	25.3%	20.1	19
2	25	8	5.7	10.7	5.0	27
3	19	6	4.3	8.0	3.7	33
4	21	6	4.8	8.0	3.2	39
5	29	4	6.6	5.3	-1.2	43
6	75	4	17.0	5.3	-11.6	47
7	13	4	2.9	5.3	2.4	51
8	41	4	9.3	5.3	-3.9	55
9	12	3	2.7	4.0	1.3	58
10	6	3	1.4	4.0	2.6	61
11	7	3	1.6	4.0	2.4	64
12	3	2	0.7	2.7	2.0	66
13	22	2	5.0	2.7	-2.3	68
14	4	2	0.9	2.7	1.8	70
15	21	2	4.8	1.3	-3.4	71
16	8	1	1.8	1.3	-0.5	72
17	2	1	0.5	1.3	0.9	73
18	5	1	1.1	1.3	0.2	74
19	5	1	1.1	1.3	0.2	75

Dos and Don'ts

- Do a follow-up analysis. Implement a feedback process by conducting an analysis after changes have been made. Make periodic analysis, as new problems manifest themselves.
- Collect data. A perception-based Pareto chart can be very misleading. What we perceive to be a major problem, upon closer inspection becomes simply another problem. Hence "pet peeves" can distort a perception- oriented Pareto analysis. Instead, collect data to support a Pareto analysis.
- Correlate data. A Pareto analysis may need to be correlated with other information in order for it make sense, as in the Example.
- Don't stop with one analysis. Further studies may refine causes, showing systemic, underlying causes

FIGURE 20–22
Pareto Graph of Example

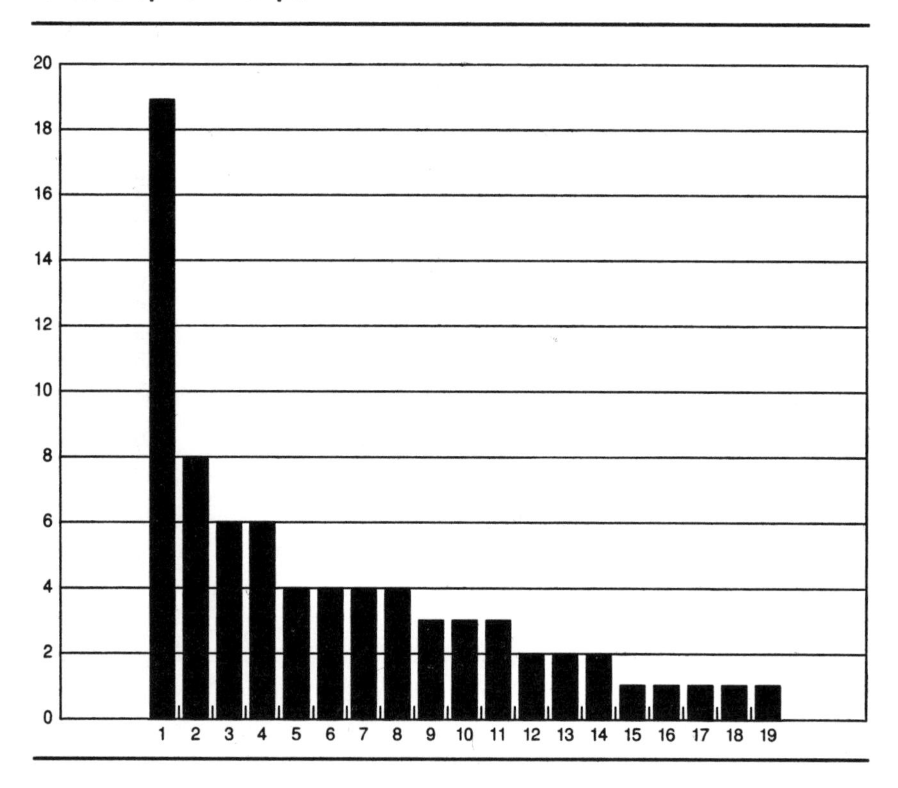

BIBLIOGRAPHY

Statistical Measures and Sampling

There are dozens of excellent works in basic statistics and in statistics applied to quality control. Look for a work that is written for the reader, and not a professor. A good place to start is with either Feigenbaum; Duncan; Grant and Leavenworth; or Gitlow, et al. If you have little background in statistics, Gitlow et al. is a good place to start.

Duncan, Acheson J. *Quality Control and Industrial Statistics*. 5th ed. Milwaukee, Wis.: ASQC Press, 1986.

Feigenbaum, Armand V. *Total Quality Control*. New York: McGraw-Hill, 1983.

Gitlow, Howard; Shelly Gitlow; Alan Oppenheim; and Rosa Oppenheim. *Tools and Methods for the Improvement of Quality*. Homewood, Ill.: Richard D. Irwin, 1989.

Grant, Eugene L., and Richard S. Leavenworth. *Statistical Quality Control*. 6th ed., New York: McGraw-Hill, 1988.

FIGURE 20–23
Plot of Delta in Example

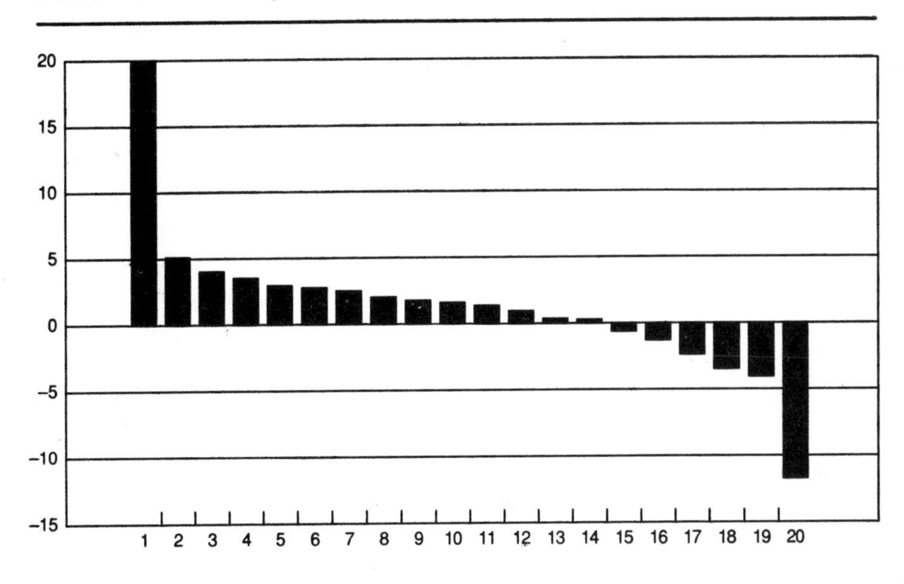

Control Charts

ASQC Automotive Division. *Statistical Process Control Manual*. Milwaukee, Wis.: ASQC Press, 1986.

AT&T Technologies. *Statistical Quality Control Handbook*, 1986.

Berger, Roger W., and Thomas H. Hart. *Statistical Process Control*. Milwaukee, Wis.: ASQC Press, 1986.

Pyzdek, Thomas. *What Every Engineer Should Know about Quality Control*. Milwaukee, Wis.: ASQC Press, 1989.

Ryan, Thomas P. *Statistical Methods for Quality Improvement*. New York: John Wiley & Sons, 1989.

Shettel-Neuber, J., and J. P. Sheposh. *Management Methods for Quality Improvement Based on Statistical Process Analysis and Control: A Literature and Field Survey*. (NPRDC Tech Note 86-21). Houston, Tex.: San Diego Navy Personnel Research and Development Center.

Shewhart, Walter A. *Economic Control of Quality of Manufactured Product*. Princeton, N.J. Van Nostrand Reinhold, 1931.

Shewhart, Walter A. *Economic Control of Manufactured Product*. (Reprint). Milwaukee, Wis.: ASQC Press, 1980.

Sorensen, S.; S. L. Dockstader; and M. J. Molof. *Developing a Statistical Process Control for Supply Operations*. (NPRDC Tech Rep 86-16). San Diego Navy Personnel Research & Development Center, 1986.

Wadsworth, Stephens, and Blan Godfrey. *Modern Methods for Quality Control and Improvement*. New York: John Wiley & Sons, 1986.

Western Electric Company. *Statistical Quality Control Handbook*. Milwaukee, Wis.: ASQC Press, 1982

Design of Experiments

Anderson, Virgil L., and Robert A. McLean. *Design of Experiments—A Practical Approach*. New York: Marcel Dekker, 1974.

Cox, D. R. *Planning of Experiments*. New York: John Wiley & Sons, 1958.

Davies, O. L., ed. *Design and Analysis of Industrial Experiments*. 2nd ed. rev. 1, New York: Hafner Publishing Company, 1956.

Hicks, Charles R. *Fundamental Concepts in the Design of Experiments*. 3rd ed. New York: Holt, Rinehart & Winston, 1982.

Menhenhall, W. *Introduction to Linear Models and the Design and Analysis of Experiments*. Belmont, Calif.: Wadsworth, 1968.

Evolutionary Operation

Box, George E. P. "Evolutionary Operation: A Method for Increasing Industrial Productivity." *Applied Statistics*, June 1957.

Box, George E. P., and Norman R. Draper. *Evolutionary Operation*. Milwaukee, Wis.: ASQC Press, 1969.

Pareto Analysis

Darling, Alan. "It Pays to Heed Pareto's Law." *Dun & Bradstreet Reports* 21, no. 1 (January/February 1983), pp. 18–21.

Goddard, Robert W. "Viewpoint: The Vital Few and the Trivial Many. *Personnel Journal*, July 1987.

Juran, J. M. *Quality Control Handbook*. 4th ed., New York: McGraw-Hill, 1988.

Scharf, Alan D. "Pareto's Law." *Industrial Business Management*. Saskatchewan Research Council-Industrial Services, May 1973.

Scharf, Alan D. "How to Use Pareto's Law. *Industrial Business Management*. Saskatchewan Research Council-Industrial Services, July 1973.

CHAPTER 21

SPECIALIZED TECHNIQUES

The techniques in this section are not applicable to all businesses, as are virtually all of the tools and techniques in the preceding chapters. Many are quite advanced, and require highly specialized engineering expertise to implement them. They are included here to acquaint the reader with techniques that may be applicable to their business. Rather than give extensive advice on how to use the tools, only a paragraph or two will be provided. Some techniques are too commodity oriented to be a general-purpose tool, and some are too advanced cognitively to be implemented without expert advice.

CONCURRENT ENGINEERING

Concurrent engineering is not so much a technique but an engineer's approach to Quality Management. Concurrent engineering attempts to link and integrate maintainability, reliability, safety, logistics, and manufacturing into the design phase through automated tools and humanistic management techniques, such as multifunction teams. This is often visually depicted as breaking down the "walls" of various engineering departments, and eliminating the old-fashioned "over-the-transom" engineering where the maintenance engineer examined a design and then passed it on to the reliability engineer, who passed it on to. . . . If you're having a hard time convincing an engineering division that Quality Management is for them, also, then give them a copy of Winner, et al.'s work on concurrent engineering. It's full of examples of incredible successes—not 5 percent per year "improvements" but 60 to 300 percent improvements—the sort of improvements that can put you out of business if you don't pay attention.

ENVIRONMENTAL STRESS SCREENING (ESS)

Sometimes facetiously referred to as *Shake and Bake,* ESS is a technique that employs thermal and/or vibrational cycling to precipitate latent defects in a product. This is somewhat like accelerated aging, but not really. The difference is that the goal is to advance the item under scrutiny into the reliability plateau—passing over the infant mortality phase prior to shipping the product. This technique has been used to great advantage in industries where infant mortalities in a product are common, such as electronic components. The objective is to stress the product close to its limit, but not to exceed this limit. The product should have a full life after and ESS; it should not be prematurely aged. Developing a screen is difficult at best, and often expensive. But once made, screening can often be incorporated into the production line.

PROTOTYPING

Prototypes can be simple cardboard items or they can be three-dimensional models on the computer. Generally it has come to mean *rapid prototyping* involving sophisticated software that provides realistic-looking computer simulations of the product or process. It has had great utility in establishing maintenance process criteria during the design phase. Computer simulations have also come a long way in the past 10 years. It is possible to model extraordinary things, such as global climate, or the human interaction in a maintenance process—thus establishing procedures and modifying design due to maintenance before production.

SNEAK CIRCUIT ANALYSIS

What is it when there is something that is not a fault, but it wasn't in the specifications either? This is a sneak—an undocumented feature. These sneaky paths in a system can unleash surprise actions at the absolutely wrong time. One sneaky circuit was responsible for unexpectedly opening the bomb bay door on a fighter aircraft. Another was responsible for a rocket rising a few feet off the ground and then abruptly stopping, and falling back into place. Another common sneak circuit was one in a 1960 automobile whereby turning off the ignition, but depressing the brake pedal, and pulling out the hazard lights turned the radio on. Sneak circuit analysis, as the name indicates, applies mainly to electronic circuits and software. It is a laborious process requiring extensive expertise on the subject. Boeing was the originator of the methods, and there are still only a few companies capable of doing a sneak circuit analysis.

JUST IN TIME

Just in time is more than just an inventory system. It is a philosophy that guides operations of an organization. The goal is to continuously improve to maximize efficiency and to eliminate waste in any form, in all parts of an organization, its suppliers, and its customers. In this context, waste is defined as any activity which does not add value to the product or service. The application of JIT requires full commitment from management above and employee involvement at all levels, particularly the line worker.

Applications of JIT are illustrated thusly:

Simplified Production or Work Processes

• Use general-purpose machines that can easily be switched from making one part to another.

• Apply group technology or cells where families of parts are produced together which can lead to reduced setup times.

• Use a U-shaped layout where material flows in on one side of the U, and comes out as finished parts on the other side of the U. This reduces material movement.

• Use blanket purchase orders which authorize a vendor to supply a certain quantity over a period of time. This reduces the individual orders processed, saving processing times, and work loads. This procedure cuts operating expenses.

Reduced Inventory Levels

• Use a "pull" system to make the materials flow as demanded by the production/work requirements. This results in less capital being tied up in inventory. The JIT system uses a *pull* system rather than a scheduled *push* system with requirements based on the units of production. Without the cushion of inventory in stock or in process, problems are immediately apparent when work centers shut down.

Improved Quality Control

• Insist on quality items from vendors or sources. With parts available only as needed, a defective part which is rejected is immediately known because of its impact on the production flow.

- Adopt a total quality control system that starts with the quality at the source for vendor-supplied items, stresses quality on the line and by the line for manufactured items, quality by the service worker as the service is performed.

Improved Quality and Reliability

- Design quality and producibility into the product. Use techniques of value engineering, design for manufacturing, and design for assembly. The goal is to eliminate all defects and scrap; therefore customers get higher quality items and fewer warranty actions.

Product Flexibility

- Less work in process gives capability to respond quickly to changes in customer demands for different items.

Volume Flexibility

- Use of pull system gives a capability to respond to surges or drops in customer demand.

Delivery Dependability

- Use of pull system and total quality control gives better response to customers in terms of delivery time, and delivery of quality products and services.

Asset Utilization

- Reduce capital investment. With reduced inventories and more efficient routing, less capital assets are required for existing processes. This provides opportunity for reduced operating expenses within existing facilities or room for expansion of the business into new ventures.

People Utilization

- Cross train employees to work in cells of production. Involved workers close to the process contribute to the continuous improvement. Workers have more sense of ownership in the product or service.

Cost Minimization

- Reduced inventory.
- Reduced material waste and lost labor due to defects.
- Balanced work load means less premium pay costs.

- Downtime costs reduced by improved and preventative maintenance.
- Simplified administrative processes reduces overhead work.

BIBLIOGRAPHY

Brazier, David, and Mike Leonard. "Concurrent Engineering: Participating in Better Designs." *Mechanical Engineering,* January 1990, pp. 52–53.

Davis, Ruth, et al. *Industrial Insights on the DoD Concurrent Engineering Program.* Roslyn, Va.: The Pymatuning Group, October 1988.

Gottesman, Ken. "JIT Manufacturing Is More than Inventory Programs and Delivery Schedules." *Industrial Engineering,* May 1991.

Hutchins, D. "Having a Hard Time with Just-in-Time." *Fortune,* June 9, 1986, pp. 64–66.

Lu, David J. *Kanban and Just-in-Time at Toyota.* Cambridge, Mass.: Productivity Press, 1985.

Miller, Jeff. Sneak circuit analysis for the common man, Rome Air Development Center: Griffiss Air Force Base, N.Y., October 1989.

Schonberger, Richard J., and James P. Gilbert. "Just-in-Time Purchasing: A Challenge for U.S. Industry." *California Management Review,* Berkeley, Calif., Fall 1983.

Vonderembse, Mark A., and Gregory P. White. *Operations Management.* St. Paul, Minn.: 1991.

Winner, Robert I.; James P. Pennell; Harold E. Bertrand; and Marko M. G. Slusarczuk. "The Role of Concurrent Engineering in Weapons System Acquisition." Alexandria, Va.: IDA, 1988.

PART 5

RESOURCES

THE ESSENTIAL TQM LIBRARY

Of the 1,000 references in the bibliography, which ones are must-buys for the TQM practitioner (aside from the volume you now hold in your hands)? The first list is a small but potent library capable of fitting in a (catalog size) briefcase. The second list is of an additional "Top 50" books that the authors have found to be of interest. Most of the works are available from bookstores, or through the ASQC Press (their address is listed in "Additional Resources"). A pleasant way to keep abreast of literature in this expanding field is to support the American Society for Quality Control as a corporate member. For a modest fee, ASQC Press will send you one copy of each new work.

THE BRIEFCASE LIBRARY

1. Deming, W. Edwards. *Out of the Crisis*. Cambridge, Mass.: Center for Advanced Engineering Study, 1986.

Deming's book is a thought-provoking work. This work will outlast the millennium as more and more companies turn from Taylor-type organizations (Theory X) to Deming organizations (an integration of Theory X and Y). Clearly expounds on Deming's 14 points.

2. Feigenbaum, Armand V. *Total Quality Control*. New York: McGraw-Hill, 1983.

Feigenbaum's tome is incredibly rich in detailed techniques. It emphasizes quality control over management techniques, but nonetheless is absolutely essential. It is an excellent place to learn more about control charts and other statistical tools in a corporate context. It is a monumental work (over 800 pages) in more than one respect. If suddenly put in charge of "Quality Control" without a lot of experience, this would be the work to read in secret every night, so one could emerge during the the day to become the amazing new quality guru.

3. Imai, Masaaki. *Kaizen: The Key to Japan's Competitive Success*. New York: Random House, 1986.

Imai's work proves to be fascinating reading on the Japanese methods for continuous improvement. If the reader prefers a different "Japanese management book" then substitute it for this selection, but be sure to peruse this one before passing it over. There are many small, compact gems to be mined in this text.

4. Juran, J. M. *Quality Control Handbook.* New York: McGraw-Hill, 1988.

Juran's work is a classic handbook on quality control. Be sure to get the most recent edition. If you bought only this work (or Feigenbaum's) you would have enough ideas and concepts to implement for the better part of a decade or more.

5. **Malcolm Baldrige National Quality Award Winners Case Studies.**

These are roadmaps of how real companies, working under real conditions, have accomplished the quality transformation. The material is generally available from the companies themselves.

6. Peters, Thomas J. *Thriving on Chaos.* New York: Alfred A. Knopf, 1988.

This is Peters's magnum opus, and is an absolutely, engaging work. He presents a scheme for "management revolution" which looks much like the principles of Deming, Juran, Crosby, and others rolled into one. He seems to pull together ideas from his previous works into a coherent methodology. Accused of being shrill in some circles, his enthusiasm is invigorating. For those of you who don't like Tom Peters, select *Quality or Else* by llyoy Dobyns as an alternate "anecdotal" classic.

7. Sun-Tzu, *The Art of War.*

A personal favorite of the authors, this work has carried many Asian warriors and leaders to success. Occasionally cryptic, always intriguing, and still applicable some 2,500 years after it was written, this work can pilot a transformation in one's self or organization. The translation by R. L. Wing is perhaps the most modern, putting the work in a strategy rather than war context (the Chinese pictogram for war can represent strategy as well). Other available translations are suitable as well; some emphasize modern warfare practices more than others.

A TQM LIBRARY

1. Aquilano, Nicholas and Richard Chase. *Fundamentals of Operations Management.* Homewood, Ill.: Richard D. Irwin, 1991.
2. Augustine, Norman. *Augustine's Laws.*: New York: Viking Penguin, 1986.
3. Bennis, Warren, and Burt Nanus. *Leaders: The Strategies for Taking Charge.* New York: Harper & Row, 1985.
4. Berry, Thomas H. *Managing the Total Quality Transformation.* New York: McGraw-Hill, 1990.
5. Crosby, Phillip B. *Quality Is Free.* New York: McGraw-Hill, 1979.

6. Defense Science Board. *Transition from Development to Production* (DoD 4245.7-M.), Department of Defense, 1985.
7. Defense Science Board. *Best Practices: How to Avoid Surprises in the World's Most Complicated Technical Environment* (NAVSOP-6071). Washington, D.C.: Department of the Navy, 1986.
8. Deming, W. Edwards. *Japanese Methods for Productivity and Quality.* Washington, D.C.: George Washington University, 1981.
9. Deming, W. Edwards. *Quality, Productivity and Competitive Position*. Cambridge, Mass.: Institute of Technology, Center for Advanced Engineering Study, 1982.
10. Dobyns, Lloyd, and Clare Crawford-Mason. *Quality or Else*. Boston: Houghton Mifflin, 1991.

11–12. Drucker, Peter. —Pick your two favorites.

13. Drucker, Peter. *New Realities*. Harper & Row: New York: Harper & Row, 1989.
14. Ernst and Young. *Total Quality: An Executive's Guide for the 1990's*. Homewood, Ill.: Richard D. Irwin, 1991.
15. Garvin, David A. *Managing Quality—The Strategic and Competitive Edge*. New York: Free Press, 1988.
16. Gitlow, Gitlow, Oppenheim, and Oppenheim. *Tools and Methods for the Improvement of Quality.* Homewood Ill.: Richard D. Irwin, 1990.
17. Goldratt, Elihu, and J. Cox. *The Goal: Excellence in Manufacturing*. Croton-on-Hudson, North River Press, N.Y.: 1984.
18. Grant, E. L., and Richard Leavenworth. *Statistical Quality Control.* New York: McGraw-Hill, 1988.
19. Harrington, H. James. *The Improvement Process*. New York: McGraw-Hill, 1987.
20. Hartley, Robert F. *Management Mistakes and Successes,* 3rd ed. New York: John Wiley & Sons, 1991.
21. Harvey, Jerry. *The Abilene Paradox*. Lexington, Mass.: Lexington Books, 1988.
22. Hatakeyomo, Yoshio. *Manager Revolution: A Guide to Survival in Today's Changing Workplace*. Cambridge, Mass.: Productivity Press, 1981.
23. Ishikawa, Kaoru. *What Is Total Quality Control?* Englewood Cliffs, N.J.: Prentice Hall, 1985.
24. Juran, J. M. *Juran on Planning for Quality.* New York: Free Press, 1988.

25. Juran, J. M. *Juran on Quality Improvement Workbook*. New York: Juran Enterprises, Inc., 1981.
26. Juran, J. M. *Managerial Breakthrough*. New York: McGraw-Hill, 1964.
27. Juran, J. M., and Frank M. Gryna, Jr. *Quality Planning and Analysis*. New York: McGraw-Hill, 1980.
28. Kanter, Rosabeth Moss. *The Change Masters*. New York: Simon & Schuster, 1983.
29. Karatsu, Hajime. *Tough Words for American Industry*. Cambridge, Mass.: Productivity Press, 1988.
30. Karatsu, Hajime. *TQC Wisdom of Japan*. Cambridge, Mass.: Productivity Press, 1988.
31. Kouzes, James, and Barry Posner. *The Leadership Challenge: How to Get Extraordinary Things Done in Organizations*. San Francisco: Jossey-Bass, 1987.
32. Mizuno, Sigeru. *Management for Quality Improvement: The Seven New Tools*. Cambridge, Mass.: Productivity Press, 1984.
33. Ozeki, Kazuo, and Tesuichi Asaka. *Handbook of Quality Tools*. Cambridge, Mass.: Productivity Press, 1990.
34. Packard Commission. *A Quest for Excellence*. Presidents' Blue Ribbon Commission on Defense Management. Washington, D.C.: U.S. Government Printing Office, 1986.
35. Peters, Thomas J., and Robert H. Waterman, Jr. *In Search of Excellence*. New York: Harper & Row, 1982.
36. Peters, Thomas J., and Nancy Austin. *A Passion for Excellence*. New York: Random House, 1985.
37. Scherkenbach, William. *The Deming Route to Quality and Productivity*. Milwaukee, Wis.: ASQC Quality Press, 1986.
38. Schonberger, Richard J. *World Class Manufacturing—The Lessons of Simplicity Applied*. New York: Free Press, 1986.
39. Shingo, Shigeo. *Sayings of Shigeo Shingo: Key Strategies for Plant Improvement*. Cambridge, Mass.: Productivity Press, 1988.
40. Shingo, Shigeo. *Study of Toyota Production System from an Industrial Engineering Viewpoint*. 1981 Japan Management Association, Tokyo, Japan: Productivity Press, 1987.
41. Shingo, Shigeo. *Zero Quality Control*. Cambridge, Mass.: Productivity Press, 1985.
42. Smith, Martin. *Maxims of Management*. Piscatway, N.J.: New Century, 1986.

43. Stahl, Michael J., and Gregory M. Bounds. *Competing Globally through Customer Value*. New York: Quorum, 1991.
44. Stogdill, Ralph. *Handbook of Leadership: A Survey of Theory and Research*. New York: Free Press, 1974.
45. Taguchi, Genichi. *Introduction to Quality Engineering*. Dearborn, Mich.: American Supplier Institute Inc., 1986.
46. Tribus, M. *Quality First: Selected Papers on Quality and Productivity*. Washington, D.C.: Improvement, National Society of Professional Engineers, March 1988.
47. Tribus, M. *Reducing Deming's 14 Points to Practice*. Cambridge, Mass.: Center for Advanced Engineering Study, MIT, 1984.
48. Western Electric. *Statistical Quality Control Handbook*. Easton, Pa.: Mack Publishing, 1977.
49. Winner, Robert I.; James P. Pennell; Harold E. Bertrand; and Marko Slusarczuk. *The Role of Concurrent Engineering in Weapon Systems Acquisition*. Alexandria, Va.: Institute of Defense Analysis, 1988.

ANNOTATED JOURNALS LIST

IEEE Engineering Management Review

Contains reprints of articles from many of the journals here, and provides a shortcut to subscribing to too many journals. The emphasis tends to be on engineering processes, but quality is given a good hearing. IEEE, P.O. Box 800, Somerset, N.J. 08875

IEEE Transactions on Engineering Management

Original contributions from the same society as above. Especially useful to those involved in manufacturing and research and development types. IEEE, P.O. Box 800, Somerset, N.J. 08875

IIE Transactions

Scholarly, but suprisingly readable. Almost every issue has one or more articles related to quality control or quality management. Institute of Industrial Engineers, 25 Technology Park/Atlanta, Norcross, Ga. 30092

Industrial Engineering

A trade magazine often containing short publicity articles on quality in manufacturing. Institute of Industrial Engineers, 25 Technology Park/Atlanta, Norcross, Ga. 30092

Journal of Management

Scholarly, but readable, and containing at least one or more articles per issue on quality management themes. The articles are generally practice oriented, and have many references. Inexpensive. Texas A & M University, College of Business Administration, College Station, Tex. 77843

Journal of Quality Technology

Essential for the practicing statistician in quality control and experimentation. American Society for Quality Control, 310 W Wisconsin Ave, Milwaukee, Wis. 53203

Management Review

A trade publication of the American Management Association. Easy to read. American Management Association, 135 W 50th St., New York, N.Y. 10020

National Productivity Review

A vital journal containing numerous articles on TQM. Expensive but worth it. Executive Enterprises, 22 W 21st St., New York, N.Y. 10010

Proceedings of the Annual Quality Congress

Presents papers from the annual symposium. The quality is spotty, but there are always numerous articles of interest. Can be difficult to obtain, unless you attend. American Society for Quality Control, 310 W Wisconsin Ave, Milwaukee, Wis. 53203

Proceedings of the Reliability Availability Maintainability Symposium

Presents papers from the annual symposium. Some are quite specialized, but there is always something interesting. A bargain as well—it comes with membership in the ASQC/IEEE reliability division. IEEE, PO Box 800, Somerset, N.J. 08875

Quality

A trade publication on quality. Often has case studies and lessons learned. Hitchcock Publishing Co., 25 W 550 Geneva Rd., Wheaton, Ill. 60188

Quality Assurance

A European source (in English) of quality assurance and control. Institute of Quality Assurance, 54 Princes Gate Exhibition Rd, London SW7 2PG, United Kingdom

Quality Control and Applied Statistics

This journal contains "book reports" of selected journal articles from scholarly and other journals (in fact every journal on this list) and can be a one stop for information on quality management and statistical quality control. Executive Sciences Institute, 1005 Mississippi Avenue, Davenport, Ia. 52803

Quality Digest

A useful compendium of reprinted and original articles on quality management. 1425 Vista Way, Red Bluff, Calif. 96080

Quality Engineering

A recent journal on quality engineering, and has provided readable case studies. Contains considerable information on standards. American Society for Quality Control, 310 W. Wisconsin Ave., Milwaukee, Wis. 53203

Quality and Reliability Engineering International

If you are an engineer, this is an essential journal. Unusual for an academic journal in that it provides news notes and a course schedule. Some articles contain many equations, but they usually do not get bogged down in proofs. John Wiley & Sons, 200 Meacham Ave., Elmont, N.Y. 11003 / Baffins Lane, Chichester West Sussex PO19 1UD, United Kingdom

Quality Progress

An essential magazine on quality control and quality management. American Society for Quality Control, 310 W. Wisconsin Ave., Milwaukee, Wis. 53203

Sloan Management Review

Fresh, enjoyable reading on the latest management concerns. Sloan School of Management, MIT, 50 Memorial Dr., Cambridge, Mass. 02139

Technometrics

For statisticians, a vital journal. American Statistical Association, 1429 Duke St., Alexandria, Va. 22314

Total Quality Management

Presents scholarly articles on the subject. Not quite as practical as other journals. Carfax Publishing Co., 35 South St., Hopkinton, Mass. 01748

ADDITIONAL RESOURCES

PRIMARY SOCIETIES

The following societies have as their major emphasis either quality and/or management, or have major divisions which do.

American Management Association
135 West 50th St.
New York, NY 10020

American Society for Quality Control
310 West Wisconsin Ave.
Milwaukee, WI 53203
1-800-248-1946, (414) 272-8575,
ASQC Quality Press, same address as the society.

European Organization for Quality Control
PO Box 2613, CH 3001
Bern, Switzerland

Institute of Industrial Engineers
25 Technology Park/Atlanta
Norcross, GA 30092
(404) 449-0460

OTHER ADDRESSES

The following include university research centers, secondary Quality Management related societies, and other useful addresses.

American Center for the Quality of Work Life
1411 K Street NW
Washington, DC 20016
(202) 338-2933

American Production and Inventory Control Society
500 West Annandale Rd.
Falls Church, VA 22046
(703) 237-8344

American Productivity and Quality Center
123 North Post Oak Lane
Houston, TX 77024
(713) 681-4020

American Productivity Management Asssociation
500 Frontage Road, Suite 395
Northfield, IL 60096
(708) 501-2650

American Society for Training and Development
1630 Duke St.
Alexandria, VA 22313
(703) 683-8100

American Statistical Association
1429 Duke St.
Alexandria, VA 22314

Association for Quality and Participation
801-B West 8th St., Suite 501
Cincinnati, OH 45023
(513) 381-1959

Center for Manufacturing Productivity and Technology Transfer, Rensselaer Polytechnic Institute
Troy, NY 12181
(518) 270-6724

Center for Quality and Productivity
University of Maryland
College Park, MD 20742
(301) 403-4535

Executive Sciences Institute
1005 Mississippi Avenue
Davenport, IA 52803
(319) 324-4463

Georgia Productivity Center
Georgia Institute of Technology
Atlanta, GA 30332
(404) 894-3404

Institute of Management Sciences
290 Westminster St.
Providence, RI 02903

Institute of Quality Assurance
London
United Kingdom

JUSE—Japanese Union of Scientists and Engineers
5-10-11 Sendagaya Shuyaku
Tokyo, Japan
FAX: 03-356-1798

Management and Behavioral Science Center, Wharton School, University of Pennsylvania
Vance Hall
3788 Spruce Street
Philadelphia, PA 19104
(215) 898-5736

Manufacturing Productivity Center
Illinois Institute of Technology
10 West 35th Street
Chicago, IL 60616
(312) 567-4800

Operations Management Association
Naman & Schneider Associates Group
6801 Sang Suite 255
Waco, TX 76710

Purdue Productivity Center
School of Industrial Engineering
CIDMAC Program
Purdue University
Grissom Hall
West Lafayette, IN 47907
(317) 494-5441

Society of Manufacturing Engineers
One SME Drive
P.O. Box 930
Dearborn, MI 48128
(313) 271-1500

University of Wisconsin—Madison
Center for Quality and Productivity Improvement
620 Walnut St.
Madison, WI 53705
(608) 263-2520

U.S. Office of Personnel Management
Office of Productivity Programs
P.O. Box 14080
Washington, DC 20044
(202) 632-5684

Work in America Institute
700 White Plains Road
Scarsdale, NY 10583
(914) 472-9600

BIBLIOGRAPHY

This is an alphabetical list of books, articles, and reports relating to Quality Management. It is by no means exhaustive, even though there are about 1,000 items listed. Entries are listed in alphabetical order by the primary author's last name. Subject listings can be found at the end of chapters.

Abbott, L. *Quality and Competition*. New York: Columbia University Press, 1955.

Abegglen, James C. *Management and Worker: Japanese Solution*. New York: Kodansha International, 1973.

Abegglen, James C., and George Stalk, Jr. *Kaisha: The Japanese Corporation*. New York: Basic Books, 1985, pp. 9–10.

Ackerman, Roger B.; Roberta Coleman; Elias Leger; and John C. MacDorman. *Process Quality Management and Improvement Guidelines*. AT&T Bell Laboratories, 1987.

Ackoff, Russell. *Redesigning the Future: A Systems Approach to Societal Problems*. New York: John Wiley & Sons, 1974.

Ackoff, Russell. "The Management of Change and the Change It Requires of Management." *Systems Practice*, 3, no. 5 pp. 427–40.

Ackoff, Russell, and F. Emery. *On Purposeful Systems*. Chicago: Aldine/Atherton: 1972.

Adam, Everett E., Jr. "Quality Circle Performance." *Journal of Management* 17, no. 1 (1991), pp. 25–39.

Adams, Walter, and James W. Brock. *The Bigness Complex: Industry, Labor and Government in the American Economy*. New York: Pantheon Books, 1986.

Aguayo, Rafael. *Dr. Deming: The American Who Taught the Japanese about Quality*. New York: Fireside, 1991.

Akao, Y. *Quality Deployment: A Series of Articles April 1986–March 1987*. Lawrence, Mass.: Growth Opportunity Alliance of Greater Lawrence, 1987.

Albert, K. J. *The Strategic Management Handbook*. New York: McGraw-Hill, 1983.

Aly, Nael; Venetta J. Maytubby; and Ahmad K. Elshennawy. "Total Quality Management: An Approach and a Case Study." *Computers and Industrial Engineering* 19, nos. 1–4 (1990), pp. 111–16.

American Productivity Center. *Allen-Bradley: First Line Supervisors Play Pivotal Role in Employee Communication Program Aimed at Boosting Productivity*. Case Study No. 49. Houston, Tex.: American Productivity Center, 1985.

Amsden, Robert T.; Howard E. Butler; and Davida M. Amsden. *SPC Simplified: Practical Steps to Quality*. White Plains, N.Y.: Unipub, 1989.

"An American Miracle that Works." *Productivity* 3, no. 11 (1982), pp. 1–5.

Anderson, Virgil L., and Robert A. McLean. *Design of Experiments, A Practical Approach*. New York: Marcel Dekker, 1974.

Ansari, A., and B. Modarress. "JIT Purchasing as a Quality and Productivity Center." *International Journal of Production Research* 26, no. 1 (1980), pp. 19–26.

Ansoff, H. Igor. "Critique of Henry Mintzberg's 'The Design School: Reconsidering the Basic Premises of Strategic Management.' " *Strategic Management Journal* 12 (1991), pp. 449–61.

Aquilano, N. J., and R. B. Chase. *Fundamentals of Operations Management*. Homewood, Ill.: Richard D. Irwin, 1991.

Aquino, Michael A. "Improvement versus Compliance: A New Look at Auditing." *Quality Progress,* October 1990, pp. 47–49.

Arter, Dennis R. *Quality Audits for Improved Performance*. Milwaukee, Wis.: ASQC Press, 1989.

ASD(A&L). *Transition from Development to Production*. DoD 4245.7-M, Department of Defense, Government Printing Office, Washington, D.C.: September 1985.

ASQC Automotive Division. *Statistical Process Control Manual*. Milwaukee, Wis.: ASQC Press, 1986.

ASQC Chemical and Process Industries Division. *Experiments in Industry: Design, Analysis, and Interpretation of Results*. Ed. Donald D. Snee, Lynne B. Hare, and J. Richard Trout. Milwaukee, Wis.: ASQC Press, 1985.

ASQC Chemical and Process Industries Division. Chemical Interest Committee. *Quality Assurance for the Chemical and Process Industries—A Manual of Good Practices*. Milwaukee, Wis.: ASQC Press, 1987.

ASQC Construction Technical Committee. *Quality Management for the Constructed Project*. Milwaukee, Wis.: ASQC Press, 1987.

ASQC Customer-Supplier Technical Committee. *Procurement Quality Control*. Ed. James L. Bossert. 4th ed. Milwaukee, Wis.: ASQC Press, 1988.

ASQC Energy Division. *Matrix of Nuclear Quality Assurance Program Requirements*. 3rd ed. Milwaukee, Wis.: ASQC Press, 1982.

ASQC Energy Division. *Nuclear Quality Systems Auditor Training Handbook*. 2nd ed. Milwaukee, Wis.: ASQC Press, 1986.

ASQC Food, Drug, and Cosmetic Division. *Food Processing Industry Quality System Guidelines*. ASQC Press, 1986.

ASQC Human Resources Division. *Human Resources Management*. Ed. Jill P. Kern; Jon J. Riley; and Louis N. Jones. Milwaukee, Wis.: ASQC Press, 1987.

ASQC Product Safety and Liability Prevention Technical Committee. *Product Recall Planning Guide*. Milwaukee, Wis.: ASQC Press, 1981.

ASQC Quality Audit Technical Committee. *How to Plan an Audit*. Ed. Charles B. Robinson. Milwaukee, Wis.: ASQC Press, 1987.

ASQC Quality Costs Committee. *The Management of Quality: Preparing for a Competitive Future*. Ed. John T. Hagan. Milwaukee, Wis.: ASQC Press, 1984.

ASQC Quality Costs Committee. *Guide for Reducing Quality Costs*. Milwaukee, Wis.: ASQC Press, 1987.

ASQC Quality Costs Committee. *Quality Costs: Ideas and Applications*. vol. 1, 2nd ed. Ed. Andrew F. Grimm. Milwaukee, Wis.: ASQC Press, 1987.

ASQC Quality Costs Committee. *Quality Costs: Ideas and Applications*. vol. 2. Ed. Jack Campanella. Milwaukee, Wis.: ASQC Press, 1989.

ASQC Quality Costs Committee. *Principles of Quality Costs: Principles, Implementation, and Use*. 2nd. ed. Ed. Jack Campanella. Milwaukee, Wis.: ASQC Press, 1990.

ASQC Statistics Division. *Glossary and Tables for Statistical Quality Control*. 2nd. ed. Milwaukee, Wis.: ASQC Press, 1983.

AT&T Technologies. *Statistical Quality Control Handbook*. Redbank, N.J.: AT&T, 1986.

Aubrey, Charles A. II. *Quality Management in Financial Services*. Milwaukee, Wis.: ASQC Press, 1985.

Aubrey, Charles A. II, and Patricia Felkins. *Teamwork: Involving People in Quality and Productivity Improvement*. White Plains, N.Y.: Unipub, 1988.

Augustine, Norman. *Augustine's Laws*. New York: Viking Penguin, 1986, p. 104.

"Automation Adds to Rolls Royce's Careful Work Study: Case Study." *Automated Factory* 50 (1986), pp. 9–11.

Axline, Larry L. "TQM: A Look in the Mirror." *Management Review* 80 (July 1991), pp. 64–71.

Badiru, Adedeji B. "A Systems Approach to Total Quality Management." *Industrial Engineering* 22, no. 3 (March 1990), pp. 33–36.

Baillie, Allan S. "Subcontracting Based on Integrated Standards: The Japanese Approach." *Journal of Purchasing and Materials Management* (Spring 1986), pp. 17–26.

Baillie, Allan S. "The Deming Approach: Being Better than the Best." *Advanced Management Journal* 51 (Autumn 1986), pp. 15–19.

Baird, John E. Jr. *Quality Circles Leader's Manual*. Milwaukee, Wis.: ASQC Press, 1982.

Baird, John E. Jr. *Quality Circles Participant's Manual*. Milwaukee, Wis.: ASQC Press, 1982.

Baird, John E. Jr., and David J. Rittof. *Quality Circles Facilitator's Manual*. Milwaukee, Wis.: ASQC Press, 1983.

Baird, Lloyd S.; Richard W. Beatty; and Craig E. Schneier. "What Performance Management Can Do for TQI. *Quality Progress* 21, no. 3 (1988), pp. 28–32.

Banks, Jerry. *Principles of Quality Control*. Milwaukee, Wis.: ASQC Press, 1989.

Bannon, L. "GE Tries New Management System at Bromont Plant." *American Metal Market/Metalworking News*, October 7, 1985, pp. 14–16.

Barker, Thomas B. *Quality by Experimental Design*. New York: Marcel Dekker, 1985.

Barr, Vilma. "Six Steps to Smoother Product Design." *Mechanical Engineering*, January 1990, pp. 48–51.

Barry, Thomas J. *Quality Circles: Proceed with Caution*. Milwaukee, Wis.: ASQC Press, 1988.

Barry, Thomas J. *Management Excellence through Quality*. Milwaukee, Wis.: ASQC Press, 1991.

Baumgarten, S., and J. S. Hensel. "Add Value to Your Service." Ed. C. Surprenant. Chicago: American Marketing Association, 1987, pp. 105–10.

Beckham, J. Daniel. "The Power of Owning a High-Quality Market Position Can Be Overwhelming." *Healthcare Forum*, March/April 1987, pp. 13–14.

Beer, Michael C. "Corporate Change and Quality." *Quality Progress*, February 1988.

Belohav, Jane. *Championship Management*. Cambridge, Mass.: Productivity Press, 1990.

Bemowski, Karen. "The Benchmarking Bandwagon." *Quality Progress* 24, no. 1 (January 1991), pp. 19–24.

Bennis, Warren. "The Coming Death of Bureaucracy." *Think* 32 (November/December 1966), pp. 30–35.

Bennis, Warren. *The Temporary Society*. New York: Harper & Row, 1968.

Bennis, Warren. *The Planning of Change*. New York: Holt, Rinehart, & Winston, 1976.

Bennis, Warren. *Why Leaders Can't Lead*. San Francisco: Jossey-Bass, 1989.

Bennis, Warren, and Burt Nanus. *Leaders: The Strategies for Taking Charge*. New York: Harper & Row, 1985.

Berger, Roger W., and Thomas H. Hart. *Statistical Process Control*. New York: Marcel Dekker, 1986.

Berger, Roger W.; David L. Shores; and Mary Thompson, eds. *Quality Circles*. New York: Marcel Dekker, 1986.

Berkman, Barbara N. "AMP Brings Quality to the Countryside." *Electronic Business*, October 16, 1989, pp. 249–50.

Berkman, Barbara N. "European Companies Join the Quality Crusade." *Electronic Business*, October 1989, pp. 263–66.

Berry, B. H. "Dana's Project 90 Aims at Manufacturing Excellence." *Iron Age*, March 1987, pp. 37–41.

Berry, Leonard L.; A. Parasuraman; and A. Zeithaml. "The Service-Quality Puzzle." *Business Horizons* 31, no. 5 (September/October 1988), pp. 35–43.

Berry, Thomas H. *Managing the Total Quality Transformation*. New York: McGraw-Hill, 1990.

Bertrand, Kate. "Marketers Discover What Quality Really Means. *Business Marketing*, April 1987.

Bessinger, R. C., and Waino Suojanen, eds. *Management and the Brain: An Integrative Approach to Organizational Behavior*. Atlanta: Georgia State University Business Publications, 1983.

Besterfield, Dale H. *Quality Control*. 3rd ed. Englewood Cliffs, N.J.: Prentice Hall, 1990.

Bharol, Chowdhry Ram. "Problems of R&D Manpower in India." *R&D Management* 19, no. 4 (October 1989), pp. 335–41.

Bhote, Keki R. "Motorola's Long March to the Malcolm Baldrige National Quality Award." *National Productivity Review* 8, no. 4 (Autumn 1989), pp. 365–76.

Bhote, Keri. "America's Quality Health Diagnosis: Strong Heart, Weak Head." *Management Review,* May 1989.

Birch, David. *Job Creation in America: How Our Smallest Companies Put the Most People to Work.* New York: Free Press, 1987, p. 3.

Bisgaard, Soren. *A Practical Aid for Experiments.* Madison, Wis.: University of Wisconsin, 1988.

Black, R. A. "Fact, Creativity, Teamwork, and Rules: Understanding Leadership Styles." *Industrial Management,* September–October 1990, pp. 17–20.

Black, Sam P. "A.Q.I. 2000." *Proceedings of IMPRO 89.* Wilton, Conn.: Juran Institute, 1989.

Blake, R. R., and J. S. Mouton. *The Managerial Grid.* Houston, Tex.: Gulf Publishing, 1964.

Blake, R. R., and J. S. Mouton. *Corporate Excellence through Grid Organization and Development.* Houston, Tex.: Gulf Publishing, 1968.

Blanchard, Kenneth, and Spencer Johnson. *The One Minute Manager.* Ed. Pat Golbitz. New York: William Morrow, 1982.

Bodensteiner, Wayne, and John W. Priest, "Designing Quality into Defense Systems." *Quality Progress* 20 (June 1987), pp. 93–96.

"Boeing Pays Now to Save Later." *BusinessWeek,* October 27, 1986, p. 50.

Bolivar, J. "Management, Workers Unite to Boost Productivity." *Production Engineering,* July 1986, pp. 14–16.

Booher, Diane, "Quality or Quantity Communication." *Quality Progress* 21, no. 6 (June 1988), pp. 65–68.

Boorstin, Daniel, and Gerald Parshall. "History's Hidden Turning Points: The True Watersheds in Human Affairs Are Seldom Spotted Quickly amid the Tumult of Headlines Broadcast on the Hour." *U.S. News & World Report* 110 (April 22, 1991), pp. 52–114.

"Boosting Productivity at American Express." *BusinessWeek,* October 5, 1981, pp. 62–66.

Bossert, James L., ed. Procurement Quality Control. 4th ed. Milwaukee, Wis.: ASQC Press, 1988.

Bossert, James L. ed. *QFD: A Practitioner's Approach.* Milwaukee, Wis.: ASQC Press, 1990.

Bothwell, L. *The Art of Leadership.* Englewood, Cliffs, N.J.: Prentice Hall, 1983.

Bottorff, Dana. "Japanese Mode of Managing Translates into Success for Management Research Group." *New England Business* 10 (April 16, 1988), pp. 49–54.

Bower Joseph L., and Thomas Hout. "Fast-Cycle Capability for Competitive Power." *Harvard Busines Review,* November/December 1988.

Box, George E. P. "Evolutionary Operation: A Method for Increasing Industrial Productivity." *Applied Statistics,* June 1957.

Box, George E. P. "Signal-to-Noise Ratios, Performance Criteria, and Transformations." *Technometrics* 30, no. 1 (February 1988), pp. 1–40.

Box, George E. P., and Norman R. Draper. *Evolutionary Operation*. New York: John Wiley, 1969.

Box, George E. P., and Norman R. Draper. *Empirical Model Building and Response Surfaces*. New York: John Wiley & Sons, 1987.

Box, George E. P.; William G. Hunter; and J. Stuart Hunter. *Statistics for Experimenters*. New York: John Wiley & Sons, 1978.

Bradford, D. L., and A. R. Cohen. *Managing for Excellence*. New York: John Wiley & Sons, 1984.

Braverman, Jerome D. *Fundamentals of Statistical Quality Control*. Milwaukee, Wis.: ASQC Press, 1981.

Brazier, David, and Mike Leonard. "Concurrent Engineering: Participating in Better Designs." *Mechanical Engineering,* January 1990, pp. 52–53.

Broh, R. A. *Managing Quality for Higher Profits*. New York: McGraw-Hill, 1982.

Brooks, George, and J. R. Linklater. "Statistical Thinking and W. Edwards Deming's Teachings in the Administrative Environment: Statistics and the Persistent Pursuit of Improvement Have Generated Many Efficiencies at Windsor Export Supply." *National Productivity Review* 5 (Summer 1986), pp. 271–310.

Brown, Donna. "10 ways to Boost Quality." *Management Review* 80 (January 1991), pp. 5–11.

Brown, J. H. U., and J. Comola. *Educating for Excellence: Improving Quality and Productivity in the 90s*. New York: Auburn House, 1991.

Brown, Larry A. "Controlling New Product Quality at Sun." *Manufacturing Systems* 8, no. 6 (June 1990), pp. 44–49.

Brown, Mark Graham. *Baldrige Award Winning Quality*. White Plains, N.Y.: Unipub, 1991.

Brush, Gary G. *Volume 12: How to Choose the Proper Sample Size*. Milwaukee, Wis. ASQC Press, 1988.

Bull, Steve. "Total Quality Management in a Redundancy Situation." *European Management Journal* 9, no. 3 (September 1991), pp. 288–94.

Burgess, John A. *Design Assurance for Engineers and Managers*. Milwaukee, Wis.: ASQC Press, 1984.

Burke, R. J. "Methods of Resolving Interpersonal Conflict." *Personnel Administration,* 1969, pp. 48–55.

Burr, Irving W. *Statistical Quality Control Methods*. New York: Marcel Dekker, 1976.

Burr, Irving W. *Elementary Statistical Quality Control*. New York: Marcel Dekker, 1979.

Burr, John T. *SPC Tools for Operators*. Milwaukee, Wis.: ASQC Press, 1989.

Burr, John T. "The Tools of Quality. Part I: Going with the Flow(chart)." *Quality Progress* 23, no. 6 (June 1990), pp. 64–67.

Burt, David N. "Managing Product Quality through Strategic Purchasing." *Sloan Management Review* 30, no. 3 (October 1989), pp. 39–48.

Butler, Arthur G., Jr. "Project Management: A Study in Organizational Conflict." *Academy of Management Journal* 16 (March 1973), pp. 84–101.

Butterfield, Ronald W. *Quality Service—Pure and Simple*. Milwaukee, Wis.: ASQC Press, 1990.

Butterfield, Ronald W. "Deming's 14 Points Applied to Service." *Training: The Magazine of Human Resources Development* 28 (March 1991), pp. 50–56.

Byham, William. *ZAPP: The Lightning of Empowerment*. New York: Crown Publishers, 1990.

Callahan, Joseph M. "Deming Involves Suppliers, Engineers (Part Three)." *Automotive Industries* 162 (January 1982), pp. 32–41.

Calvin, Thomas W. *Volume 7: How and When to Perform Bayesian Acceptance Sampling*. rev. ed., Milwaukee, Wis.: ASQC Press, 1990.

Camp, Robert. "Benchmarking: The Search for the Best Practices that Lead to Superior Performance. Parts 1–5." *Quality Progress,* January–May 1989.

"Campbell Soup Adopts Total Systems Approach." *Productivity Letter* 6, no. 5 (1986), pp. 1–2.

"Can a Company Spend Too Much on Quality?" *Electronic Business,* May 1, 1988.

Caplan, Frank. *The Quality System: A Sourcebook for Managers and Engineers*. 2nd ed. Radnor, Penn.: Chilton, 1990.

Carlsen, Robert D.; JoAnn Gerber; and James F. McHugh. *Manual of Quality Assurance Procedures and Forms*. Englewood Cliffs, N.J.: Prentice Hall, 1981.

Carr, David K., and Ian D. Littman. *Excellence in Government*. Milwaukee, Wis.: ASQC Press, 1990.

Carrubba, Eugene R., and Ronald D. Gordon. *Product Assurance Principles: Integrating Design Assurance and Quality Assurance*. New York: McGraw-Hill, 1988.

Carter, C. L. Jr. *Quality Assurance, Quality Control, and Inspection Handbook*. 4th ed. Milwaukee, Wis.: ASQC Press, 1984.

Carter, C. L., Jr. *The Control and Assurance of Quality, Reliability, and Safety*. Milwaukee, Wis.: ASQC Press, 1990.

Casalou, Robert F. "Total Quality Management in Health Care." *Hospital & Health Services Administration* 36 (Spring 1991), pp. 134–213.

Case, John. "Zero-Defect Management." *Inc*. 9 (February 1987), pp. 17–22.

Cassidy, Robert. "Corporation of the Year: A Perpetual Idea Machine." *Research & Development* 31, no. 11 (November 1989), pp. 54–64.

Castro, Janice. "Making It Better." *Time,* November 13, 1989, pp. 78–81.

Charbonneau, Harvey C., and Gordon L. Webster. *Industrial Quality Control*. Milwaukee, Wis.: ASQC Press, 1978.

Christiansen, Donald. *Engineering Excellence Cultural and Organizational Factors*. New York: IEEE Press, 1987.

Christopher, William. *Productivity Measurement Handbook*. Cambridge, Mass.: Productivity Press, 1985.

Ciampa, D. *Manufacturing's New Mandate: The Tools for Leadership*. New York: John Wiley & Sons, 1988.

Claussen, Ronald L. "Annual Quality Improvement: A Manufacturing Perspective." *Proceedings of IMPRO 88*. Wilton, Conn.: Juran Institute, Inc., 1988.

Clemens, John K., and Douglas F. Mayer. *The Classic Touch: Leadership Lessons from Homer to Hemingway*. Homewood, Ill.: Richard D. Irwin, 1987.

Clements, Richard Barrett. *Creating and Assuring Quality*. Milwaukee, Wis.: ASQC Press, 1990.

Cochran, William G. *Sampling Techniques*. 3rd ed. Milwaukee, Wis.: ASQC Press, 1977.

Cohen, Louis. "Quality Function Deployment: An Application Perspective from Digital Equipment Corp." *National Productivity Review*, Summer 1988.

Cole, R. E. *Work, Mobility, and Participation: A Comparative Study of American and Japanese Industry*. Berkeley: University of California Press, 1979.

Collins, Frank C., Jr. *Quality: The Ball in Your Court*. Milwaukee, Wis.: ASQC Press, 1987.

Commins, Kevin. "U.S. Companies Urged to Adopt Holistic Thinking." *Journal of Commerce and Commercial* 384 (June 25, 1990), pp. 4–11.

Cooper, Kenneth. *Aerobics*. New York: M. Evans, 1967.

Copp, Richard. *Report on Japanese Quality Engineering Using Designed Experiments—The Taguchi Method*. Dearborn, Mich.: American Supplier Institute, July 1984.

Coppola, Anthony. *RADC Guide to Basic Training in TQM Analysis Techniques*. Rome, N.Y.: RADC, 1989.

Cornell, John A. *Experiments with Mixtures: Designs, Models, and the Analysis of Mixture Data*. 2nd ed. New York: John Wiley, 1990.

Cornell, John A. *Volume 5: How to Run Mixture Experiments for Product Quality*. rev. ed. ASQC Press, 1990.

Cornell, John A. *Volume 8: How to Apply Response Surface Methodology*. rev. ed. Milwaukee, Wis.: ASQC Press, 1990.

Couger, J. D., and R. W. Knapp. *System Analysis Techniques*. New York: John Wiley & Sons, 1974.

Cound, Dana. *A Leader's Journey to Quality*. New York: Marcel Dekker, 1991.

Cousins, Norman. *Anatomy of an Illness*. New York: Bantam Books, 1981.

Cox, D. R. *Planning of Experiments*. New York: John Wiley & Sons, 1958.

Crosby, Phillip B. *Quality Is Free*. New York: McGraw-Hill, 1979.

Crosby, Philip B. *Quality without Tears*. New York: McGraw-Hill, 1984.

Crosby, Philip B. *Running Things—The Art of Making Things Happen*. New York: McGraw-Hill, 1986.

Crosby, Philip B. "Quality—Management's Choice." *Quality,* Anniversary Issue, 1987.

Crosby, Philip B. *The Eternally Successful Organization*. New York: McGraw-Hill, 1988.

Crosby, Philip B.; W. Edwards Deming; and J. M. Juran. "Three Preachers, One Religion." *Quality,* September 1986, pp. 22–25.

Cross, Lisa. "The Changing Workforce." *Graphic Arts Monthly,* June 1989, pp. 38–45.

Crossfield, R. T., and B. G. Dale. "Mapping Quality Assurance Systems: A Methodology." *Quality and Reliability Engineering International* 6, no. 3 (June–August 1990), pp. 167–78.

Crossfield, R. T., and B. G. Dale. "The Use of Expert Systems in Total Quality Management: An Exploratory Study." *Quality and Reliability Engineering International* 7 (1991), pp. 19–26.

"Customer Satisfaction Powers Everything Motorola Does." *National Productivity Report* 19, no. 6 (March 31, 1990), pp. 1–4.

Dale, B. G., and P. Shaw. "Failure Mode and Effects Analysis in the U.K. Motor Industry: A State-of-the-Art Study." *Quality and Reliability Engineering International,* 6, no. 3 (June–August 1990), pp. 179–88.

Dalkey, N. C. "The Delphi Method: An Experimental Study of Group Opinion." *Research Paper RM-5888-PR,* June 1969. The RAND Corporation, Santa Monica, California.

Daniels, Aubrey C. *Performance Management: Improving Quality Productivity through Positive Reinforcement*. 3rd ed. Tucker, Ga.: Performance Management Publications, 1990.

Danzer, H. H. "Audit—A Center of Future Quality Assurance." *EOQC Quality* 32, no. 1 (March 1988), pp. 13–17.

Darling, Alan. "It Pays to Heed Pareto's Law." *Dun & Bradstreet Reports* 21, no. 1 (January/February 1983), pp. 18–21.

David, A. J., et al. *Quality by Design: A Quality Manual for the AT&T R&D Community.* Holmdel, N.J.: AT&T Bell Laboratories Quality Assurance Center, 1987.

Davidow W., and B. Uttal. *Total Customer Service—The Ultimate Weapon*. New York: Harper & Row, 1989.

Davies, O. L., ed. *Design and Analysis of Industrial Experiments*. 2nd ed., rev. 1. New York: Hafner Publishing, 1956.

Davis, J. H. *Group Performance*. Reading, Mass.: Addison-Wesley Publishing, 1969.

Davis, Ruth, et al. *Industrial Insights on the DoD Concurrent Engineering Program.* Roslyn, Va.: The Pymatuning Group, October 1988.

Davis, Stanley M. *Managing Corporate Culture*. New York: Harper & Row, 1984.

de Vries, J., and H. van de Water. "Quality Circles and Quality of Working Life: Results of a Study in Seven Large Organizations." *EOQ Quality* 2, no. 2 (July 1990), pp. 4–10.

Deal, T. E., and A. A. Kennedy. *Corporate Cultures*. Reading, Mass.: Addison-Wesley Publishing, 1982.

DeBono, E. *Lateral Thinking for Management*. New York: McGraw-Hill, 1971.

DeBono, Edward. *Tactics: The Art and Science of Success*. Boston: Little, Brown, 1984.

Defense Science Board. *Best Practices*. (NAVSOP-6071), 1986.

Defense Science Board. *Transition from Development to Production*. (DoD 4245.7-M). Department of Defense, Government Printing Office, Washington, D.C.: 1985.

Dehnad, K., ed. *Quality Control, Robust Design, and the Taguchi Method*. Pacific Grove, Calif.: Brooks/Cole Publishing, 1988.

Delbecq, Andre L., et al. *Group Techniques for Program Planning: Guide to Nominal Group and Delphi Processes*. Cleveland, Ohio: Greenbriar Books, 1985.

Deming Library series of videotapes, Films, Inc., Chicago, Ill.

Deming, W. Edwards. *Japanese Methods for Productivity and Quality.* Washington, D.C.: George Washington University, 1981.

Deming, W. Edwards. *Quality, Productivity and Competitive Position*. Cambridge, Mass.: Center for Advanced Engineering Study, MIT Press, 1982.

Deming, W. Edwards. "Quality: Management's Commitment to Quality." *Business* 35 (January 1985), pp. 50–55.

Deming, W. Edwards. *Out of the Crisis*. Cambridge, Mass.: Center for Advanced Engineering Study, MIT Press, 1986.

Deming, W. Edwards. "New Principles of Leadership." *Modern Materials Handling* 42 (October 1987), pp. 37–41.

Deming, W. Edwards, "Out of the Crisis." *Journal of Organizational Behavior Management* 10 (Spring 1989), pp. 205–13.

Dempsey, W. A. "Vendor Selection and the Buying Process." *Industrial Marketing Management* 7 (1978), pp. 257–67.

Department of the Army. Memorandum, Implementation of Total Quality Management, August 1989.

Department of Defense. *Total Quality Management Guide. Volumes I and II*. DoD 5000.510G, Washington, D.C.: Department of Defense, 1990.

Desatnick, Robert L. *Managing to Keep the Customer.* Delran, N.J.: MacMillan, 1987.

Desatnick, Robert L. "Long Live the King." *Quality Progress,* April 1989.

DeToro, Irving J. *Doing It Right: Quality through Employee Involvement*. Milwaukee, Wis.: ASQC Press, 1990.

DeYoung, H. Garrett. "Back from the Brink: Xerox Redefines Its Notion of Quality." *Electronic Business,* October 16, 1989, pp. 50–54.

DeYoung, H. Garrett. "Preachings of Quality Gurus: Do It Right the First Time. *Electronic Business,* October 16, 1989, pp. 88–94.

Diamond, William J. *Practical Experiment Designs*. Milwaukee, Wis.: ASQC Press, 1981.

Diller, Wendy. "Ford's Better Idea for Cutting Supplier Costs." *Dun's Business Month* 121 (June 1983), pp. 75–81.

Dingus, Victor, and William Golomski, eds. *A Quality Revolution in Manufacturing*. Atlantic, Ga.: IIE Press, 1991.

DiPrimio, Anthony. *Quality Assurance in Service Organizations*. Milwaukee, Wis.: ASQC Press, 1987.

Dobyns, Lloyd. "Ed Deming Wants Big Changes, and He Wants Them Fast." *Smithsonian* 21 (August 1990), pp. 74–79.

Dobyns, Lloyd, and Clare Crawford-Mason. *Quality or Else*. Boston: Houghton Mifflin, 1991.

Docstader, S. L. "Managing TQM Implementations: A Matrix Approach." Unpublished manuscript. San Diego: Navy Personnel Research & Development Center, 1987.

Dodge, H. F. "A Sampling Inspection Plan for Continuous Production." *Annals of Mathematical Statistics* 14 (1942), pp. 264–79.

Dodge, H. F., and H. G. Romig. *Sampling Inspection Tables—Single and Double Sampling*, 2nd ed. New York: John Wiley & Sons, 1959.

Dodge, H. F., and M. N. Torrey. "Additional Continuous Sampling Plans." *Industrial Quality Control* 7 (1951), pp. 7–12.

Doering, Robert D. "An Approach toward Improving the Creative Output of Scientific Task Teams." *IEEE Transactions on Engineering Management* 20, no. 1 (February 1973), pp. 29–31.

Donis, Peter P. "The Genesis of Annual Quality Improvement at Caterpillar." *Proceedings of IMPRO 88*. Wilton, Conn.: Juran Institute, 1988.

Doyukai, Kigyo. *Minshuka Shian (Tentative View of Democratization of Business Enterprises)*. Tokyo: Doyukai, 1947.

Doyukai, Kigyo. *Nendai no Kigyo Keiei (Management for Eighties)*. Tokyo: Doyukai, 1980.

"Dr. Deming Shows Pontiac the Way." *Fortune* 107 (April 18, 1983), pp. 66–71.

Dreyfus, Joel. "Victories in the Quality Crusade." *Fortune*, October 10, 1988, pp. 80–84.

Drucker, Peter F. *The Changing World of the Executive*. New York: Times Books, 1985.

Drucker, Peter F. "The Coming of the New Organization." *Harvard Business Review*, January/February 1988, pp. 45–53.

Drucker, Peter F. *The Effective Executive*. New York: Harper & Row, 1967.

Drucker, Peter F. *Managing in Turbulent Times*. New York: Harper & Row, 1980.

Drucker, Peter. *The New Realities*. New York: Harper & Row, 1989.

Drucker, Peter F. *Technology, Management and Society*. New York: Harper & Row, 1970.

Dumaine, Brian. "A Humble Hero Drives Ford to the Top." *Fortune*, January 4, 1988.

Dumaine, Brian. "How Managers Can Succeed through Speed." *Fortune*, February 13, 1989.

Dumaine, Brian. "Who Needs a Boss?" *Fortune* (May 1990), p. 7.

Duncan, Acheson J. *Quality Control and Industrial Statistics.* 5th ed. Homewood, Ill.: Richard D. Irwin, 1986.

Duncan, W. Jack, and Joseph G. Van Matre. "The Gospel According to Deming: Is It Really New?" *Business Horizons* (July/August 1990), pp. 3–17.

Dutton, Barbara. "Switching to Quality Excellence." *Manufacturing Systems,* March 1990, pp. 51–53.

Dyer, Constance. *Canon Production System: Creative Involvement of the Total Workforce.* Cambridge, Mass.: Productivity Press, 1984.

Eastman Kodak Company. *Keeping the Customer Satisfied–A Guide to Field Service.* Milwaukee, Wis.: ASQC Press, 1989.

Ebrahimpour, Maling, and Sang M. Lee. "Quality Management Practices of American and Japanese Electronic Firms in the United States." *Production and Inventory Management Journal* 29, no. 4 (Fourth Quarter 1988), pp. 28–31.

Edosomwan, Johnson. "The Baldrige Award: Focus on Total Customer Satisfaction." *Industrial Engineering,* July 1991, pp. 24ff.

Edosomwan, Johnson A., and Arvind Ballakur, eds. *Productivity and Quality Improvement in Electronics Assembly.* New York: McGraw-Hill, 1991.

Edosomwan, Johnson, and Wanda Savage-Moore. "Assess Your Organization's TQM Posture and Readiness to Successfully Compete for the Malcolm Baldrige Award." *Industrial Engineering,* February 1991, pp. 22–24.

Edwards, C. D. "The Meaning of Quality." *Quality Progress,* October 1968, pp. 36–39.

Electronics System Division. *ESD Process Improvement Guide.* ESD: Hanscom AFB Mass., 1991.

Ernst & Young Quality Improvement Consulting Group. "Total Quality: An Executive's Guide for the 1990s." Homewood, Ill.: Richard D. Irwin, 1990.

Ertel, Danny. "How to Design a Conflict Management Procedure That Fits Your Dispute." *Sloan Management Review* 32, no. 4 (Summer 1991), pp. 29–42.

Ettlie, J. E. "Technology Transfer—From Innovators to Users." *Industrial Engineering,* June 1973, pp. 16–23.

Ettlie, J. E. *Taking Charge of Manufacturing.* San Francisco: Jossey-Bass, 1989.

Failure Mode Effects and Criticality Analysis, MIL-STD-1629. Washington, D.C.: U.S. Government Printing Office.

Fargo, Francis T. *Handbook of Dimensional Measurement.* 2nd ed. New York: Industrial Press, 1982.

Farrow, John. "Quality Audits: An Invitation to Managers." *Quality Progress,* January 1987.

Fechter, William F., and Renee B. Horowitz. "Visionary Leadership Needed by all Managers." Industrial Management 33, no. 4 (July/August 1991), pp. 2–5.

Feenstra, Robert C. "Quality Change under Trade Restraints in Japanese Autos." *The Quarterly Journal of Economics* 103, no. 1 (February 1988), pp. 131–46.

Feher, Bela, and Mark F. Levine. *Organization Redesign for Productivity Improvement: Method Case Study.* San Diego: Navy Personnel and Research Development Center, 1985.

Feigenbaum, Armand V. *Total Quality Control.* New York: McGraw-Hill, 1983.

Feigenbaum, Armand V. "Total Quality Leadership." *Quality* 25, no. 4 (April 1986), pp. 18–22.

Feigenbaum, Armand V. "Quality Is Universal." *Quality,* Anniversary Issue, 1987.

Feigenbaum, Armand V. "America on the Threshold of Quality." *Quality* 29, no. 1 (January 1990), pp. 16–18.

Feigenbaum, Ed; Pamela McCorduck; and Penny Nii. *The Rise of the Expert Company.* New York: Times Books, 1988.

Feldman, Raymond G. "Achieving World Class." *P&IM Review with APICS News,* April 1989, pp. 40–42.

Ferguson, Gary A. "Printer Incorporates Deming—Reduces Errors, Increases Productivity." *Industrial Engineering* 22 (August 1990), pp. 32–33.

Fine, Charles, and Evan L. Porteus. "Dynamic Process Improvement." *Operations Research* 37, no. 4 (July/August 1989), pp. 580–91.

Finkelman, Daniel. "If the Customer Has an Itch, Scratch It." *The New York Times,* May 14, 1989.

Finkelstein, Marvin S.; Edward J. Harrick; and Paul E. Sultan. "Sharing Information Spawns Trust, Productivity, and Quality." *National Productivity Review* 10 (Summer 1991), pp. 295–304.

Fisher, B. A. *Small Group Decision Making: Communication and the Group Process.* New York: McGraw-Hill, 1974.

Flaherty, Gerald S. "The Total Quality Team." *Proceedings of IMPRO 88.* Wilton, Conn.: Juran Institute, 1988.

Flood, Robert L., and Michael C. Jackson. *Creative Problem Solving: Total Systems Intervention.* New York: John Wiley & Sons, 1991.

Ford, Henry. *Today and Tomorrow.* Cambridge, Mass.: Productivity Press, 1988.

"Formulating a Quality Improvement Strategy." PIMSLETTER 31, p. 31.

Foster, Lowell W. *Modern Geometric Dimensioning and Tolerancing.* 2nd ed. Milwaukee, Wis.: ASQC Press, 1982.

Foster, Richard. *Innovation: The Attacker's Advantage.* New York: Summit Books, 1986, pp. 121–35.

Fox, Robert E. "MRP, Kanban or OPT—What's Best?" *Inventories and Production,* July/August 1982.

Fox, Robert E. "OPT: An Answer for America, Leapfrogging the Japanese." Part IV. *Inventories and Production* 3 (March/April 1983), p. 24.

Frankwicz, Michael. "A Study of Project Management Techniques." *Journal of Systems Management* 24 (October 1973), pp. 18–22.

Freeman, Nancy Brooke. "Harley Davidson's Face for Survival." *American Machinist,* January 1985, pp. 71–75.

Frehr, U. "The Rough Road from QC to TQC." *EOQC Quality* 31, no. 3 (September 1987), pp. 8–10.

Fried, Louis. "Don't Smother Your Project in People." *Management Advisor* 9, no. 3 (March 1972), pp. 46–49.

Friedman, M., and D. Ulmer. *Treating Type A Behavior and Your Heart.* New York: Alfred A. Knopf, 1984.

Fukuda, Ryuji. *Managerial Engineering: Techniques for Improving Quality, and Productivity in the Workplace.* Stamford, Conn.: Productivity Press, 1984.

Fuller, F. Timothy. "Eliminating Complexity from Work: Improving Productivity by Enhancing Quality." *National Productivity Review,* Autumn 1985.

Gabor, Andrea. "The Leading Light of Quality: An Innovative Florida Utility Borrows a Page from Japan, Inc." *U.S. News & World Report* 105 (November 28, 1988), pp. 53–63.

Gabor, Andrea. "The Man Who Changed the World of Quality." *International Management* 43 (March 1988), pp. 42–44.

Gabor, Andrea. *The Man Who Discovered Quality: The Management Genius of W. Edwards Deming.* Milwaukee, Wis.: ASQC Press, 1991.

Gabriele, M. C. "Sikorsky Offers Free Consultation Plan for Suppliers." *American Metal Market/Metalworking News,* June 2, 1986, p. 17.

Gale, Bradley T. "How Quality Drives Market Share." *The Quality Review,* Summer 1987.

Gale, Bradley T., and Robert D. Buzzell. "Market Perceived Quality: Key Strategic Concept, Planning Review." *International Society for Planning and Strategic Management,* March/April 1989.

Gale, Bradley T., and Robert D. Buzzell. "Market Perceived Quality: Key Strategic Concept." *Planning Review,* March/April 1989, pp. 6–16.

Garvin, David A. "Quality on the Line." *Harvard Business Review,* September/October 1983.

Garvin, David A. "What Does 'Product Quality' Really Mean?" *Sloan Management Review* 26, no. 2 (1985), pp. 25–43.

Garvin, David A. "Competing on the Eight Dimensions of Quality." *Harvard Business Review,* November/December 1987, pp. 101–8.

Garvin, David A. *Managing Quality: The Strategic and Competitive Edge.* New York: Free Press, 1987.

Garvin, David A. *Managing Quality Edge.* New York: Free Press, 1988.

Gause, Donald C., and Gerald M. Weinberg. *Exploring Requirements: Quality before Design.* New York: Dorset House, 1989.

Gebhart, Fred. "Putting Deming to Work in Pharmacy Education." *Drug Topics* 134 (April 9, 1990), pp. 64–71.

Ghiselin, B. ed. *The Creative Process.* New York: Doubleday, 1966.

Gibson, P. *Quality Circles: An Approach to Productivity Improvement*. Elmsford, N.Y.: Pergamon Press, 1983.

Gilbert, G. Ronald. "Jump-Start Your Team for Quality." *Government Executive*, November 1990, p. 54.

Gilks, John F. "Total Quality: Wave of the Future." *Canadian Business Review*, Spring 1990, pp. 17–20.

Ginzberg, Eli, and George Vojta. *Beyond Human Scale: The Large Corporation at Risk*. New York: Basic Books, 1985, pp. 218–19.

Gitlow, Howard S., and Paul T. Hertz. "Product Defects and Productivity." *Harvard Business Review* 61 (September/October 1983), pp. 131–211.

Gitlow, Howard S., and Process Management International. *Planning for Quality, Productivity, and Competitive Position*. Milwaukee, Wis.: ASQC Press, 1990.

Gitlow, Howard S., and Shelly Gitlow. *The Deming Guide to Quality and Competitive Position*. Englewood Cliffs, N.J.: Prentice Hall, 1986.

Gitlow, Howard; Shelly Gitlow; Alan Oppenheim; and Rosa Oppenheim. *Tools and Methods for the Improvement of Quality*. Homewood, Ill.: Richard D. Irwin, 1989.

Gluck, F. W.; S. P. Kaufman; and S. A. Walleck. "Strategic Management for Competitive Advantage." *Harvard Business Review* 58, no. 4 (July/August 1980), pp. 154–61.

GOAL/QPC. *Memory Jogger*. Lawrence, Mass.: GOAL/QPC, 1988.

Goddard, Robert W. "Viewpoint—The Vital Few and the Trivial Many." *Personnel Journal*, July 1987.

Godfrey, A. Blanton. "Strategic Quality Management, Part I." *Quality* 29, no. 3 (March 1990), pp. 17–22.

Gold, Philip. "Table Turning Is the Lesson." *Insight*, September 30, 1991, pp. 36–38.

Goldberg, E.; V. Hulton; P. Konoske; and M. Monda. *Introduction to Total Quality Management: Selected Readings*. (NPRDC Tech Note 87-23). San Diego: Navy Personnel Research and Development Center, ADA181 325, 1987, p. 325.

Goldratt, Elihu, and J. Cox. *The Goal: Excellence in Manufacturing*. Croton-on-Hudson, N.Y.: North River Press, 1984.

Goldratt, Elihu, and R. Fox. *The Race for a Competitive Edge*. Milford, Conn.: Creative Output, 1986.

Goodman, John; Arlene Malech; and Theodore Marra. "I Can't Get No Satisfaction." *The Quality Review*, Winter 1987.

Gordon, Thomas. *Leader Effectiveness Training*. New York: Bantam Books, 1977.

Gottesman, Ken. "JIT Manufacturing Is More than Inventory Programs and Delivery Schedules." *Industrial Engineering*, May 1991.

Gottinger, Hans W. *Coping with Complexity*. Dordrecht, Neth.: D. Reidel Publishing, 1990.

Grahn, Dennis. "The Buchanan Scale Identifies the Best and the Worst." *Quality Progress*, September 1989, p. 120.

Grant, Eugene L., and Richard S. Leavenworth. *Statistical Quality Control*. 6th ed. New York: McGraw-Hill, 1988.

Grant, Philip C. *The Effort-Net Return Model of Employee Motivation*. Westport, Conn.: Quorum, 1990.

Gray, Jeffrey. *The Psychology of Fear and Stress*. New York: McGraw-Hill, 1971.

Grayson, J., and C. O'Dell. "American Business—A Two-Minute Warning." *Ten Tough Issues Managers Must Face*. New York: Free Press, 1988.

Greene, Tony. "Design Automation Vendors Define Quality on Own Terms." *Electronic Business,* October 16, 1989, pp. 209–10.

Greenwood, Frank. "How to Survive by Raising Quality While Dropping Costs." *Journal of Systems Management* 39 (September 1988), pp. 36–43.

Greenwood, Frank; Mary M. Greenwood; and W. Edwards Deming. "How to Raise Quality and Productivity While Cutting Costs." *Records Management Quarterly* 24 (July 1990), pp. 8–17.

Gregerman, Ira. *Knowledge Worker Productivity.* New York: American Management Association, 1981.

Grieco, Peter L., Jr., and Michael W. Gozzo. *Made in America*. Milwaukee, Wis.: ASQC Press, 1987.

Griffin, Gerald R. *Machiavelli on Management*. New York: Praeger Publishers, 1991.

Griffith, Gary K. *Quality Technician's Handbook*. New York: John Wiley, 1986.

Griffith, Gary K. *Statistical Process Control Methods—For Long and Short Runs*. Milwaukee, Wis.: ASQC, 1989.

Griffiths, David N. *Implementing Quality with a Customer Focus*. Milwaukee, Wis.: ASQC Press, 1991.

Groocock, John. *The Chain of Quality.* New York: John Wiley & Sons, 1986.

Grove, A. S. *High Output Management*. New York: Random House, 1983.

Gryna, F. *Quality Circles: A Team Approach to Problem Solving*. New York: American Management Association (AMACOM), 1982.

Guaspari, John. "You Want to Buy into Quality? Then You've Got to Sell It." *Management Review,* January 1988.

Gunter, Berton. "A Perspective on the Taguchi Methods." *Quality Progress* 20, no. 6 (June 1987), pp. 44–52.

Gutierrez, Genaro J., and Panagiotis Kouvelis. "Parkinson's Law and Its Implications for Project Management." *Management Science* 37, no. 8 (August 1991), pp. 990–1001.

Haavind, Robert. "Hewlett-Packard Unravels the Mysteries of Quality." *Electronic Business,* October 16, 1989, pp. 101–5.

Haavind, Robert. "Motorola's Unique Problem: What to Do for an Encore." *Electronic Business,* October 16, 1989, pp. 60–66.

Haavind, Robert. "New Thinking on Quality." *Electronic Business* 15 no. 20 (October 16, 1989), pp. 24–27.

Haavind, Robert. "Solectron's Chen Crusades for U.S. Manufacturing Excellence." *Electronic Business,* October 16, 1989, pp. 72–74.

Hahn, Gerald J. "Some Things Engineers Should Know about Experimental Design." *Journal of Quality Technology* 9, no. 1 (January 1977), pp. 13–20.

Hahn, Gerald J. "Experimental Design in the Complex World." *Technometrics,* 26, no. 1 (February 1984), pp. 19–31.

Hale, Roger L.; Douglas R. Hoelscher; and Ronald E. Kowal. *Quest for Quality.* Minneapolis, Minn.: Tennant Company, 1987, pp. 11–12.

Hall, D. M. *Management of Human Systems.* Cleveland, O.: Association for Systems Management, 1971.

Hall, Robert W. *Kawasaki U.S.A.: Transferring Japanese Production Methods to the United States—A Case Study.* Falls Church, Va.: American Production and Inventory Control Society, 1982.

Hall, Robert W. *Zero Inventories.* Homewood, Ill.: Richard D. Irwin, 1983.

Hall, Stephen S.J. *Quality Assurance in the Hospitality Industry.* White Plains, N.Y.: Unipub, 1990.

Hamlin, Jerry L., ed. *Success Stories in Productivity Improvement.* Atlanta, Ga.: IIE Press, 1991.

Hammond, Joshua. "Claude I. Taylor—Quality Is the Measure of Value." *Quality Progress,* August 1986.

Hammons, Charles, and Gary A. Maddus. "An Obligation to Improve." *Management Decision* 27 (November 1989), pp. 5–14.

Hardaker, Maurice, and Bryan K. Ward. "Getting Things Done, How to Make a Team Work." *Harvard Business Review,* November 1987.

Harman, Willis, and John Hormann. *Creative Work: The Constructive Role of Business in a Transformational Society.* Indianapolis, Ind.: Knowledge Systems, 1991.

Harrington, H. James. *The Improvement Process: How America's Leading Companies Improve Quality.* New York: McGraw-Hill, 1986.

Harrington, H. James. *Excellence—The IBM Way.* Milwaukee, Wis.: ASQC Quality Press, 1988.

Harrington, H. James. *The Quality/Profit Connection.* Milwaukee, Wis.: ASQC Press, 1989.

Harrington, H. James. *Business Improvement Process.* New York: McGraw-Hill, 1991.

Harrington, H. James and Jack B. ReVelle. "Hughes Aircraft Manages Total Quality Control in a DOD Environment." *Industrial Engineering* 21, no. 12 (December 1989).

Hart, Marilyn K., and Robert F. Hart. *Quantitative Methods for Quality and Productivity Improvement.* Milwaukee, Wis.: ASQC Press, 1989.

Hartley, Robert F. *Management Mistakes and Successes.* 3rd ed. New York: John Wiley & Sons, 1991.

Harvey, Jerry. *The Abilene Paradox.* Lexington, Mass.: Lexington Books, 1988.

Hatakeyomo, Yoshio. *Manager Revolution: A Guide to Survival in Today's Changing Workplace*. Cambridge, Mass.: Productivity Press, 1981.

Hauser, John R., and Don Clausing. "The House of Quality." *Harvard Business Review*, May/June 1988, pp. 63–73.

Hawken, Paul. *The Next Economy*. New York: Holt, Rinehart & Winston, 1983, pp. 172–73.

Hayes, Glenn E. *Quality and Productivity: The New Challenge*. Milwaukee, Wis.: ASQC Press, 1985.

Hayes, Glenn E. *Quality Assurance: Management and Technology*, rev. ed. Milwaukee, Wis.: ASQC, 1990.

Hayes, R. H. "Why Japanese Factories Work." *Harvard Business Review*, July/August 1981.

Hayes, Robert H., and Steven C. Wheelwright. *Restoring Our Competitive Edge: Competing through Manufacturing*. New York: John Wiley & Sons, 1984.

Hayward, Gordon P. *Introduction to Nondestructive Testing*. Milwaukee, Wis.: ASQC Press, 1978.

Hegland, Don. "Stop Burning the Toast." *Production Engineering* 30 (April 1983), pp. 7–11.

Heiberger, Richard M. *Computation for the Analysis of Designed Experiments*. Milwaukee, Wis.: ASQC Press, 1989.

Helfgott, Roy B. "On the Demise of the Long Run." *Journal of Economic Perspectives* 3, no. 4 (Fall 1989), pp. 149–52.

Heller, R. *The Supermanagers*. New York: E. P. Dutton, 1984.

Heller, Robert. "Anachronistic Attitudes: There's a Little Hope of a Renaissance in Manufacturing When Only Lip Service Are Paid to Issues Like Quality." *Management Today*, April 1991, pp. 28–31.

Henkoff, Ronald. "What Motorola Learns from Japan." *Fortune*, April 24, 1989.

Hensel, James S. "Service Quality Improvement and Control: A Customer-based Approach." *Journal of Business Research* 20, no. 1 (January 1990), pp. 43–54.

Herrmann, John L. *Leadership and Wealth*. Milwaukee, Wis.: ASQC Press, 1989.

Hersey, Paul, and K. H. Blanchard. "The Management of Change." *Training and Development Journal* 26, no. 1 (January 1972).

Hersey, Paul, and K. H. Blanchard. "The Management of Change." *Training and Development Journal* 26, no. 2 (February 1972).

Hersey, Paul, and K. H. Blanchard. "The Management of Change." *Training and Development Journal* 26, no. 3 (March 1972).

Hersey, Paul. *The Situational Leader*. New York: Warner Books, 1984.

Herzberg, F.; B. Mausner; and B. Snyderman. *The Motivation to Work*. New York: John Wiley & Sons, 1959.

Heskett, James L. *Managing in the Service Economy*. Boston: Harvard Business School Press, 1986, p. 67.

Heslop, Jon M. "The New Philosophy." *Rubber World* 193 (March 1986), pp. 21–23.

Hickman, Craig, and Michael Silva. *Creating Excellence*. New York: New American Library, 1984.

Hicks, Charles R. *Fundamental Concepts in the Design of Experiments*. 3rd ed. New York: Holt, Rinehart & Winston, 1982.

Hildebrand, Carol. "The Deming Prize: No Longer a Stranger at Home." *Computerworld* 23 (December 11, 1989), pp. 100–01.

Hill, Terry. *Manufacturing Strategy: Text and Cases*. Homewood, Ill.: Richard D. Irwin, 1989.

Hintze, Larry L. "Improving Productivity through the Use of Pareto's Principle." *United States Army Management Engineering College*, September 1988, 1–11.

Hodgson. "Deming's Never-Ending Road to Quality." *Personnel Management* 19 (July 1987), pp. 40–45.

Hodlin, Steven F. "Tap into Quality." *Quality Engineering* 1, no. 2 (1988–89), pp. 127–34.

Hoff, Benjamin. *The Tao of Pooh*. New York: Penguin Books, 1983.

Hohner, Gregory J. "TQC: Method of Champions—Part I." *Manufacturing Systems* 8, no. 3 (March 1989), pp. 62–64.

Hohner, Gregory J. "TQC: Method of Champions—Part II." *Manufacturing Systems* 8, no. 4 (April 1989), pp. 22–32.

Hohner, Gregory J. "TQC: Method of Champions—Part III." *Manufacturing Systems* 7, no. 5 (May 1989), pp. 51–53.

Holmes, Donald S., and A. Erhan Mergen. "The Dynamic Histogram Chart." *Quality and Reliability Engineering International* 6, no. 2 (April–May 1990), pp. 107–11.

Hopkins, Shirley A. "Have U.S. Financial Institutions Really Embraced Quality Control?" *National Productivity Review* 8, no. 4 (Autumn 1989), pp. 407–20.

Horton, Thomas R. "Taking Personal Stock." *Quality*, January 1990, p. 72.

Howell, Vincent. *Quality Improvement through Continuing Education: An Engineer's Guide for Life-Long Learning*. Milwaukee, Wis.: ASQC Press, 1986.

Hudiberg, John J. *Winning with Quality: The FPL Story*. White Plains, N.Y.: Unipub, 1991.

Huge, Ernest C. *Total Quality*. Milwaukee, Wis.: ASQC Press, 1990.

Hunt, Daniel V. *Quality in America: How to Implement a Competitive Quality Program*. Homewood, Ill.: Richard D. Irwin, 1991.

Hutchins, D. "Having a Hard Time with Just-in-Time." *Fortune*, June 9, 1986, pp. 64–66.

Huthwaite, Bart. "How to Recognize a Great Design." *Design News* 47 (January 7, 1991), pp. 122–31.

Hyde, Bill. "Using Cultural Strengths Can Win Back Manufacturing." *Industrial Management* 33, no. 4 (July/August 1991), pp. 12–16.

"If Japan Can, Why Can't We?" videotape Films, Inc., 1980.

Iglewicz, Boris, and David C. Hoaglin. "Use of Boxplots for Process Evaluation. *Journal of Quality Technology* 19, no. 4 (October 1987), pp. 180–90.

Imai, Masaaki. *Kaizen: The Key to Japan's Competitive Success*. New York: Random House, 1986.

Ingle, S. *Quality Circle Master Guide*. Englewood Cliffs, N.J.: Prentice Hall, 1981.

Inglesby, Tom. "Do We Know Too Much or Too Little?" *Manufacturing Systems*, May 1989, p. 4.

Inglesby, Tom. "The 'Standard of the World' Makes a Comeback." *Manufacturing Systems*, April 1991, pp. 16–24.

Ingman, Lars C. "Buying 'right': Pushing Upstream." *Pulp & Paper* 65 (April 1991), pp. 175–82.

"Inside Xerox: Moving in a Quality Direction." *Electronics Business*, April 1, 1987.

International Association of Quality Circles. *QC Sources: Selected Writing on Quality Circles*. IAQC, 1984.

"Is America Turning It Around in Quality?" *Iron Age* 226 (July 22, 1983), pp. 40–44.

Ishikawa, Kaoru. *Guide to Quality Control*. rev. ed. White Plains, N.Y.: Kraus International Publications, 1982.

Ishikawa, Kaoru. "How to Apply Companywide Quality Control in Foreign Countries." *Quality Progress*, 22, no. 9 (September 1989), pp. 70–74.

Ishikawa, Kaoru. *Quality Control Circles at Work*. Tokyo: JUSE, 1984.

Ishikawa, Kaoru. *What Is Total Quality Control? The Japanese Way*. Englewood Cliffs, N.J.: Prentice Hall, 1985.

Itami, H. *Keiei Senryaku no Ronri* ("The Theory of Corporate Strategy"). Tokyo: Nihon Keizai Shinbun, 1981.

Iwata, R. *Nihonteki Keiei no Henseigenri* ("Fundamental Framework of Japanese Management"). Tokyo: Bunshin Do, 1978.

Jacobson, G., and J. Hillkirk. *Xerox: American Samurai*. New York: Macmillan, 1986.

Jamieson, Archibald. *Introduction to Quality Control*. Milwaukee, Wis.: ASQC Press, 1982.

Jans, N. A. "Organizational Commitment, Career Factors and Career/Life Stage." *Journal of Organizational Behavior* 10 (1989), pp. 247–66.

Japan Management Association. *ZD no Shintenkai* ("New development of ZD"). Tokyo: Nihon Noritsu Kyokai, 1978.

Johnson, L. Marvin. *Quality Assurance Program Evaluation*. rev. ed. Milwaukee, Wis.: ASQC Press, 1990.

Johnson, Ross H., and Richard T. Weber. *Buying Quality*. Milwaukee, Wis.: ASQC Press, 1985.

Johnson, Ross H., and William O. Winchell. "Educating for Quality: A Different Approach." *Quality Progress* 21, no. 9 (September 1988), pp. 48–50.

Joiner, Brian, and Peter Scholtes. "The Quality Manager's New Job." *Quality Progress*, October 1986.

Jonason, Per. "Project Management, Swedish Style." *Harvard Business Review*, November/December 1971, pp. 104–09.

Jones, Louis, and Ronald McBride. *An Introduction to Team-Approach Problem Solving*. Milwaukee, Wis.: ASQC Press, 1990.

Jones, Peter. "The American Who Saved Japan." *Scholastic Update* 119 (April 6, 1987), pp. 8–11.

Jones, Steven D.; Randy Powell; and Scott Roberts, "Comprehensive Measurement to Improve Assembly-Line Work Group Effectiveness." *National Productivity Review* 10, no. 1 (Winter 1990/91), pp. 45–55.

Juran, J. M. "China's Ancient History of Managing for Quality." *Quality Progress* 23, no. 7 (July 1990), pp. 31–35.

Juran, J. M. *Juran on Quality Improvement Workbook*. New York: Juran Enterprises, 1981.

Juran, J. M. *Juran on Planning for Quality*. New York: Free Press, 1988.

Juran, J. M. *Juran's New Quality Road Map*. New York: Free Press, 1991.

Juran, J. M. *Managerial Breakthrough*. New York: McGraw-Hill, 1964.

Juran, J. M. *Quality Control Handbook*. New York: Mc-Graw Hill, 1988.

Juran, J. M. "The Quality Trilogy." *Quality Progress*, August 1986, pp. 19–24.

Juran, J. M. and Frank M. Gryna, Jr. *Quality Planning and Analysis*. New York: McGraw-Hill, 1980.

Juran Institute. "The Tools of Quality. Part V: Check Sheets." *Quality Progress* 23, no. 10 (October 1990), pp. 51–56.

Jurgen, Ronald K. *Computers and Manufacturing Productivity*. New York: IEEE Press, 1986.

Kackar, Raghu N. "Off-Line Quality Control, Parameter Design, and the Taguchi Method." *Journal of Quality Technology* 17, no. 4 (October 1985), pp. 176–209.

Kackar, Raghu N. "Taguchi's Quality Philosophy: Analysis and Commentary." *Quality Progress* 19, no. 12 (December 1986), pp. 21–29.

Kackar, Raghu N. "Quality Planning for Service Industries." *Quality Progress* 21, no. 8 (August 1988), pp. 39–42.

Kagaku, Gijyutsu-Cho. *Kagaku Gijyutsu Hakusho* ("The White Paper of Science and Technology of Japan"). Tokyo: The Ministry of Science and Technology, 1982.

Kagono, T.; Y. Nonaka; K. Sakakibara; and A. Okumura. *Nichibei Kigyo no Keiei Hikaku* ("Business Comparison between Japanese and American Firms") Tokyo: Nihon Keizai Shinbun, 1983.

Kahn, R. L.; D M. Wolfe; R. P. Quinn; J. D. Snock; and R. A. Rosenthal. *Organizational Stress: Studies in Role Conflict and Ambiguity*. New York: John Wiley & Sons, 1964.

Kane, Edward J. "IBM's Quality Focus on the Business Process." *Quality Progress*, April 1986.

Kane, Victor E. *Defect Prevention: Use of Simple Statistical Tools*. Milwaukee, Wis.: ASQC Press, 1989.

Kanter, Rosabeth Moss. *The Change Masters*. New York: Simon & Schuster, 1983.

Kanter, Rosabeth Moss. "Managing the Human Side of Change." *Management Review,* April 1985, pp. 52–56.

Kanter, Rosabeth Moss. "Quality Leadership and Change." *Quality Progress,* February 1987, pp. 45–51.

Kantrow, Alan M. "Wide-Open Management at Chaparrel Steel." *Harvard Business Review,* May/June 1986, pp. 99–101.

Kaplan, Robert S. "Yesterday's Accounting Undermines Production." *Harvard Business Review,* July/August 1984, pp. 95–102.

Kaplan, Robert S. "Measuring Manufacturing Performance: A New Challenge for Mangerial Accounting Research." *The Accounting Review,* February 1985.

Karabatsos, Nancy. "Absolutely, Postively Quality." *Quality Progress* 23, no. 5 (May 1990), pp. 24–28.

Karatsu, Hajime. *Mastering the Tools of QC*. Tokyo: JUSE, 1987.

Karatsu, Hajime. *An Invitation to QC*. Tokyo: JUSE, 1988.

Karatsu, Hajime. *TQC Wisdom of Japan: Managing for Total Quality Control*. Cambridge, Mass.: Productivity Press, 1988.

Karatsu, Hajime, and Karen Jones. *Tough Words for American Industry.* Cambridge, Mass.: Productivity Press, 1988.

Kast, D. "The Motivational Basis of Organizational Behavior." *Behavioral Science* 9, no. 2 (1964), pp. 131–43.

Katz, Donald R. "Coming Home." *Business Month* 132 (October 1988), pp. 56–65.

Kaydos, Will. *Measuring, Managing and Maximizing Performance*. Cambridge, Mass.: Productivity Press, 1991.

Kazdin, A. E. *Behavior Modification in Applied Settings*. Homewood, Ill.: Richard D. Irwin, 1980.

Kearns, David T. "A Corporate Response." *Quality Progress,* February 1988.

Kenworthy, Harry W. "Total Quality Concept: A Proven Path to Success." *Quality Progress,* July 1986, pp. 21–24.

Kepner, Charles H., and Benjamin B. Trigoe. *The New Rational Manager.* Princeton, N.J.: Princeton Research Press, 1981.

Kerr, John. "For Bourns, 'acceptable' Is Not Good Enough." *Electronic Business,* October 16, 1989, pp. 235–36.

Kerr, John. "These Days, Intel Thinks Impatience Is a Virtue." *Electronic Business,* October 16, 1989, pp. 111–12.

Kerzner, Harold. *Project Management*. New York: Van Nostrand Reinhold, 1984.

Khalil, Tarek M., and Bulent A. Bayraktar, eds. *Management of Technology II*. Atlanta, Ga.: IIE Press, 1991.

Khorramshahgol, Reza; A. Ason Okoruwa; and Hossein Azani. "Capital Budgeting in a Multiple Objective Environment." *Journal of Information and Optimization Sciences* 10, no. 3 (1989), pp. 567–77.

Khuri, Andre L., and John A. Cornell. *Response Surfaces: Designs and Analyses*. New York: Marcel Dekker, 1987.

Kiechel, Walter, III. "Executives Ought to Be Funnier." *Fortune*, December 12, 1983, p. 208.

Kiechel, Walter, III. "Visionary Leadership and Beyond." *Fortune*, July 21, 1986, pp. 127–28.

Kiechel, Walter, III. "Corporate Strategy for the 1990s." *Fortune*, February 29, 1988, p. 34.

King, Bob. *Better Designs in Half the Time: Implementing QFD Quality Function Deployment in America*. Methuen, Mass.: GOAL/QPC, 1987.

King, Carol. "A Framework for a Service Quality Assurance System." *Quality Progress*, September 1987.

King, Robert E. "Hoshin Planning, The Foundation of Total Quality Management." *ASQC Quality Congress Transactions*, 1989.

Kinlaw, Dennis C. *Developing Superior Work Teams*. New York: Free Press, 1990.

Kinsella, Bridget. "What 'Training' Means: According to Those that Have Done It, Setting the Groundwork for Quality Improvement Is a Slow and Often Painstaking Process." *The Printing Industry* 62 (November 1990), pp. 122–32.

Kirkham, Roger L. "How to Manage Changes Coming from Tight Times." *Federal Manager's Quarterly*, January 1987, pp. 30–37.

Kirschmann, John D. *The Nutrition Almanac*. New York: McGraw-Hill, 1984.

Klein, Janice. *Revitalizing Manufacturing*. Homewood, Ill.: Richard D. Irwin, 1990.

Knowlton, Christopher. "What America Makes Best." *Fortune*, March 1988, pp. 40–48.

Kobayashi, S. *Sony wa Hito o Ikasu* ("Sony Makes the Best Use of Its Human Resources"). Tokyo: Nippon Keiei Shuppan-kai, 1965.

Kobayashi, S. *Creative Management*. New York: American Management Associations, 1971.

Kodansha Ltd. *The Best of Japan*. Cambridge, Mass.: Productivity Press, 1988.

Koska, Mary T. "Adopting Deming's Quality Improvement Ideas: A Case Study." *Hospitals* 64 (July 5, 1990), pp. 58–65.

Kotter, J. *The General Managers*. New York: Macmillan, 1982.

Kouzes, James, and Barry Posner. *The Leadership Challenge: How to Get Extraordinary Things Done in Organizations*. San Francisco: Jossey-Bass, 1987.

Kouzes, James M., and Barry Z. Posner. "The Credibility Factor: What Followers Expect from Their Leaders." *Management Review* 79, no. 1 (January 1990), pp. 29–33.

Kraar, Louis. "25 Who Help the U.S. Win: Innovators Everywhere Are Generating Ideas to Make America a Stronger Competitor. They Range from a Boss Who Demands the Impossible to a Mathematician with a Mop." *Fortune* 123 (Spring/Summer 1991), pp. 34–112.

Kraljic, P. "Purchasing Must Become Supply Management." *Harvard Business Review,* 1983, pp. 109–17.

Krishnaiah, P. R., and C. R. Rao. *Handbook of Statistics 7: Quality Control and Reliability.* Amsterdam, Neth.: North-Holland, 1988.

Krishnamoorthi, K. S. *Quality Control for Operators and Foremen.* Milwaukee, Wis.: ASQC Press, 1989.

Kume, H. *Statistical Methods for Quality Improvement.* Tokyo: JUSE, 1985.

Kurokawa, Kaneyuki. "Quality and Innovation." *IEEE Circuits and Devices,* July 1988, pp. 3–8.

Kusaba, Ikuro. "Statistical Methods in Japanese Quality Control." *Societas Qualitatis* 2, no. 2 (May/June 1988).

Kuzela, Lad. "Deming's Delight: Ford, GM Make Front-end Gains." *Industry Week* 221 (May 14, 1984), pp. 86–92.

Labell, Fran, and Dean Duxbury. "Campbell's Worldwide QA Program Yields Quality, Safety, Savings; Goal: Consumer Satisfaction." *Food Processing* 46 (August 1985), pp. 60–65.

Laford, Richard J. *Ship-to-Stock: An Alternative to Incoming Inspection.* Milwaukee, Wis.: ASQC Press, 1986.

Lammermeyr, Horst U. *Human Relations—The Key to Quality.* White Plains, N.Y.: Unipub, 1990.

Landwehr, James M., and Ann E. Watkins. *Exploring Data.* Palo Alto, Calif.: Seymour Publications, 1987.

Lange, Christian. "Ritual in Business: Building a Corporate Culture through Symbolic Management." IM 33, no. 4 (July/August 1991), pp. 21–23.

Lascelles, D. M., and B. G. Dale. "A Study of the Quality Management Methods Employed by UK Automotive Suppliers." *Quality and Reliabaility Engineering International* 4, no. 4 (October–December 1988), pp. 301–09.

Lascelles, David, and Barrie Dale. "Quality Management: The Chief Executive's Perception and Role." *Journal of European Management* 8, no. 1 (March 1990), pp. 67–75.

Lash, Linda M. *Complete Guide to Customer Service.* New York: John Wiley, 1989.

Latack, J. C.; H. J. Joseph; and R. J. Aldag. "Job Stress: Determinants and Consequences of Coping Behaviors." Working paper, Graduate School of Business, University of Wisconsin, Madison, Wis., 1985.

Latham, G. P., and E. A. Locke. "Goal-Setting—A Motivational Technique that Works." *Organizational Dynamics* 8 (Autumn 1979), pp. 68–80.

Latzko, William J. *Quality and Productivity for Bankers and Financial Managers.* New York: Marcel Dekker, 1986.

Lawler, Edward E. III. *High Involvement Management: Participative Strategies for Improving Organizational Performance.* San Francisco: Jossey-Bass, 1986.

Lawrence, P. R., and D. Dyer. *Renewing American Industry.* New York: Macmillan, 1983.

Lawton, Robin L. *Creating a Customer-Centered Culture in a Service Environment.* Innovative Management Technologies, 1988.

Lawton, Robin L. "Creating a Customer-Centered Culture for Service Quality." *Quality Progress* 22, no. 5 (May 1989), pp. 34–36.

Leavitt, H. J. *Managerial Psychology.* 4th ed. Chicago: University of Chicago Press, 1978.

Leddick, Susan. "Teaching Managers to Support Quality Improvement Efforts." *National Productivity Review* 10 (Winter 1991), pp. 69–76.

Lefevre, Henry L. *Quality Service Pays.* White Plains, N.Y.: Unipub, 1989.

Leibowitz, Michael R. "Baldrige Winners Start Ahead of the Pack." *Electronic Business,* October 16, 1989, pp. 80–82.

Lenz, J. "What Happens When You Don't Stimulate." *Proceedings of the First International Conference on Simulation in Manufacturing.* Stratford-upon-Avon, England, March 5–7, 1985.

Leon, Ramon V.; Anne C. Shoemaker; and Raghu N. Kacker. "Performance Measures Independent of Adjustment: An Explanation and Extension of Taguchi's Signal-to-Noise Ratios." *Technometrics* 29, no. 3 (August 1987), pp. 253–85.

Leonard, Frank S., and W. Earl Sasser. "The Incline of Quality." *Harvard Business Review,* September/October 1982.

Levering, R.; M. Moskowitz; and M. Katz. *The 100 Best Companies to Work for in America.* Reading, Mass.: Addison-Wesley Publishing, 1984.

Levinson, H. *The Exceptional Executive.* New York: The New American Library, 1971.

Levinson, H., and S. Rosenthal. *CEO: Corporate Leadership in Action.* New York: Basic Books, 1984.

Lewin, Kurt. *The Conceptual Representation and the Measurement of Psychological Forces.* Durham, N.C.: Duke University Press, 1938.

Lewin, Kurt. "Frontiers in Group Dynamics." *Human Relations* 1, no. 1 (1947).

Lewin, Kurt. "Group Decision and Social Change." *Readings in Social Psychology.* New York: Holt, Rinehart & Winston, 1958, pp. 197–211.

Lewis, Wayne M., and R. Michael Jens. "Project Management Lessons from the Past Decade of Mega-Projects." *Project Management Journal* 18, no. 5 (December 1987), pp. 69–74.

Liker, Jeffrey K.; David B. Roitman; and Ethel Roskies. "Changing Everything All at Once: Work Life and Technological Change." *Sloan Management Review,* Summer 1987, pp. 29–47.

Likert, Rensis. *New Patterns of Management.* New York: McGraw-Hill, 1961.

Lindenmeyer, Carl, and Lawrence J. Caldwell. "Packaging IE Student Field Projects." *IIE Focus/Industrial Engineering,* 1990, pp. 5–8.

Li-Ping Tang; Peggy Smith Tollison Thomas; and Harold D. Whiteside. "Quality Circle Productivity as Related to Upper-Management Attendance, Circle Initiation, and Collar Color." *Journal of Management* 15, no. 1 (March 1989), pp. 101–13.

Liswood, Laura A. *Serving Them Right: Innovative and Powerful Customer Retention Strategies*. Milwaukee, Wis.: ASQC Press, 1990.

"Living the American dream." *Design News* 47 (January 7, 1991), pp. 120–21.

Lodge, Charles. "Six Gurus Show the Way to Improved Product Quality." *Plastics World,* August 1989, pp. 29–40.

Loehr, Dr. James E., and Peter J. McLaughlin. *Mentally Tough*. New York: M. Evans, 1986.

Loehr, Dr. James E., and Dr. Jeffrey Migdow. *Take a Deep Breath*. New York: Villard Books, 1986.

Longdorf, Bob. "Deming on Quality." *Business* 35 (November 1984), pp. 44–111.

Lu, David J. *Kanban and Just-in-Time at Toyota*. Cambridge, Mass.: Productivity Press, 1985.

Lu, David J. *Inside Corporate Japan: The Art of Fumble Free Management*. Cambridge, Mass.: Productivity Press, 1987.

Lubben, Richard T. *Just-In-Time Manufacturing: An Aggressive Manufacturing Strategy.* New York: McGraw-Hill, 1988.

Luthans, Fred. *Organizational Behavior.* New York: McGraw-Hill, 1973.

Luthans, Fred, and R. Kreitner. *Organizational Behavior Modification*. Glenview, Ill.: Scott, Foresman, 1978.

Lynn, Monty, and David P. Osborn. "Deming's Quality Principles: A Health Care Application." *Hospital & Health Services Administration* 36 (Spring 1991), pp. 111–210.

Maass, Richard A., and the ASQC Customer-Supplier Technical Committee. *World Class Quality—An Innovative Prescription for Survival*. Milwaukee, Wis.: ASQC Press, 1988.

Maccoby, M. *The Gamesman*. New York: Simon & Schuster, 1976.

Maccoby, M. *The Leader: A New Face for American Management*. New York: Ballantine Books, 1983.

Maccoby, Michael. "Deming Critiques American Management." *Research-Technology Management* 33 (May/June 1990), pp. 43–52.

Maccoby, Michael. "Productivity with a Human Face. Long Practiced in Japan, the Management Ideas of Edward Deming Are Finally Starting to Catch on Here, Too." *Washington Monthly* 23 (March 1991), pp. 55–63.

MacKenzie, R. Alex. *The Time Trap*. New York: McGraw-Hill, 1972.

Maddi, Salvatore R., and Suzanne C. Kobasa. *The Hardy Executive: Health Under Stress*. Homewood, Ill.: Richard D. Irwin, 1984.

Main, Jeremy. "The Curmudgeon Who Talks Tough on Quality." *Fortune,* June 25, 1984, pp. 118–22.

Main, Jeremy. "Detroit Is Trying Harder for Quality." *BusinessWeek,* November 1, 1982.

Main, Jeremy. "Detroit's Cars Really Are Getting Better." *Fortune,* February 2, 1987, p. 95.

Main, Jeremy. "How to Win the Baldrige Award." *Fortune,* April 23, 1990, pp. 101–16.

Main, Jeremy. "Manufacturing the Right Way." *Fortune,* May 21, 1990, pp. 54–64.

Main, Jeremy. "Under the Spell of the Quality Gurus: These Consultants Get up to $10,000 a Day to Help Companies Improve Their Products." *Fortune* 114 (April 18, 1986), pp. 30–34.

Mainstone, Larry E., and Ariel S. Levi. "Fundamentals of Statistical Process Control." *Journal of Organizational Behavior Management* 9, no. 5 (Spring 1987), p. 117.

"Management Discovers the Human Side of Automation." *BusinessWeek,* September 29, 1986, pp. 70–79.

Mann, Nancy R. "Dr. W. Edwards Deming: Back to Basics: Satisfying the Customer." *Road & Track* 35 (February 1984), pp. 14–22.

Mann, Nancy R. *The Keys to Excellence.* Santa Monica, Calif.: Prestwick Books, 1985.

Mann, Nancy R. "Why It Happened in Japan and Not in the U.S." *Chance* 1, no. 3 (Summer 1988), pp. 8–15.

Mans, Jack. "World Class Manufacturing Comes to Modesto." *Prepared Foods* 155 (September 1986), pp. 80–89.

Mansir, Brian E., and Nicholas R. Schacht. *Total Quality Management: A Guide to Implementation.* Bethesda, Md.: 1989.

Manual on Quality Control of Materials. Philadelphia, Pa.: American Society for Testing and Materials, 1951.

Manz, Charles, and H. P. Sims. "Leading Workers to Lead Themselves: The External Leadership of Self-managing Work Teams. *Administrative Science Quarterly* 32, no. 1 (1990), pp. 106–09.

Marlow, Edward, and Richard Schilhavy. "Expectation Issues in Management by Objectives Programs." *IM,* January/February 1991, pp. 29–32.

Marquardt, Donald W. "Youden Address: Quality Audits in Relation to International Business Strategy—What Is Our National Posture?" *Statistics Division Newsletter,* Winter 1989, pp. 10–13.

Marquardt, Donald W. "Meeting the Worldwide Quality Challenge." *Quality Progress,* August 1988, pp. 34–37.

Martin Marietta Corporation. *Quality Engineering Workmanship Standards Manual.* Milwaukee, Wis.: ASQC Press, 1988.

Maslow, Abraham H. *Motivation and Personality.* New York: Harper & Row, 1954.

Mason, Robert L.; Richard F. Gunst; and James L. Hess. *Statistical Designs & Analysis of Experiments with Applications to Engineering and Science.* Milwaukee, Wis.: ASQC Press, 1989.

Matsushita, K. *Keiei wa Kachidakai Itonami* ("Management as Engagement of Great Value"). *Zaikajin Shiso Zenshu,* vol. 2. Tokyo: Daiyamondo-sha, 1968.

Matsushita, K. *Jissen Keiei Tetsugaku* ("Practical Management Philosophy"). Kyoto: PHP Kenkyusho, 1978.

Matsushita, Konosuke. "Off the Cuff." *Business Policy.* Tokyo, 1979.

Matsushita, K. *Not for Bread Alone.* Kyoto: PHP Institute, 1984.

Mawhinney, Thomas C. "OBM, SPC, and Theory D: A Brief Introduction." *Journal of Organizational Behavior Management* 8 (Spring/Summer 1986), pp. 89–117.

McCarthy, Laurie R. "Catch Molding Flaws with High-Tech Measuring Tools." *Plastics World,* August 1989, pp. 43–47.

McCreadie, John. "Quantum Rebuilds Profits with the Help of a Friend." *Electronic Business,* October 16, 1989, pp. 241–42.

McElroy, John. "Deming Was Right." *Automotive Industries* 170 (April 1990), pp. 5–11.

McGrath, J. *Social and Psychological Factors in Stress.* New York: Holt, Rinehart & Winston, 1970.

McGregor, Douglas. *Leadership and Motivation.* Cambridge, Mass.: MIT Press, 1983.

McIlvaine, Paul J. "Pearls of Wisdom." *Program Management,* November/December 1989, p. 15.

McKay, Harvey. *Swim with the Sharks.* New York: William Morrow, 1988.

McLean, Gary N., and Sam Parkenham-Walsh. "An In-Process Model for Improving Quality Management Processes. *Consultation* 6, no. 3 (Fall 1987).

Melan, Eugene H. "Process Management in Service and Administrative Operations." *Quality for Progress,* June 1985.

Melan, Eugene H. "Process Management: A Unifying Framework for Improvement." *National Productivity Review* 8, no. 4 (Autumn 1989), pp. 305–406.

Menhenhall, W. *Introduction to Linear Models and the Design and Analysis of Experiments.* Belmont, Calif.: Wadsworth, 1968.

Meyer, Michael, and Jennifer Meyer. "The Myth of German Efficiency." *Newsweek,* July 30, 1990, p. 36.

Michaelson, Gerald A. "The Turning Point of the Quality Revolution." *Across the Board* 27 (December 1990), pp. 40–46.

Mickelson, Elliot S. *Construction Quality Program Handbook.* Milwaukee, Wis.: ASQC Press, 1986.

MIL-STD-105D: Sampling Procedures and Tables for Inspection by Attributes, Superintendent of Documents, Washington, D.C.: U.S. Government Printing Office, 1963.

MIL-STD-414: Sampling Procedures and Tables for Inspection by Variables for Percent Defective, Superintendent of Documents, Washington, D.C.: U.S. Government Printing Office, 1957.

MIL-STD-1235A (MU): Single and Multilevel Continuous Sampling Procedures and Tables for Inspection by Attributes, Superintendent of Documents, Washington, D.C.: U.S. Government Printing Office, 1974.

MIL-STD-1235 A-1: Functional Curves of the Continuous Sampling Plans, Superintendent of Documents, Washington, D.C.: U.S. Government Printing Office, 1975.

MIL-STD-1235B: Single- and Multi-Level Continuous Sampling Procedures and Tables for Inspection by Attributes, Superintendent of Documents, Washington, D.C.: U.S. Government Printing Office, 1981.

Miller, Jeff. *Sneak Circuit Analysis for the Common Man*. Griffiss Air Force Base, N.Y.: Rome Air Development Center, October 1989.

Miller, Jeffery G., and Thomas E. Vollmann. "The Hidden Factory." *Harvard Business Review*, September/October 1985.

Miller, Lawrence M. *Barbarians to Bureaucrats: Corporate Life Cycle Strategies*. New York: Potter, 1989.

Miller, Rock. "Continuing the Taguchi Tradition." *Managing Automation* 2 (February 1988), pp. 34–36.

Mills, Charles A. *The Quality Audit: A Management Evaluation Tool*. New York: McGraw-Hill, 1989.

Mills, P. K., and B. Z. Posner. "The Relationships among Self-Supervision, Structure and Technology in Professional Service Organizations." *Academy of Management Journal* 25 (1982), pp. 437–42.

Mindess, Harvey. *Laughter and Liberation*. Los Angeles: Nash, 1971.

Mirvis, Philip H., and Donald L. Kanter. "Combatting Cynicism in the Workplace." *National Productivity Review* 8, no. 4 (Autumn 1989), pp. 377–94.

Mitzenberg, H. *The Nature of Managerial Work*. New York: Harper & Row, 1973.

Mizuno, Shigeru, ed. *Management for Quality Improvement: The Seven New QC Tools*. Cambridge, Mass.: Productivity Press, 1988.

Mizuno, Shigeru. *Company-wide Total Quality Control*. White Plains, N.Y.: UNIPUB-Kraus International, 1987.

Mizuno, Shigeru. *Management for Quality Improvement: The Seven New QC Tools*. Cambridge, Mass.: Productivity Press, 1988.

Modic, Stanley J. "What Makes Deming Run?" *Industry Week* 236 (June 20, 1988), pp. 84–94.

Moen, Ronald D., and Thomas W. Nolan. "Process Improvement." *Quality Progress*, September 1987.

Mohr, W., and H. Mohr. *Quality Circles: Changing Images of People at Work*. Reading, Mass.: Addison-Wesley Publishing, 1983.

Monda, M.; P. Thrapp; and E. L. Goldbert. *Quality Management: An Annotated Bibliography*. (NPRDC Tech Note 72-86-07). San Diego Navy Personnel Research Center, 1986.

Monden, Y. *Toyota Production System*. Industrial Engineering and Management Press, Institute of Industrial Engineers: Atlanta, Ga, 1983.

Monden, Yasuhiro; Rinya Shibakawa; Saturo Takayanagi; and Teruya Nagao. *Innovations in Management: The Japanese Corporation*. Atlanta, Ga.: Institute of Industrial Engineers, 1991.

Montgomery, Douglas C. *Design and Analysis of Experiments*. New York: John Wiley, 1984.

Montgomery, Douglas C. *Introduction to Statistical Quality Control*. 2nd ed. Milwaukee, Wis.: ASQC Press, 1990.

Mooney, Marta. "Process Management Technology." *National Productivity Review* 5 (Autumn 1986), pp. 366–76.

Moran, John W.; Richard P. Talbot; and Russell M. Benson. *A Guide to Graphical Problem-Solving Processes*. Milwaukee, Wis.: ASQC Press, 1990.

Moreau, Dan, and Carol Chapman. "Change Agents: W. Edwards Deming Is the American Who Taught the Japanese How to Compete." *Changing Times* 43 (September 1989), pp. 132–41.

Morse, Wayne J.; Harold P. Roth; and Kay M. Poston. *Measuring, Planning, and Controlling Quality Costs*. Montvale, N.J.: National Association of Accountants, 1987.

Mosbacher, Robert A. "U.S. Must Produce Quality Products." *American Metal Market* 97 (August 30, 1989), pp. 14–21.

Motiska, Paul J., and Karl A. Shilliff. "10 Precepts of Quality." *Quality Progress* 23, no. 2 (February 1990), pp. 27–28.

Motoiu, Radu. "Evaluating the Quality Systems by Psychological Investigation." *EOQ Quality* 2 (1990), pp. 11–12.

Multi-Level Continuous Sampling Procedures and Tables for Inspection by Attributes, Inspection, and Quality Control Handbook (Interim) H106, Superintendent of Documents, Washington, D.C.: U.S. Government Printing Office, 1958.

Naden, Jim. "The SDA Strategy for Total Quality." *The Accountant's Magazine* 94 (June 1990), pp. 20–22.

Nadler, Gerald, and Shozo Hibino. *Breakthrough Thinking*. New York: Prima Publishing, 1991.

Nakajima, Seiichi. *Terotekunology* ("Terotechnology"). Tokyo: Nihon Noritsu Kyokai (Japan Management Association), 1981.

Nakajima, Seiichi. *Seisan Kanri no tameno TPM Tenkai Puroguramu Nyumon* ("Introduction to TPM Development Program for Production Management"). Tokyo: Nihon Noritsu Kyokai (Japan Management Association), 1983.

Nakajima, Seiichi. "TQC to TPM wa dokoga chigauka" ("What Differences Exist Between TQC and TPM?"). *Plant Engineer* 1 (1984), pp. 20–24.

Nakajima, Seiichi. *Introduction to TPM: Total Productive Maintenance*. Cambridge, Mass.: Productivity Press, 1991.

Nasar, Sylvia. "Competitiveness: Getting It Back." *Fortune,* April 27, 1987, p. 223.

National Tooling and Machining Association, "Measuring and Gaging in the Machine Shop," ASQC, 1981.

Nemoto, Masao, and David Lu, transl. and ed. *Total Quality Control for Management: Strategies and Techniques from Toyota and Toyoda Gosei*. Englewood Cliffs, N. J.: Prentice Hall, 1987.

"New Quality Data Says Cars Built Better." *Automotive Marketing* 18 (June 1969), pp. 16–23.

Newman, R. G. "Insuring Quality: Purchasing's Role." *Journal of Purchasing and Materials Management,* 1988, pp. 14–20.

Newman, R. G. "The Buyer-Supplier Relationship under Just-in-Time. *Production and Inventory Management Journal,* 1988, pp. 45–50.

Newsletter for Continuous Improvement. Milwaukee, Wis.: ASQC Press, March 1991.

Nishiyama, K. *Jissen teki Kanri Kaikei no Hoko* ("Direction of Practical Management Accounting"), *Keiei Jitsumu,* February 1983, pp. 20–26.

Nora, John; O. Raymond Rogers; and Robert Stramy. *Transforming the Workplace.* Princeton, N.J.: Princeton Research Press, 1985.

Novi, Tom. "Using the Quality Improvement Process to Improve Your Mail Distribution System." *MAIL: The Journal of Communication Distribution* 3, no. 2 (February/March 1991), pp. 35–43.

"Now Dr. Deming Is Lecturing Automakers." *Industry Week* 210 (August 24, 1981), pp. 28–34.

O'Guin, Michael. "Focus the Factory with Activity-Based Accounting." *Management Accounting,* February 1990, pp. 36–41.

Oakland, John S., and Ric Grayson. "Quality Assurance Education and Training in the U.K." *Quality and Reliability Engineering International* 3 (1987), pp. 169–75.

Oberle, Joseph. "Quality Gurus: The Men and Their Message." *Training: The Magazine of Human Resources Development* 27 (January 1990), pp. 47–56.

Ohno, Taiichi. *Toyota Production System: Beyond Large Scale Production.* Cambridge, Mass.: Productivity Press, 1978.

Ohno, Taiichi. *Workplace Management.* Cambridge, Mass.: Productivity Press, 1982.

Ohno, Taiichi, and Setsuo Mito. *Just-in-Time for Today and Tomorrow.* Cambridge, Mass.: Productivity Press, 1986.

Oneal, M. "Harley-Davidson: Ready to Hit the Road Again." *BusinessWeek* 70 (July 21, 1986).

Osanaiye, P. A., and S. A. Alebiosu. "Effects of Industrial Inspection Errors on Some Plans that Utilize the Surrounding Lot Information." *Journal of Applied Statistics* 15, no. 3 (1988), pp. 295–304.

Ott, Ellis R., and Edward G. Schilling. *Process Quality Control: Troubleshooting and Interpretation of Data.* 2nd ed. New York: Marcel Dekker, 1990.

Ouchi, William G. "A Conceptual Framework for the Design of Organizational Control Mechanisms." *Management Science* 25 (September 1979), pp. 833–848.

Ouchi, William G. *Theory Z: How American Business Can Meet the Japanese Challenge.* Reading, Mass.: Addison-Wesley Publishing, 1982.

Ouchi, William G., and M. A. McGuire. "Organizational Control: Two Functions." *Administrative Quarterly* 20 (December 1975), pp. 559–69.

Overman, Stephenie. "Teamwork Boosts Quality at Wallace." *HR Magazine* 36 (May 1991), pp. 30–35.

Owen, D. B. *Beating Your Competition through Quality.* Milwaukee, Wis.: ASQC Press, 1989.

Owen, Jean V. "Total Quality at Westinghouse." *Manufacturing Engineering*, July 1989, pp. 48–49.

Ozawa, M. *Total Quality Control and Management*. Tokyo: JUSE, 1988.

Ozeki, Kazuo, and Asaka Tesuichi. *Handbook of Quality Tools*. Cambridge, Mass.: Productivity Press, 1990.

Packard, David A. *A Quest for Excellence: Final Report to the President by the President's Blue Ribbon Commission on Defense Management*. Washington, D.C.: U.S. Government Printing Office, June 1986.

Page, E. S. "Continuous Inspection Schemes." *Biometrika* 41 (1954), pp. 100–15.

Parasuraman, A.; V. A. Zeithaml; and L. L. Berry. "A Conceptual Model of Service Quality and Its Implications for Future Research." *Journal of Marketing* 49 (Fall 1985), pp. 41–50.

Parker, Clare O. "IPC Applies SCAT to Chronic Quality Control Problems." *Rubber World* 191 (January 1985), pp. 26–33.

Parker, Mike, and Jane Slaughter. "Management by Stress." *Technology Review*, October 1988, pp. 37–44.

Pascale, R. T., and A. G. Athos. *The Art of Japanese Management*. New York: Penguin Books, 1982.

Paton, Scott M. "Force Field Analysis Proves Money Isn't Everything." *Quality Digest*, April 1989, pp. 21–25.

Patterson, Douglas O. "Saying Is One Thing, Doing Is Another!" *Journal of the Institute of Environmental Sciences*, January/February 1991, pp. 17–20.

Persico, John Jr. "Team up for Quality Improvement." *Quality Progress*, January 1989.

Peter, Lawrence J., and Raymond Hull. *The Peter Principle: Why Things Always Go Wrong*. New York: William Morrow, 1969.

Peters, Thomas J. "It's Time to Get Back to Basics." *Quality*, May 1986, pp. 14–20.

Peters, Thomas J. *Thriving on Chaos: Handbook for Measurement Revolution*. New York: Alfred A. Knopf, 1987.

Peters, Thomas J., and Nancy Austin. *A Passion for Excellence*. New York: Random House, 1985.

Peters, Thomas J., and Robert H. Waterman, Jr. *In Search of Excellence*. New York: Harper & Row, 1982.

Pfau, Loren D. "Total Quality Management Gives Companies a Way to Enhance Position in Global Marketplace." *Industrial Engineering*, April 1989, pp. 17–20.

Phillips, Julien R., and Allan A. Kennedy. *The Leader-Manager*. Ed. John N. Williamson. New York: John Wiley & Sons, 1984, p. 198.

Pierce, Richard J. *Involvement Engineering: Engaging Employees in Quality and Productivity*. Milwaukee, Wis.: ASQC Press, 1986.

Pierce, Richard J. *Leadership, Perspective and Restructuring for Total Quality*. Milwaukee, Wis.: ASQC Press, 1991.

Pines, Ellis. "The Gurus of TQM: After Years of Neglect, TQM Thinkers Now Have an Audience." *Aviation Week & Space Technology* 132 (May 21, 1990), pp. 529–534.

Piotrowski, John L. "The R&M 2000 Initiative Air Force R&M Policy Letters." *IEEE Transactions on Reliability* 36, no. 3 (August 1987), pp. 278–80.

Pittenger, Oliver W. *So You Are the Supervisor.* Milwaukee, Wis.: ASQC Press, 1986.

Pondy, L. R.; R. J. Boland, Jr.; and H. Thomas. *Managing Ambiguity and Change.* New York: John Wiley & Sons, 1988.

Port, Otis. "How to Make It Right the First Time." *BusinessWeek,* June 8, 1987, pp. 74–75.

Port, Otis; Zachary Shiller; Gregory L. Miles; and Amy Schulman. "Smart Factories: America's Turn?" *BusinessWeek,* May 8, 1989, pp. 142–46.

Port, Otis; Resa King; and William J. Hampton. "How the New Math of Productivity Adds Up." *BusinessWeek,* June 6, 1988, pp. 103–13.

Price, Frank. *Right Every Time—Using the Deming Approach.* New York: Marcel Dekker, 1990.

Pritchard, Robert D.; Philip L. Roth; Steven D. Jones; and Patricia Galgay Roth. "Implementing Feedback Systems to Enhance Productivity: A Practical Guide." *National Productivity Review* 10, no. 1 (Winter 1990/91), pp. 57–67.

"Productivity Case Study: An American Miracle That Works." *Productivity* 3, no. 11 (1982), pp. 1–5.

Pryor, Lawrence S. "Benchmarking: A Self-Improvement Strategy." *The Journal of Business Strategy,* November/December 1989, pp. 28–32.

Pyzdek, Thomas. *An SPC Primer.* Milwaukee, Wis.: ASQC Press, 1984.

Pyzdek, Thomas. *Pyzdek's Guide to SPC, Volume I: Fundamentals.* Milwaukee, Wis.: ASQC Press, 1989.

Pyzdek, Thomas. *What Every Engineer Should Know about Quality Control.* New York: Marcel Dekker, 1989.

Pyzdek, Thomas. *What Every Manager Should Know about Quality.* New York: Marcel Dekker, 1991.

QC Circle Headquarters. *How to Operate QC Circle Activities.* Tokyo: JUSE, 1985.

QC Circle Headquarters. "QC Circle Koryo—General Principles of the QC Circle." Tokyo: JUSE, 1980.

"Quality Awards: The OEM Way of Saying Thanks." *Electronic Business,* March 15, 1988.

"Quality Control: An International Concept?, The Evolution of Japanese Management." *Japan Economic Journal,* January 14, 1989.

Quality Function Deployment. Detroit, Mich.: American Supplier Institute, 1989.

Quinn, James Brian, and Christopher Gagnon. "Will Service Follow Manufacturing into Decline?" *Harvard Business Review,* November/December 1986, p. 95.

Ralston, David. "The Benefits of Flextime: Real or Imagined?" *Journal of Organizational Behavior* 10, no. 4 (October 1989), pp. 369–373.

Rao, Vittal. "Total Quality: A Commitment to Excellence." *APICS, Readings in Productivity Improvement,* 1985, pp. 18–23.

Rayner, Bruce C. P. "Avnet, Other Distributors Getting into the Groove." *Electronic Business,* October 16, 1989, pp. 199–202.

Rayner, Bruce C. P. "Education: At the Heart of Quality." *Electronic Business,* October 16, 1989, pp. 222–30.

Rayner, Bruce C. P. "Rockwell Commits to Organizational Excellence." *Electronic Business,* October 16, 1989, pp. 161–64.

Rayner, Bruce C. P. "The Pentagon Revives Its Interest in Quality." *Electronic Business,* October 16, 1989, pp. 149–52.

Raz, Tzvi, and Marlin U. Thomas, eds. *Design of Inspection Systems—Selected Readings.* Milwaukee, Wis.: ASQC Press, 1990.

Recardo, Ronald. "The What, Why and How of Change Management." *Manufacturing Systems,* May 1991, pp. 52–58.

Reidenbach, R. Eric; Troy A. Festervand; and Michael MacWilliam. "Effective Corporate Response to Negative Publicity." *Business* 37, no. 4 (October–December 1987), pp. 9–17.

Reynolds, Edward A. "The Science (Art?) of Quality Audit and Evaluation." *Quality Progress* 23, no. 7 (July 1990), pp. 55–56.

Rice, Valerie. "Financial Analysts Figure the Bottom Line on Quality." *Electronic Business,* October 16, 1989, p. 131.

Rice, Valerie. "Signetics Brings Quality to the Old World." *Electronic Business,* October 16, 1989, pp. 257–58.

Rice, Valerie. "Spreading the Gospel: Quality Is Everybody's Business at TI." *Electronic Business,* October 16, 1989, pp. 121–25.

Rice, Valerie, and John McCreadie. "Credo at Small Companies: Quality Equals Simplicity." *Electronic Business,* October 16, 1989, pp. 136–42.

Richman, Louis S. "The Coming World Labor Shortage." *Fortune,* April 9, 1990, pp. 69–77.

Rickards, Tudor; Simon Aldridge; and Kevin Gaston. "Factors Affecting Brainstorming: Towards the Development of Diagnostic Tools for Assessment of Creative Performance." *R&D Management* 18, no. 4 (October 1988), pp. 309–20.

Riley, Frank. "Don't Curse the Darkness." *Manufacturing Engineering,* February 1990, p. 7.

Ringle, William M. "The American Who Remade 'made in Japan.' " *Nation's Business,* (February 1981), pp. 67–74.

Riordan, John J., and William Cotliar. *How to Develop Your GMP/QA Manual.* 3rd ed. Milwaukee, Wis.: ASQC Press, 1983.

Roberts, Wess. *Leadership Secrets of Attila the Hun.* New York: Warner, 1989.

Robinson, Charles B. "Auditing a Quality System. Part 2: Audit Policy and Protocol." *Quality Progress* 23, no. 2 (February 1990), pp. 54–58.

Robinson, Charles B. *Auditing a Quality System for the Defense Industry.* Milwaukee, Wis.: ASQC Press, 1990.

Robinson, Stanley L., and Richard K. Miller. *Automated Inspection and Quality Assurance*. New York: Marcel Dekker, 1989.

Rodgers, Buck. *Getting the Most Out of Yourself and Others*. New York: Harper & Row; 1987.

Rohan, Thomas M. "Quality Picks Up: Everybody Wants to Get into the Act." *Industry Week* 227 (October 28, 1985), pp. 20–22.

Rohan, Thomas M. "New Crisis in Quality: Programs Are Impaired by Isolation, Teams Lack Corporate Direction, and the Top Boss Doesn't Always Participate." *Industry Week* 239 (October 15, 1990), pp. 11–24.

Rooney, Charles. "Measuring Quality Progress." *American Paint & Coatings Journal* 75, January 21, 1991 pp. 36–48.

Rosander, A. C. *Applications for Quality Control in the Service Industries*. New York: Marcel Dekker, 1985.

Rosander, A. C. *The Quest for Quality in Services*. White Plains, N.Y.: Unipub, 1989.

Rosander, A. C. *Deming's 14 Points Applied to Services*. New York: Marcel Dekker, 1991.

Rose, Frank. "Now Quality Means Service Too." *Fortune,* April 22, 1991, pp. 97–111.

Rosenberg, Jim. "Philosophy and Practice of Quality: International Newspaper Group Hears a New Management Gospel." *Editor & Publisher* 123 (December 1, 1990), pp. 32–43.

Ross, H. Terrance, and Marion M. Ormsby. "Teamwork Breeds Quality at Hearing Technology, Inc." *National Productivity Review* 9, no. 3 (Summer 1990), pp. 321–27.

Ross, Joel E., and David E. Wegman. "Quality Management and the Role of the Accountant." *Industrial Management* 32, no. 4 (July/August 1990), pp. 21–23.

Ross, Philip J. *Taguchi Techniques for Quality Engineering*. New York: McGraw-Hill, 1988.

Rossier, Paul E., and D. Scott Sink. "What's Ahead for Productivity and Quality Improvement?" *Industrial Engineering* 22, no. 3 (March 1990), pp. 25–31.

Rossler, Paul E. "Challenging Mainstream IE Thinking: The Human Element." *Industrial Engineering,* September 1991, pp. 53–58.

Roth, Glenn. "Management, Measurement, and Analysis of the Supplier Base." *Quality Engineering* 1, no. 1 (1988–89), pp. 55–62.

Rout, L. "Hyatt Hotel's Gripe Sessions Help Chief Maintain Communication with Workers." *The Wall Street Journal* (July 16, 1981).

Rubinstein, Sidney P. *Participative Systems at Work: Creating Quality and Employment Security.* Milwaukee, Wis.: ASQC Press, 1987.

Russell, J. P. *The Quality Master Plan*. Milwaukee, Wis.: ASQC Press, 1990.

Ryan, Thomas P. *Statistical Methods for Quality Improvement*. New York: John Wiley & Sons, 1989.

Safizadeb, M. Hossein. "The Case of Work Groups in Manufacturing Operations." *California Management Review* 33, no. 6 (Summer 1991), pp. 61–82.

Salm, James L. "Examining the Costs of Quality." *Manufacturing Systems*, April 1991, pp. 48–50.

Sammons, Donna. "Driving a Hard Bargain." *Carpenter* 7 (May 1985), pp. 165–73.

Samuels, Mike, and Nancy Samuels. *Seeing with the Mind's Eye*. New York: Random House, 1975.

Sandholm, L. "Management Training—A Prerequisite of TQC." *EOQC* 23, no. 4 (December 1989), pp. 5–10.

Saraph, Jayant V.; P. George Benson; and Roger G. Schroeder. "An Instrument for Measuring the Critical Factors of Quality Management." *Decision Sciences* 20, no. 4 (Fall 1989), pp. 810–29.

Sarazen, J. Stephen. "Schools Build in Quality." *Quality Progress* 22, no. 1 (January 1989), pp. 38–41.

Sarazen, J. Stephen. "The Tools of Quality—Part II; Cause-and-Effect Diagrams." *Quality Progress* 23, no. 7 (July 1990), pp. 59–62.

Sashkin, Marshall. "A Theory of Organizational Leadership: Vision, Culture and Charisma." *Proceedings of Symposium on Charismatic Leadership in Management*. Montreal, Que.: McGill University, 1987.

Sayle, Allan J. *Management Audits: The Assessment of Quality Management Systems*. 2nd ed. Milwaukee, Wis.: ASQC Press, 1988.

Scharf, Alan D. "How to Use Pareto's Law." *Industrial Business Management*. Saskatchewan Research Council-Industrial Services, July 1973.

Scharf, Alan D. "Pareto's Law." *Industrial Business Management*. Saskatchewan Research Council-Industrial Services, May 1973.

Schein, Edgar. *Organizational Culture and Leadership*. San Francisco: Jossey-Bass, 1985.

Scherkenbach, William. *The Deming Route to Quality and Productivity: Road Maps and Roadblocks*. Rockville, Md.: Mercury Press, 1986.

Schilling, Edward G. *Acceptance Sampling in Quality Control*. New York: Marcel Dekker, 1982.

Schlenker, Emily C. *An Organizational Stress Inventory*. Rock Island, Ill.: 1991.

Schmidt, Michael S., and Larry C. Meile. "Taguchi Designs and Linear Programming Speed New Product Formulation." *Interfaces* 19, no. 5 (September/October 1989), pp. 49–56.

Schmidt, Stephen R., and Robert G. Launsby. *Understanding Industrial Designed Experiments*. Milwaukee, Wis.: ASQC Press, 1988.

Scholtes, Peter R. *An Elaboration of Deming's Teachings on Performance Appraisal*. Madison, Wis.: Joiner Associates, 1987.

Scholtes, Peter R. *The Team Handbook*. Madison, Wis.: Joiner Associates, 1988.

Scholtes, Peter R., and Heero Hacquebord. "Beginning the Quality Transformation, Part I; and Six Strategies for Beginning the Quality Transformation, Part II." *Quality Progress*, July/August 1988.

Schonberger, Richard J. *Building a Chain of Customers—Linking Business Functions to Create the World Class Company*. New York: Free Press, 1990.

Schonberger, Richard J. *Japanese Manufacturing Techniques: Nine Hidden Lessons in Simplicity*. New York: Free Press, 1982.

Schonberger, Richard J. *World Class Manufacturing—The Lessons of Simplicity Applied*. New York: Free Press, 1987.

Schonberger, Richard J. *World Class Manufacturing Casebook: Implementing JIT and TQC*. New York: Free Press, 1987.

Schonberger, Richard J., and James P. Gilbert. "Just-in-Time Purchasing: A Challenge for U.S. Industry." *California Management Review*, Fall 1983.

Schrader, Lawrence J. "An Engineering Organization's Cost of Quality Program." *Quality Progress*, January 1986, pp. 29–33.

Schrantz, Joe. "IQ Is for Real at Glidden." *Industrial Finishing* 66 (December 1990), pp. 12–24.

Schrock, Edward M., and Henry L. Lefevre. *The Good and the Bad News about Quality*. New York: Marcel Dekker, 1988.

Schultz, Louis E., and Darrell R. Schroeder. "Pathway to Continuous Process Improvement." *Process Management Institute*, pp. 1–11.

Schultz, Louis E. *The Role of Top Management in Effecting Change to Improve Quality and Productivity*. Minneapolis, Minn.: Process Management Institute, 1985, pp. 1–8.

Schultz, Louis E. *Overview of Quality Management Philosophies*. Minneapolis, Minn.: Process Management Institute, 1986.

Schultz, Louis E. *Pathway to Continuous Improvement*. Bloomington, Minn.: Process Management Institute.

Schultz, Louis E. "Creating a Vision for Strategy and Quality: A Way to Help Management Assume Leadership." *Concepts in Quality Proceedings*. November 1988.

Schuster, Michael. "Gain Sharing: Do It Right the First Time." *Sloan Management Review* 28, no. 2 (Winter 1987), pp. 17–25.

Schwartz, Karen D. "DOD Hopes Quality Control Can Save Time, Money." *Government Computer News*, January 8, 1990, p. 40.

Schwarz, Robert A. *Midland City: Recovering Prosperity through Quality*. Milwaukee, Wis.: ASQC Press, 1989.

Sease, D. R. "How U.S. Companies Devise Ways to Meet Challenge from Japan." *The Wall Street Journal*, September 16, 1986, pp. 1, 23.

Seemer, Robert H. "Keeping in Step with the Environment: Applying TQC to Energy Supply." *National Productivity Review* 9, no. 4 (Autumn 1990), pp. 439–55.

Sellers, Patricia. "Getting Customers to Love You." *Fortune*, March 13, 1989.

Selye, Hans. *Stress without Distress*. Philadelphia: J. B. Lippincott, 1974.

Sepehri, Mehran, ed. *Quest for Quality: Managing the Total System*. Atlanta, Ga.: IIE Press, 1991.

Shah, Syed, and George Woelki. "Aerospace Industry Finds TQM Essential for TQS." *Quality*, March 1991, pp. 14–24.

Shainin, Dorian, and Peter D. Shainin. "The Issues in Quality—The Next 10 Years." *Quality*, Anniversary Issue, 1987.

Shapiro, Benson. "Manage Customers for Profits (Not Just for Sales)." *Harvard Business Review*. Boston: September/October 1987.

Shapiro, Benson P. "What the Hell Is Market Oriented?" *Harvard Business Review*, November/December 1988.

Shaw, M. E. *Group Dynamics: The Psychology of Small Group Behavior*. New York: McGraw-Hill, 1976.

Sheehy, Barry. "Hitting the Wall: How to Survive Your Quality Program's First Crisis." *National Productivity Review* 9, no. 3 (Summer 1990), pp. 329–35.

Sherwin, D. S. "Management of Objectives." *Harvard Business Review* 54 (May/June 1976), pp. 149–60.

Shettel-Neuber, J., and J. P. Sheposh. *Case Study of a Quality Management Effort at the Naval Air Rework Facility*. (NPRDC Tech Note 72-86-09). San Diego: Navy Personnel Research and Development Center, 1986.

Shettel-Neuber, J., and J. P. Sheposh. *Management Methods for Quality Improvement Based on Statistical Process Analysis and Control: A Literature and Field Survey*. (NPRDC Tech Note 86-21). San Diego Navy Personnel Research and Development Center, 1986.

Shewhart, Walter A. *Economic Control of Quality of Manufactured Product*. Princeton, N.J.: Van Nostrand Reinhold, 1931.

Shewhart, Walter A. *Economic Control of Manufactured Product*. Milwaukee, Wis.: ASQC Press, 1980. (Reprint.)

Shimoyamada, Kaoru. "The President's Audit: QC Audits at Komatsu." *Quality Progress*, January 1987, pp. 44–49.

Shingo, Shigeo. *A Revolution in Manufacturing: The SMED System*. Stamford, Conn.: Productivity Press, 1985.

Shingo, Shigeo. *Zero Quality Control: Source Inspection and the Poka-Yoke System*. Cambridge, Mass.: Productivity Press, 1986.

Shingo, Shigeo. *Study of Toyota Production System from an Industrial Engineering Viewpoint*. Cambridge, Mass.: Productivity Press, 1987.

Shingo, Shigeo. *Non-Stock Production: The Shingo System for Continuous Improvement*. Cambridge, Mass.: Productivity Press, 1988.

Shingo, Shigeo. *Poka-Yoke: Improving Product Quality by Preventing Defects*. Cambridge, Mass.: Productivity Press, 1989.

Shingo, Shigeo. *Sayings of Shigeo Shingo: Key Strategies for Plant Improvement*. Cambridge, Mass.: Productivity Press, 1988.

Shinohara, Isao, ed. *New Production System: JIT—Crossing Industry Boundaries*. Cambridge, Mass.: Productivity Press, 1988.

Shores, Dick. "TQC: Science, Not Witchcraft." *Quality Progress* 22, no. 4 (April 1989), pp. 42–45.

Shores, Richard A. *Survival of the Fittest*. Milwaukee, Wis.: ASQC Press, 1986.

Singer, A. J.; G. F. Churchill; and B. G. Dale. "Some Supplier Quality Assurance Assessment Issues." *Quality and Reliability Engineering International* 5, no. 5 (April–June 1989), pp. 101–11.

Single Level Continuous Sampling Procedures for Inspection by Attributes, Inspection and Quality Control Handbook (Interim) H107. Superintendent of Documents. Washington, D.C.: U.S. Government Printing Office, 1959.

Sink, D. Scott, and Thomas C. Tuttle. *Planning and Measurement in Your Organization of the Future*. Atlanta, Ga.: IIE Press, 1991.

Sirgy, M. Joseph. "Can Business and Government Help Enhance the Quality of Life of Workers and Consumers?" *Journal of Business Research* 22 (1991), pp. 327–33.

Slass, Robert. "Turning Deming's Points into Action." *Industry Week* 236 (June 20, 1988), pp. 14–21.

Sloan, Alfred P., Jr. *My Years with General Motors*. New York: Doubleday. (Reprint 1991.)

Sloan, David, and Scott Weiss. *Supplier Improvement Process Handbook*. Milwaukee, Wis.: ASQC Press, 1987.

"Smart Design: Quality Is the New Style." *BusinessWeek,* April 11, 1988, pp. 102–17.

Smith, G. F. "Managerial Problem Identification." *OMEGA* 17, no. 1 (1989), pp. 27–36.

Smith, Jason. *Learning Curve for Cost Control*. Atlanta, Ga.: IIE Press, 1991.

Smith, Kevin. "Ten Best Friends of the Automobile." *Car and Driver* 36 (January 1991), pp. 96–105.

Smith, M. T. "Air Force Systems Command Approach to R&M." *IEEE Transactions on Reliability* 36, no. 3 (August 1987), pp. 291–94.

Smith, Martin. *Maxims of Management*. Piscatway, N.J.: New Century, 1986.

Smock, Doug. "What Really Is the Cost of Quality?" *Plastics World,* August 1989, p. 11.

Snee, Ronald D. "Graphical Analysis of Process Variation Studies." *Journal of Quality Technology* 15, no. 2 (April 1983), pp. 76–88.

Snee, Ronald D. "Statistical Thinking and Its Contribution to Total Quality." *The American Statistician,* May 1990.

Snee, Ronald D.; Lynne B. Hare; and J. Richard Trout. *Experiments in Industry: Design, Analysis, and Interpretation of Results*. Milwaukee, Wis.: ASQC Press, 1985.

Snodgrass, Thomas J., and Muthiah Kase. *Function Analysis*. Madison, Wis.: University of Wisconsin Press, 1986.

Sontag, Harvey. *Corporate Perceptions: A Quality Primer.* Milwaukee, Wis.: ASQC Press, 1989.

Sorensen, S.; S. L. Dockstader; and M. J. Molof. *Developing a Statistical Process Control for Supply Operations*. (NPRDC Tech Rep 86-16). Houston, Tex.: San Diego Navy Personnel Research & Development Center, 1986.

Spechler, Jay W. ed. and co-author. *When America Does It Right: Case Studies in Service Quality*. Atlanta, Ga.: IIE Press, 1991.

Squeglia, Nicholas L. *Zero Acceptance Number (C=10) Sampling Plans,* 3rd ed. Milwaukee, Wis.: ASQC Press, 1986.

Squires, Frank H. *Successful Quality Management*. Wheaton, Ill.: Hitchcock Publishing, 1980.

Squires, Frank H. "Skunk Works." *Quality,* November 1987, p. 64.

Squires, Frank H. "Who Is Responsible for Quality?," *Quality,* December 1987, p. 73.

Stahl, Michael J., and Gregory M. Bounds. *Competing Globally through Customer Value*. New York: Quorum Books, 1991.

Starr, M. *Production Management—Systems and Synthesis*. Englewood Cliffs, N.J.: Prentice Hall, 1972.

Statistical Procedures for Determining Validity of Suppliers' Attributes Inspection, Quality Control and Reliability Handbook (Interim) H109, Superintendent of Documents, Washington, D.C.: U.S. Government Printing Office, 1960.

Staveley, J. C. and B. G. Dale. "Some Factors to Consider in Developing a Quality-Related Feedback System." *Quality and Reliability Engineering International* 3, no. 4 (1987), pp. 265–71.

Stebbing, Lionel. *Quality Assurance: The Route to Efficiency and Competitiveness*. New York: Halstead Press, 1986.

Stephens, Kenneth S. *How to Perform Continuous Sampling (CSP)*. vol. 2. Milwaukee, Wis.: ASQC Press, 1979.

Stephens, Kenneth S. *How to Perform Skip-Lot and Chain Sampling,* vol. 4. Milwaukee, Wis.: ASQC Press, 1982.

Stephenson, Stan. "Are You Up to Delivering Quality, Deming style?" *Motor Age* 103 (April 1984), pp. 5–11.

Stitt, John. *Managing for Excellence*. Milwaukee, Wis.: ASQC Press, 1990.

Stogdill, Ralph. *Handbook of Leadership: A Survey of Theory and Research*. New York: Free Press, 1974.

Stowell, Daniel M. "Quality in the Marketing Process." *Quality Progress* 22, no. 10 (October 1989), pp. 57–62.

Stratton, A. Donald. "Kaizen and Variability." *Quality Progress* 23, no. 4 (April 1990), pp. 44–46.

Stratton, A. Donald. *An Approach to Quality Improvement That Works*. Milwaukee, Wis.: ASQC Press, 1991.

Stratton, Brad. "Payment in Kind." *Quality Progress,* April 1989.

Strickland, Jack. "Total Quality Management." *Army Research, Development & Acquisition Bulletin,* March/April 1988, pp. 1–4.

Strickland, Jack. "Key Ingredients to Total Quality Management." *Defense,* March/April 1989, pp. 17–21.

Strickland, Jack. "Total Quality Management: Linking Together People and Processes for Mission Excellence." Army Research, *Development & Acquisition Bulletin,* May/June 1989, pp. 9–12.

Strickland, Jack, and Peter Angiola. *Total Quality Management in the Department of Defense*. Washington, D.C.: U.S. Government Printing Office, 1989.

Stuelpnagel, Thomas R. "Total Quality Management." *National Defense,* November 1988, pp. 57–62.

Sullivan, Edward. "OPTIM: Linking Cost, Time and Quality." *Quality Progress,* April 1986.

Sullivan, Lawrence P. "Quality Function Deployment." *Quality Progress,* June 1986, pp. 39–50.

Sullivan, Lawrence P. "Policy Management through Quality Function Deployment." *Quality Progress* 21, no. 6 (June 1988), pp. 18–20.

Sullivan, Lawrence P. "The Power of Taguchi Methods." *Quality Progress,* June 1987, pp. 76–79.

Sumanth, David J.; Johnson Edosomwan; D. Scott Sink; and William B. Werther. eds. *Productivity Management Frontiers,* vol. III. Atlanta, Ga.: IIE Press, 1991.

Sun-Tzu. *The Art of Strategy.* trans. R. L. Wing. New York: Dolphin/Doubleday, 1988.

Sun-Tzu. *The Art of War.* trans. Samuel B. Griffith. London: Oxford University Press, 1963.

Suzaki, Kiyoshi. "Work-in-Process Management: An Illustrated Guide to Productivity Management." *Production and Inventory Management,* Third Quarter 1985.

Suzaki, Kiyoshi. *The New Manufacturing Challenge: Techniques for Continuous Improvement*. New York: Free Press, 1987.

Suzuki, N. "Japanese Management in the U.S.: Its Facts and Fallacies." Mimeograph, 1983.

Suzuki, N. "Japanese Catch-up Effort of Market Research Skills with U.S.: But for Whom?" *Proceedings on Research Developments in International Marketing,* University of Manchester Institute of Science and Technology, Manchester, England, 1984.

Swift, Jill A., and Timothy J. Flynn. "Methodology for Developing a Quality Plan within a Manufacturing Company." *Quality Engineering* 1, no. 4 (1989), pp. 467–86.

Taguchi, Genichi. *Introduction to Quality Engineering*. Dearborn, Mich.: American Supplier Institute, 1986.

Taguchi, Genichi. *System of Experimental Design*. Detroit, Mich.: American Supplier Institute, 1988.

Taguchi, Genichi, and Don Clausing. "Robust Quality." *Harvard Business Review,* January/February 1990, pp. 65–75.

Taguchi, Genichi; Don Clausing; Connie Dyer; and A. Lance Ealey. "Robust Quality; Design Products Not to Fall in the Field." *Harvard Business Review,* January/February 1990, pp. 65–76.

Taguchi, Genichi, and Yu-In Wu. *Introduction to Off-Line Quality Control Systems*. Central Quality Control Association, 1980.

Talley, Dorsey J. *Management Audits for Excellence*. Milwaukee, Wis.: ASQC Press, 1988.

Talley, Dorsey J. *Total Quality Management: Performance and Cost Measures*. Milwaukee, Wis.: ASQC Press, 1991.

Tannenbaum, Robert, and Warren H. Schmidt. "How to Choose a Leadership Pattern." *Harvard Business Review*. Classic, May/June 1973, pp. 162–80.

Tapiero, Charles S. "Production Learning and Quality Control." *IIE Transactions* 19, no. 4 (December 1987), pp. 362–69.

Taylor, Alex III. "Why Toyota Keeps Getting Better and Better and Better." *Fortune*, November 19, 1990, pp. 66–79.

Taylor, Frederick W. *The Principle of Scientific Management*. New York: Harper & Row, 1911.

Taylor, James R. *Quality Control Systems: Procedures for Planning Quality Programs*. New York: McGraw-Hill, 1989.

"Ten Molders of Modern Japan." *Scholastic Update* 116 (November 11, 1983), pp. 32–42.

Teresko, John. "Quality." *Industry Week* 215 (October 16, 1982), pp. 54–64.

Thamhain, H.J., and D. L. Wilemon. "The Effective Management of Conflict in Project-Oriented Work Environments." *Defense Management Journal* 11, no. 3 (July 1971), p. 975.

"The New American Corporation," *Information Week*, January 15, 1990, pp. 36–38.

"The Push for Quality." *BusinessWeek*, June 8, 1987, pp. 130–44.

"The Quality Revolution: Internally First, Externally Second." *Vital Speeches* (August 1, 1990), pp. 625–34.

"The Search for Quality: An Endless Marathon." *Design News* 47 (March 25, 1991), pp. 396–401.

"The Tools of Quality, Part V: Check Sheets." *Quality Progress*, October 1990, pp. 51–56.

Thompson, P.; G. DeSouza: and B. T. Gale. "The Strategic Management of Service Quality." *PIMSLETTER, No. 33*. Cambridge, Mass.: Strategic Planning Institute, 1985.

Thompson, P. *Quality Circles: How to Make Them Work in America*. New York: American Management Association (AMACOM), 1983.

Thor, Carl G. "Getting the Most from Productivity Statistics." *National Productivity Review* 9, no. 4 (Autumn 1990), pp. 457–66.

Tichy, Noel M., and Mary Anne Devanna. *The Transformational Leader*. New York: John Wiley & Sons, 1986.

Tomlinson, William H.; Steven K. Paulson; Douglas H. Briggs; and Joji Arai. "Company Identity, Quality Improvement, and Labor Management Relations in Danish,

Italian, Japanese, Scottish, and U.S. Firms.'' *National Productivity Review* 10 (Spring 1991), pp. 129–216.

Tompkins, James A. *Winning Manufacturing: The How-To-Book of Successful Manufacturing*. Raleigh, N.C.: Tompkins Associates, 1991.

''Total Quality Management: A New Look at a Basic Issue.'' *Vital Speeches* 57 (April 15, 1991), pp. 415–22.

Total Quality Management: Selected Readings and Resources. San Diego: Navy Personnel Research and Development Center, 1986.

Townsend, Patrick L., and Joan A. Gebhardt. *Commit to Quality*. New York: John Wiley & Sons, 1990.

Trachtenberg, Jeffrey A. ''How Do We Confuse Thee? Let Us Count the Ways.'' *Forbes*, March 21, 1988, pp. 156–60.

Treleven, M. ''Single Sourcing: A Management Tool for the Quality Supplier.'' *Journal of Purchasing and Materials Management*, 1987, pp. 19–24.

Trepo, Georges X. ''Introduction and Diffusion of Management Tools: The Example of Quality Circles and Total Quality Control.'' *European Management Journal* 5, no. 4 (Winter 1987), pp. 287–293.

Trevor, Malcolm. *The Japanese Management Development System*. Wolfeboro, N.H.: Frances Pinter Ltd., 1986, pp. 5–6.

Tribus, M. *Deming's Redefinition of Management*. Cambridge, Mass.: Center for Advanced Engineering Study, MIT Press, 1985.

Tribus, M. *Managing to Survive in a Competitive World*. Cambridge, Mass.: Center for Advanced Engineering Study, MIT Press, 1983.

Tribus, M. *Reducing Deming's 14 Points to Practice*. Cambridge, Mass.: Center for Advanced Engineering Study, MIT Press, 1984.

Tribus, M. ''Quality First: Selected Papers on Quality and Productivity Improvement.'' Washington, D.C.: National Society of Professional Engineers, March 1988.

Tribus, Myron and Geza Szonyi. ''An Alternative View of the Taguchi Approach.'' *Quality Progress* 22, no. 5 (May 1989), pp. 46–52.

Tribus, M., and Y. Tsuda. *Creating the Quality Company*. Cambridge, Mass.: Center for Advanced Engineering Study, MIT Press, 1983.

Trice, Harrison M., and Janice Beyer. ''Cultural Leadership in Organizations.'' *Organization Science* 2, no. 2 (1991), pp. 149–69.

Tsuda, M. *Jinji Kanri no Gendaiteki Kadai* (''Contemporary Problems of Japanese Human Resource Management''). Tokyo: Zeimu Keiri Kyokai, 1981.

Tsuda, M. *Nihonteki Keiei no Ronri* (''Logic of Japanese Management''). Tokyo: Chuo Keizai Co. 1977.

Tucker, Frances Gaither; Seymour Zivan; and Robert Camp. ''How to Measure Yourself against the Best.'' *Harvard Business Review*. January/February 1987.

Tunner, Joseph R. ''A Quality Technology Primer for Managers.'' Milwaukee, Wis.: ASQC Press, 1990.

"U.S. Electronics in Japan? Not with That Quality." *Electronic Business,* April 17, 1989.

U.S. Department of Commerce. *Application Guidelines for the Malcom Baldrige National Quality Award.* Milwaukee, Wis.: Baldrige Award Consortium, ANNUA.

U.S. Department of Defense. *TQM Implementation Guide,* Volumes I and II. February 1990.

U.S. Government Accounting Office. NSIAD91-190 as reported in *On Q,* September 1991, pp. 24 and 416.

Uttal, B. "The Corporate Culture Vultures." *Fortune* 108, no. 8 (October 17, 1983), pp. 66–72.

Vanisina, Leopold S. "Total Quality Control: An Overall Organizational Improvement Strategy." *National Productivity Review* 9, no. 1 (Winter 1989/90), pp. 59–73.

Vasilash, Gary S. "Hearing the Voide of the Customer." *Production,* February 1989.

Vendor-Vendee Technical Committee. *Procurement Quality Control.* Milwaukee, Wis.: ASQC Press, 1985.

Vetschera, R. "Group Decision and Negotiation Support—A Methodological Survey." *OR Spektrum* 12, no. 2, 1990, pp. 67–77.

Vonderembse, Mark A., and Gregory P. White. *Operations Management.* St. Paul, Minn.: 1991.

Von Oech, Roger. *A Whack on the Side of the Head: How to Unlock Your Mind for Innovation.* New York: Warner Books, 1983.

Vroom, V. *Work and Motivation.* New York: John Wiley & Sons, 1964.

"W. Edwards Deming: Shogun of Quality Control." *FE: The Magazine for Financial Executives* 2 (February 1986), pp. 24–28.

Wadsworth, Harrison M.; Kenneth S. Stephens; and A. Blanton Godfrey. *Modern Methods for Quality Control and Improvement.* New York: John Wiley & Sons, 1986.

Wadsworth, Stephens, and Blan Godfrey. *Modern Methods for Quality Control and Improvement.* New York: John Wiley & Sons, 1986.

Waite, Charles L., Jr. "Timing Is Everything." *Quality Progress,* April 1989.

Wall, T. D.; N. J. Kemp; P. R. Jackson; and C. W. Clegg. "Outcomes of Autonomous Work Groups: A Long-Term Field Experiment." *Academy of Management Journal* 29, no. 2 (1986), pp. 286–304.

Walleigh, Richard C. "What's Your Excuse for Not Using JIT?" *Harvard Business Review,* March/April 1986.

Walsh, Loren M.; Ralph Wurster; and Raymond J. Kimber. eds. *Quality Management Handbook.* New York: Marcel Dekker, 1986.

Walton, Mary. "The Deming Management Method." *The New York Times Book Review,* October 26, 1986, p. 38.

Walton, Mary. *The Deming Management Method.* New York: Putnam, 1986.

Walton, Mary. *The Deming Management Method.* New York: Dodd, Mead & Company, 1986.

Walton, Mary. *Deming Management at Work.* New York: Putnam, 1990.

Walton, Mary. "Deming's Parable of the Red Beads." *The New York Times Book Review,* October 26, 1986, p. 38.

Walton, R. E. "Work Innovations in the United States." *Harvard Business Review,* July/August 1979, pp. 88–98.

Waterman, Robert H. *The Renewal Factor.* New York: E. P. Putnam, 1987.

Weaver, Charles N. *TQM: A Step-by-Step Guide to Implementation.* Milwaukee, Wis.: ASQC Press, 1991.

Webb, Janette, and Patrick Dawson. "Measure for Measure: Strategic Change in an Electronic Instruments Corporation." *Journal of Management Studies* 28, no. 2 (March 1991), pp. 191–206.

Webster, Cynthia. "Toward the Measurement of the Marketing Culture of a Service Firm." *Journal of Business Research* 21, no. 4 (December 1990), pp. 345–362.

Weisberg, R. W. *Creativity: Genius and Other Myths.* New York: Freeman, 1986.

Wellins, Richard S., et al. *Empowered Teams.* San Francisco: Jossey-Bass, 1991.

West, Michael A., and James L. Farr. *Innovation and Creativity at Work.* New York: John Wiley & Sons, 1990.

Western Electric Company. *Statistical Quality Control Handbook.* Milwaukee, Wis.: ASQC Press, 1982.

Westland, Cynthia Lane. *Quality: The Myth and the Magic.* Milwaukee, Wis.: ASQC Press, 1990.

Westland, J. Christopher. "Assessing the Economic Benefits of Information Systems Auditing." *Information Systems Research* 1, no. 3 (September 1990), pp. 309–24.

Westley, Frances, and Henry Mintzburg. "Visionary Leadership and Strategic Management." *Strategic Management Journal* 10, no. 2 (1989), pp. 17–32.

Wetherill, G. Barrie. *Sampling Inspection and Quality Control.* Milwaukee, Wis.: ASQC Press, 1982.

"What Makes Sammy Run?" *OMEGA* 18, no. 4 (1990), pp. 339–53.

Wheeler, Donald J., and David S. Chambers. *Understanding Statistical Process Control.* Knoxville, Tenn.: Statistical Process Controls, 1986.

Whiteside, D.; R. Brandt; Z. Schiller; and A. Gabor. "How GM's Saturn Could Run Rings around Old Style Car Makers." *BusinessWeek,* January 28, 1985, pp. 126–28.

Whiting, Rick. "At Compaq, Reaching the Top Begins at the Bottom." *Electronic Business,* October 16, 1989, pp. 191–92.

Willborn, Walter. *Quality Management System: A Planning and Auditing Guide.* New York: Industrial Press, 1989.

Willborn, Walter. "Registration of Quality Programs." *Quality Progress* 21, no. 9 (September 1988), pp. 56–58.

Willborn, Walter, and the ASQC Quality Audit Technical Committee. *Audit Standards: A Comparative Analysis.* Milwaukee, Wis.: ASQC Press, 1987.

Willborn, Walter O., and Madhav N. Sinha. *The Management of Quality Assurance.* New York: John Wiley & Sons, 1985.

Willoughby, W. J. *Best Practices: How to Avoid Surprises in the World's Most Complicated Technical Environment*. Department of the Navy. Washington, D.C.: U.S. March 1986.

Winchell, William. *Continuous Quality Improvement: A Manufacturing Professional's Guide*. Dearborn, Mich.: Society of Manufacturing Engineers, 1991.

Winner, Robert I.; James P. Pennell; Harold E. Bertrand; and Marko M. G. Slusarczuk. "The Role of Concurrent Engineering in Weapons System Acquisition." Alexandria, Va.: IDA, 1988.

Withey, Michael J., and William H. Cooper. "Predicting Exit, Voice, Loyalty and Neglect." *Administrative Science Quarterly* 34 (1989), pp. 521–39.

Wolak, Jerry. "An Interview with Dr. A. Gunneson." *Quality,* May 1987, pp. 50–54.

Wollschlager, Lester Jay. *The Quality Promise*. New York: Marcel Dekker, 1990.

Wood, Lamont. "Rapid Prototyping—Uphill, but Moving." *Manufacturing Systems,* December 1990, pp. 14–18.

Wood, Robert Chapman. "The Prophets of Quality." *The Quality Review,* Winter 1988, pp. 18–25.

Xerox Corporation. "Baldrige Quality Award Winner." *National Productivity Report* 18, no. 24 (December 31, 1989), pp. 1–4.

Yamashiro, A. *Gendai no Keieirinen* ("Contemporary Management Philosophy"). Tokyo: Hakuto Shobo, 1972.

Yoshida, Kosaku. "Deming Management Philosophy: Does It Work in the U.S. as Well as in Japan?" *Columbia Journal of World Business* 24 (Fall 1989), pp. 10–18.

Zaleznik, A. "Managers and Leaders: Are They Different?" *Harvard Business Review* 55, no. 3 (May/June 1977), pp. 67–78.

Zaleznik, A. *The Managerial Mystique*. New York: Harper & Row, 1989.

Zaremba, Alan. *Management in a New Key: Communication in the Modern Organization*. Atlanta, Ga.: IIE Press, 1991.

Zeithaml, Valarie, A.; L. L. Berry; and A. Parasuraman. "Communication and Control Processes in the Delivery of Service Quality." *Journal of Marketing* 52 (April 1988), pp. 35–48.

Zeithaml, Valarie; A. Parasuraman; and Leonard L. Berry. *Delivering Quality Service—Balancing Customer Perceptions and Expectations*. New York: Free Press, 1990.

Ziemke, M. Carl, and Mary S. Spann. "Warning: Don't Be Half-hearted in Your Efforts to Employ Concurrent Engineering." *IE* February 1991, pp. 45–49.

Zurn, James T. "New Product Introduction and Quality Program Management." *Quality Engineering* 1, no. 1 (1988–89), pp. 29–43.

Zygmont, J. "Flexible Manufacturing Systems: Curing the Cure-all." *High Technology,* October 1966, pp. 23–24.

APPENDIX A:

WHEN ALL ELSE FAILS . . . 75 QUICK START IDEAS

1. Have a blue-collar worker visit a customer. Let them be the factory floor rep.
2. Play with your product. Try to use as though it were the first time you have seen it.
3. What does your kid (or other family member) think of your product?
4. Take a customer to lunch.
5. Visit where your product is displayed or used.
6. Take a vacation. Forget to tell anyone exactly where you're going. Leave anything work related behind.
7. Have a one-on-one talk with the company clown. Ask only those questions needed to prompt conversation. Listen to him/her. Take notes.
8. Always carry a small notepad and pen.
9. Ask individuals what they would like as a reward. Have a tiger team session assess what rewards would be appropriate and appreciated.
10. Memorize spouse and children names (and ages) of your division—even if you're not yet the division manager. If you want a challenge, remember their birthdays and send them all a card.
11. Require everyone to submit a one-half page (or use 3x5 cards) per week or a two-week summary of what they have done in their department. On another half page or card have them identify problems or roadblocks that management should be aware of.
12. Have an improvement party once a quarter, or year. *Everyone* in the company must submit ideas for improvement. Implement as many of them as you can. If they fail, go back to the *old* way.
13. Rotate job positions for one week. Have the stock boy run the register and meet the public. Have the cashier stock the shelves or run a forklift (yes, this may require training). Do this on a regular basis.

14. Remove all reserved parking spaces, except for the handicapped and the mail courier. This means *your* parking space, too.
15. Create a new job position, at least temporarily. Have someone greet people at the door in a retail establishment. Create a guru to whom people can go in the office for advice on computer problems.
16. Throw away personnel records older than three years. All records. Forgive your employees. Do you *really* still care that they dropped the ball once back in 1973?
17. Eliminate the performance review system altogether. Who needs it? The employees know how they are doing, and if they don't, a performance review will shock them into poorer performance, not better. They only serve to perpetuate us-them mentalities of the 19th century.
18. Constantly ask stupid questions: Why do I have to get up to change the channel? Why do I have to dry the clothes on a clothesline? Is there anything faster than a microwave?
19. Stop any meetings that occur on a regular basis. Why don't you know what's going on without them?
20. Get out of your office at least 50 percent of the time: On the shop floor, on the sales floor, in among the cubicles, or at civic functions.
21. Eliminate your office altogether. If the paperwork won't fit in a briefcase, delegate it.
22. Take a visit (preferably with children—they ask the best questions) to an offbeat (relative to your business) place such as police station, bread factory, publisher, archaeological dig, or whatever appeals to you.
23. Don't take "Get it right the first time" too far. Failures *are* OK, valuable even. How else can we learn except by making mistakes and correcting them. Television, radio, and the telephone of just 15 years ago are abysmal concepts compared to today's products. Constantly *improve*.
24. Steal from anyone and anything you can. Not physical things—ideas! Scholars do this regularly with great benefit; they examine another field altogether for ideas that can be adapted from other fields and then be "innovative." It's not unethical—it's being smart.
25. Subscribe to an unusual (but easy to read) magazine—whatever is interesting but different from your present hobby. You don't even have to like the subject.
26. Develop a hobby that you enjoy and spend several hours per week on it. If it requires traveling on occasion, so much the better.
27. Listen to your workers, especially the frustrated new worker. They are gold mines of ideas and your company's future. Ignore youth at extreme corporate peril.
28. Listen to people who've seen it all such as World War II, the Depression, and the roaring twenties. The younger you are the more beneficial

it is. Imagine being 60 in the year 2030 but having heard first person accounts of what daily life was like in 1919?

29. Spend a day or afternoon once a week solely devoted to your children. Or to someone else's.
30. Post financial information on an employee bulletin board or in a newsletter. They deserve to have some idea of the financial health of the company.
31. Eliminate the hierarchy. Form a self-managing team. Start with only one if the idea scares you, but start.
32. Ride a bike to work. No, it is not ridiculous, ladies (and men). Perhaps our work attire is, and our planet destroying machines.
33. Install a bike rack next to the door. Buy the best. Better still, have a locked shed for the bicycles. Give bicyclists (and walkers) a nominal bonus ($1 per week for every week they ride their bike or walk everyday).
34. Hand out rewards for the most number of miles run/cycled/walked per year. Recognize everyone who achieves an average of so many per day. Reward those who are most improved, lost the most weight, by pounds and by percentage.
35. Start each day with a 20-minute session of Tai Chi, yoga, or some other slow-moving stretching exercise—for *all* employees.
36. Read a religious text. Not the one you grew up with or currently practice, unless you haven't read it in a long time.
37. Learn a language. Learn a dead one, such as Old Norse or Anglo-Saxon. If this doesn't have appeal then read what's available in translations from a specific period and place of literature.
38. Tulip bulbs need a rest every five years. Why should we be any different? Take a sabbatical and replant yourself when you get back. Give other employees sabbaticals. They don't have to be a year long—try three or four months. Allow absolutely no correspondence or phone calls. Not even insurance forms.
39. Go to somebody else's trade association meetings. If you're an engineer, go to an accounting function. Send the accountant to a meeting of the American Production and Inventory Society. Send all of them to the seminar on knowledge-based systems.
40. Train anyone in anything. Give everyone a personal not to exceed training budget. Disallow no training, no matter how unrelated—even parachute lessons for your safety engineer. Set guidelines on how many hours of work time can be used for this purpose. Bear in mind this is irrelevant if the team is properly self-managed and/or employees are properly motivated and tasked.
41. Ferret out and destroy all bureaucratic barriers and silly hierarchies. Eliminate all internal forms—they make bureaucracy's job much easier.

42. Observe the banter your employees have with customers. Customers can be sensitive (embarrassed) about making some purchases. Making remarks on a purchase will cost you a customer. It is amazing that self-service shoe stores are gobbling up the market share of service-oriented stores. This is because there is no pressure to buy useless "accessories" or to receive a snide remark such as "we don't carry shoes that large."
43. Eliminate computerized-performance monitors. If you need them that badly, you should resign your managerial post.
44. Get fired for being too creative and innovative.
45. Start a company newsletter. Let the employees edit and produce it. If you have more than four employees, one at least harbors the desire to write. Include birthdays, graduations, everything. Make sure everyone is mentioned at least two times a year.
46. Hire your friends to be customers. Ask them to relate their experiences with the clerk. How did the store look? What were the other clerks doing? What did the carts look like? The floors? This works even better if you can hire someone who doesn't like you. They will be brutally honest.
47. Constantly ask, "Why is that the manager's job?— You may do something about it." Why should the boss decide what radio station is broadcast over the Muzak system, or what should be done about people wearing shorts to work in the summer?
48. Count how many customers balk or renege—those who enter the store and then leave if the checkout lines are too long.
49. Form a "critique club" whose members offer suggestions for personal self-improvement. None of the critiques may be antagonistic.
50. Take a bus to work.
51. Bring your children in to work so they can see what you do for a living.
52. If you still have a performance appraisal system, give underachievers somewhat more praise than you think they deserve, and observe their performance as your expectation level of them rises.
53. Take your best business techniques back to school. Schools need more involvement, not more money.
54. Keep a meticulous time log for one week. Account for every minute.
55. Send out two or more congratulatory letters per week to people who have made a difference in your organization or community.
56. Read a book a month from the humanities: art, history, literature, and philosophy.
57. Go to the bookstore and order (and read) the "Briefcase" set of books listed in the Resources.
58. Explain to a child under 10 what you do or what your company makes.
59. Join the American Society for Quality Control or another society.

60. Become active in that society. Plan meetings, give a presentation, submit a paper, edit the newsletter.
61. Present an in-house seminar on specific TQM techniques.
62. Attend a seminar put on by a Baldrige Award winner, or request information about how they did it.
63. Take the organizational stress test in Chapter 15. Is it time to do something about organizational stress?
64. Write down your impression of the culture of your organization or profession. What should be changed?
65. Besides money, what motivates you? Make a list of "perks" that would be possible with your present job. If you supervise, have your employees make out a list.
66. Change leadership styles. Try a different one on for size. (You may want to warn your colleagues ahead of time.)
67. Watch a video tape of Tom Peters or some other motivational, business excellence-related speaker. Leave plenty of time (30–90 minutes) for discussion among attendees.
68. Make a worker king/queen-for-the-day, at least theoretically. What would they do to improve work processes, quality of work life, and so on.
69. Have regular (required) improvement suggestions from all workers. Implement as many of them as practicable.
70. Cultivate the five elements of leadership: intelligence, credibility, humanity, courage, and discipline.
71. Rename your employees. Call them associates. Call them stakeholders. Call them partners. Whatever you do, consider them to be invaluable business partners.
72. Start a quality education initiative in your local school or school system. Focus on developing practical tools that teachers can implement with minimal disruption of their current practices. Consider that schools are the beginning training ground of future employees.
73. Try a different culture or ethnic group on for size. How would someone from a different socioeconomic background view the problem you're having?
74. Go back in the past. Read about the everyday life of the Romans, or the Egyptians. Talk to someone who can remember seeing a car for the first time, or when radio or television was brand new. What would their views be on current business conditions.
75. Read R. L. Wing's translation of Sun-Tzu's *The Art of Strategy.* Then remember it was written 2,500 years ago.

APPENDIX B:

1992 EXAMINATION ITEMS AND POINT VALUES

1992 Examination Categories/Items	Point Values
1.0 Leadership	90
1.1 Senior Executive Leadership ... 45	
1.2 Management for Quality ... 25	
1.3 Public Responsibility ... 20	
2.0 Information and Analysis	80
2.1 Scope and Management of Quality and Performance Data and Information ... 15	
2.2 Competitive Comparisons and Benchmarks ... 25	
2.3 Analysis and Uses of Company-Level Data ... 40	
3.0 Strategic Quality Planning	60
3.1 Strategic Quality and Company Performance Planning Process ... 35	
3.2 Quality and Performance Plans ... 25	
4.0 Human Resource Development and Management	150
4.1 Human Resource Management ... 20	
4.2 Employee Involvement ... 40	
4.3 Employee Education and Training ... 40	
4.4 Employee Performance and Recognition ... 25	
4.5 Employee Well-Being and Morale ... 25	
5.0 Management of Process Quality	140
5.1 Design and Introduction of Quality Products and Services ... 40	
5.2 Process Management—Product and Service Production and Delivery Processes ... 35	

5.3	Process Management—Business Processes and Support Services	30
5.4	Supplier Quality	20
5.5	Quality Assessment	15

6.0 Quality and Operational Results 180

6.1	Product and Service Quality Results	75
6.2	Company Operational Results	45
6.3	Business Process and Support Service Results	25
6.4	Supplier Quality Results	35

7.0 Customer Focus and Satisfaction 300

7.1	Customer Relationship Management	65
7.2	Commitment to Customers	15
7.3	Customer Satisfaction Determination	35
7.4	Customer Satisfaction Results	75
7.5	Customer Satisfaction Comparison	75
7.6	Future Requirements and Expectations of Customers	35

TOTAL POINTS 1000

1.0 LEADERSHIP (*90 pts.*)

The ***Leadership*** Category examines senior executives' *personal* leadership and involvement in creating and sustaining a customer focus and clear and visible quality values. Also examined is how the quality values are integrated into the company's management system and reflected in the manner in which the company addresses its public responsibilities.

1.1 Senior Executive Leadership (*45 pts.*)
Describe the senior executives' leadership, personal involvement, and visibility in developing and maintaining a customer focus and an environment for quality excellence.

AREAS TO ADDRESS

a. senior executives' leadership, personal involvement, and visibility in quality-related activities of the company. Include: (1) reinforcing a customer focus; (2) creating quality values and setting expectations; (3) planning and reviewing progress toward quality and performance objectives; (4) recognizing employee contributions; and (5) communicating quality values outside the company

b. brief summary of the company's quality values and how the values serve as a basis for consistent communication within and outside the company

c. personal actions of senior executives to regularly demonstrate, communicate, and reinforce the company's customer orientation and quality values through all levels of management and supervision

d. how senior executives evaluate and improve the effectiveness of their personal leadership and involvement

Notes:

(1) The term "senior executives" refers to the highest-ranking official of the organization applying for the Award and those reporting directly to that official.

(2) Activities of senior executives might also include leading and/or receiving training, benchmarking, customer visits, and mentoring other executives, managers, and supervisors.

(3) Communication outside the company might involve: national, state, and community groups; trade, business, and professional organizations; and education, health care, government, and standards groups. It might also involve the company's stockholders and board of directors.

1.2 Management for Quality (*25 pts.*) Describe how the company's focus and quality values are integrated into day-to-day leadership, management, and supervision of all company units.

AREAS TO ADDRESS

a. how the company's customer focus and quality values are translated into requirements for all levels of management and supervision. Include principal roles and responsibilities of each level: (1) within their units; and (2) cooperation with other units.

b. how the company's organizational structure is analyzed to ensure that it most effectively and efficiently serves the accomplishment of the company's customer, quality, innovation, and cycle time objectives. Describe indicators, benchmarks, or other bases for evaluating and improving organizational structure.

c. types, frequency, and content of reviews of company and work unit quality plans and performance. Describe types of actions taken to assist units which are not performing according to plans.

d. key methods and key indicators the company uses to evaluate and improve awareness and integration of quality values at all levels of management and supervision

1.3 Public Responsibility (*20 pts.*) Describe how the company includes its responsibilities to the public for health, safety, environmental protection, and ethical business practices in its quality policies and improvement activities, and how it provides leadership in external groups.

AREAS TO ADDRESS

a. how the company includes its public responsibilities, such as business ethics, public health and safety, environmental protection, and waste management in its quality policies and practices. For each area *relevant and important* to the company's business, briefly summarize: (1) how potential risks are identified, analyzed, and minimized; (2) principal quality improvement goals and how they are set; (3) principal improvement methods; (4) principal quality indicators used in each area; and (5) how and when other progress is reviewed.

b. how the company promotes quality awareness and sharing with external groups

Notes:

(1) Health, safety, environmental, and waste management issues addressed in this item are those associated with the company's operations. Such issues that arise in connection with the use of products and services or disposal of products are addressed in Item 5.1.

(2) Health and safety of employees are not covered in this item. These are addressed in Item 4.5.
(3) Trends in indicators of quality improvement in 1.3a should be reported in Item 6.2.
(4) External groups may include those listed in Item 1.1, Note 3.

2.0 INFORMATION AND ANALYSIS (*80 pts.*)

The ***Information and Analysis*** Category examines the scope, validity, analysis, management, and use of data and information to drive quality excellence and improve competitive performance. Also examined is the adequacy of the company's data, information, and analysis system to support improvement of the company's customer focus, products, services, and internal operations.

2.1 Scope and Management of Quality and Performance Data and Information (***15 pts.***) Describe the company's base of data and information used for planning, day-to-day management, and evaluation of quality. Describe also how data and information reliability, timeliness, and access are assured.

AREAS TO ADDRESS

a. criteria for selecting types of data and information to be included in the quality-related data and information base. List key types included and very briefly describe how each supports quality improvement. Types of data and information should include: (1) customer-related; (2) internal operations; (3) company performance; and (4) cost and financial.

b. how the company ensures reliability, consistency, standardization, review, timely update, and rapid access to data and information throughout the company. If applicable, describe how software quality is assured.

c. key methods and key indicators the company uses to evaluate and improve the scope and quality of its data and information and how it shortens the cycle from data gathering to access. Describe efforts to broaden company units' access to data and information.

Notes:

(1) This Item permits the applicant to demonstrate the *breadth and depth* of its quality-related data. Applicants should give brief descriptions of the data under major headings such as "company performance" and subheadings such as "product and service quality" and "cycle time." Note that information on the scope and management of competitive and benchmark data is requested in Item 2.2.

(2) Actual data should not be reported in this Item. Such data are requested in other Items. Accordingly, all data reported in other Items, such as 6.1, 6.2, 6.3, 6.4, 7.4, and 7.5, are part of the base of data and information to be described in Item 2.1.

2.2 Competitive Comparisons and Benchmarks (*25 pts.*) Describe the company's approach to selecting data and information for competitive comparisons and world-class benchmarks to support quality and performance planning, evaluation, and improvement.

AREAS TO ADDRESS

a. criteria the company uses for seeking competitive comparisons and benchmarks: (1) key company requirements and priorities; and (2) with whom to compare—within and outside the company's industry

b. current scope, sources, and uses of competitive and benchmark data, including company and independent testing or evaluation: (1) product and service quality; (2) customer satisfaction and other customer data; (3) internal operations, including business processes, support services, and employee-related; and (4) supplier performance

c. how competitive and benchmark data are used to encourage new ideas and improve understanding of processes

d. how the company evaluates and improves the scope, sources, and uses of competitive and benchmark data

2.3 Analysis and Uses of Company-Level Data (*40 pts.*) Describe how quality- and performance-related data and information are analyzed and used to support the company's overall operational and planning objectives.

AREAS TO ADDRESS

a. how customer-related data (Category 7.0) are aggregated, analyzed, and translated into actionable information to support: (1) developing priorities for prompt solutions to customer-related problems; (2) determining relationships between the company's product and service quality performance and key customer indicators, such as customer satisfaction, customer retention, and market share; and (3) developing key trends in customer-related performance for review and planning

b. how company operational performance data (Category 6.0) are aggregated, analyzed, and translated into actionable information to support: (1) developing priorities for short-term improvements in company operations, including improved cycle time, productivity, and waste reduction; and (2) developing key trends in company operational performance for review and planning

c. how key cost, financial, and market data are aggregated, analyzed, and translated into actionable information to support improved customer-related and company operational performance

d. key methods and key indicators the company uses to evaluate and improve its analysis. Improvement should address: (1) how the company shortens the cycle of analysis and access to results; and (2) how company analysis strengthens integration of customer, performance, financial, market, and cost data for improved decision making.

Notes:

(1) This Item focuses primarily on analysis for company-level strategies, decision making, and evaluation. Usually, data for these analyses come from or affect a variety of company operations. Some other Items in the Criteria involve analyses of specific sets of data for special purposes such as evaluation of training. Such special-purpose analyses should be part of the information base of Items 2.1 and 2.3 so that this information can be included in larger, company-level analyses.

(2) Analyses involving cost, financial, and market data vary widely in types of data used and purposes. Examples include: relationships between customer satisfaction and market share; relationships between quality and costs; relationships between quality and revenues

and profits; consequences and costs associated with losses of customers and diminished reputation resulting from dissatisfied customers; relationships between customer retention, costs, and profits; priorities for company resource allocation and action based upon costs and impacts of alternative courses of action; improvements in productivity and resource use; and improvements in asset utilization.

3.0 STRATEGIC QUALITY PLANNING (*60 pts.*)

The ***Strategic Quality Planning*** Category examines the company's planning process and how all key quality requirements are integrated into overall business planning. Also examined are the company's short- and longer-term plans and how quality and performance requirements are deployed to all work units.

3.1 Strategic Quality and Company Performance Planning Process (*35 pts.*) Describe the company's strategic planning process for the short term (1–2 years) and longer term (3 years or more) for quality and customer satisfaction leadership. Include how this process integrates quality and company performance requirements and how plans are deployed.

AREAS TO ADDRESS

a. how the company develops plans and strategies for the short term and longer term. Describe data and analysis results used in developing business plans and priorities, and how they consider: (1) customer requirements and the expected evolution of these requirements; (2) projections of the competitive environment; (3) risks: financial, market, and societal; (4) company capabilities, including research and development to address key new requirements or technology leadership opportunity; and (5) supplier capabilities.

b. how plans are implemented. Describe: (1) the method the company uses to deploy overall plan requirements to all work units and to suppliers, and how it ensures alignment of work unit activities; and (2) how resources are committed to meet the plan requirements.

c. how the company evaluates and improves its planning process, including improvements in: (1) determining company quality and overall performance requirements; (2) deployment of requirements to work units; and (3) input from all levels of the company

Note: Review of performance relative to plans is addressed in Item 1.2.

3.2 Quality and Performance Plans (*25 pts.*)
Summarize the company's quality and performance plans and goals for the short term (1–2 years) and the longer term (3 years or more).

AREAS TO ADDRESS

a. for the company's chosen directions, including planned products and services, markets, or market niches, summarize: (1) key quality factors and quality requirements to achieve leadership; and (2) key company performance requirements

b. outline of the company's principal short-term quality and company performance plans and goals: (1) summary of key requirements and key performance indicators deployed to work units and suppliers; and (2) resources committed for key requirements such as capital equipment, facilities, education and training, and personnel

c. principal longer-term quality and company performance plans and goals, including key requirements and how they will be addressed

d. two-to-five year projection of significant improvements using the most important quality and company performance indicators. Describe how quality and company performance might be expected to compare with competitors and key benchmarks over this time period. Briefly explain the comparison.

4.0 HUMAN RESOURCE DEVELOPMENT AND MANAGEMENT (*150 pts.*)

The ***Human Resource Development and Management*** Category examines the key elements of how the company develops and realizes the full potential of the work force to pursue the company's quality and performance objectives. Also examined are the company's efforts to build and maintain an environment for quality excellence conducive to full participation and personal and organizational growth.

4.1 Human Resource Management (*20 pts.*) Describe how the company's overall human resource development and management plans and practices support its quality and company performance plans and address all categories and types of employees.

AREAS TO ADDRESS

a. how human resource plans derive from quality and company performance plans (Item 3.2). Briefly describe major human resource development initiatives or plans affecting: (1) education, training, and related skill development; (2) recruitment; (3) involvement; (4) empowerment; and (5) recognition. Distinguish between the short-term (1–2 years) and the longer-term (3 years or more) plans as appropriate.

b. key quality, cycle time, and other performance goals and improvement methods for personnel practices such as recruitment, hiring, personnel actions, and services to employees. Describe key performance indicators used in the improvement of these personnel practices.

c. how the company evaluates and uses all employee-related data to improve the development and effectiveness of the entire work force. Describe how the company's evaluation and improvement processes address all types of employees.

Notes:

(1) Human resource plans might include the following: mechanisms for promoting cooperation such as internal customer/supplier techniques or other internal partnerships; initiatives to promote labor-management cooperation, such as partnerships with unions; creation of modifications to recognition systems; mechanisms for increasing or broadening employee responsibilities; permitting employees to learn and use skills that go beyond current job assignments through redesign of work processes; creation of high performance work teams; and education and training initiatives. Plans might also include forming partnerships with educational institutions to develop employees or to help ensure the future supply of well-prepared employees.

(2) "Categories of employees" refers to the company's classification system used in its personnel practices and/or work assignments. "Types of employees" takes into account other factors, such as bargaining unit membership and demographic makeup. This includes gender, age, minorities, and the disabled.

(3) All employee-related data refers to data contained in personnel records as well as data described in Items 4.2, 4.3, 4.4, and 4.5.

4.2 Employee Involvement (*40 pts.*) Describe the means available for all employees to contribute effectively to meeting the company's quality and performance objectives; summarize trends in involvement.

AREAS TO ADDRESS

a. management practices and specific mechanisms the company uses to promote employee contributions, individually and in groups, to quality and company performance objectives. Describe how and how quickly the company gives feedback to contributors.

b. company actions to increase employee empowerment, responsibility, and innovation. Briefly summarize principal goals for all categories of employees, based upon the most important requirements for each category.

c. key methods and key indicators the company uses to evaluate and improve the extent and effectiveness of involvement of all categories and types of employees

d. trends in percent involvement for each category of employee. Use the most important indicator(s) of effective employee involvement for each category.

Note: Different involvement goals and indicators may be set for different categories of employees, depending on company needs and on the types of responsibilities of each employee category.

4.3 Employee Education and Training (*40 pts.*) Describe how the company determines what quality and related education and training is needed by employees and how the company utilizes the knowledge and skills acquired; summarize the types of quality and related education and training received by employees in all categories.

AREAS TO ADDRESS

a. (1) how the company determines needs for the types and amounts of quality and related education and training to be received by all categories and types of employees. Address: (a) relevance of education and training to company plans; (b) needs of individual employees; and (c) all work units having access to skills in problem analysis, problem solving, and process simplification; (2) methods for the delivery of education and training; and (3) how the company ensures on-the-job use and reinforcement of knowledge and skills.

b. summary and trends in quality and related education and training received by employees. The summary and trends should address: (1) quality orientation of new employees; (2) percent of employees receiving quality and related education and training in each employee category annually; (3) average hours of quality education and training per employee annually; (4) percent of current employees who have received quality and related education and training; and (5) percent of employees who have received education and training in specialized areas such as design quality, statistical, and other quantitative problem-solving methods.

c. key methods and key indicators the company uses to evaluate and improve the effectiveness of its quality and related education and training for all categories and types of employees. Describe how the indicators take into account: (1) education and training delivery effectiveness; (2) on-the-job performance improvement; and (3) employee growth.

Note: Quality and related education and training address the knowledge and skills employees need to meet their objectives as part of the company's quality and performance plans. This may include quality awareness, leadership, problem solving, meeting customer requirements, process analysis, process simplification, waste reduction, cycle time reduction, and other training that affects employee effectiveness and efficiency.

4.4 Employee Performance and Recognition (*25 pts.*) Describe how the company's employee performance, recognition, promotion, compensation, reward, and feedback processes support the attainment of the company's quality and performance objectives.

AREAS TO ADDRESS

a. how the company's performance, recognition, promotion, compensation, reward, and feedback approaches for individuals and groups, including managers, support the company's quality and performance objectives. Address: (1) how the approaches ensure that quality is reinforced relative to short-term financial considerations; and (2) how employees contribute to the company's performance and recognition approaches.

b. trends in reward and recognition, by employee category, for contributions to the company's quality and performance objectives

c. key methods and key indicators the company uses to evaluate and improve its performance and recognition processes. Describe how the evaluation takes into account cooperation, participation by all categories and types of employees, and employee satisfaction.

4.5 Employee Well-Being and Morale (*25 pts.*) Describe how the company maintains a work environment conducive to the well-being and growth of all employees; summarize trends and levels in key indicators of well-being and morale.

AREAS TO ADDRESS

a. how well-being and morale factors such as health, safety, satisfaction, and ergonomics are included in quality improvement activities. Summarize principal improvement goals, methods, and indicators for each factor relevant and important to the company's work environment. For accidents and work-related health problems, describe how root causes are determined and how adverse conditions are prevented.

b. mobility, flexibility, and retraining in job assignments to support employee development and/or to accommodate changes in technology, improved productivity, changes in work processes, or company restructuring

c. special services, facilities, and opportunities the company makes available to employees. These might include one or more of the following: counseling, assistance, recreational or cultural, non-work-related education, and outplacement

d. how and how often employee satisfaction is determined

e. trends in key indicators of well-being and morale. This should address, as appropriate: satisfaction, safety, absenteeism, turnover, attrition rate for customer-contact personnel, grievances, strikes, and worker compensation. Explain important adverse results, if any. For such adverse results, describe how root causes were determined and corrected, or give current status. Compare results on the most significant indicators with those of industry averages, industry leaders, and other key benchmarks.

5.0 MANAGEMENT OF PROCESS QUALITY (*140 pts.*)

The ***Management of Process Quality*** Category examines the systematic processes the company uses to pursue ever-higher quality and company performance. Examined are the key elements of process management including design, management of process quality for all work units and suppliers, systematic quality improvement, and quality assessment.

5.1 Design and Introduction of Quality Products and Services (*40 pts.*) Describe how new and/or improved products and services are designed and introduced and how processes are designed to meet key product and service quality requirements and company performance requirements.

AREAS TO ADDRESS

a. how designs of products, services, and processes are developed so that: (1) customer requirements are translated into design requirements; (2) all quality requirements are addressed early in the overall design process by appropriate company units; (3) designs are coordinated and integrated to include all phases of production and delivery; and (4) a process control plan that involves selecting, setting, and monitoring key process characteristics is developed.

b. how designs are reviewed and validated, taking into account key factors: (1) product and service performance; (2) process capability and future requirements; and (3) supplier capability and future requirements

c. how the company evaluates and improves the effectiveness of its designs and design processes so that new product and service introductions progressively improve in quality and cycle time

Notes:

(1) Design and introduction may include modification and variance of existing products and services and/or new products and services emerging from research and development.

(2) Applicant response should reflect the key requirements of the products and services they deliver. Factors that may need to be considered in design include: health, safety, long-term performance, environment, measurement capability, process capability, maintainability, and supplier capability.

(3) Service and manufacturing businesses should interpret product and service requirements to include all product- and service-related requirements at all stages of production, delivery, and use.

5.2 Process Management—Product and Service Production and Delivery Processes (*35 pts.*)
Describe how the company's product and service production and delivery processes are managed so that current quality requirements are met and quality and performance are continuously improved.

AREAS TO ADDRESS

a. how the company maintains the quality of processes in accord with product and service design requirements. Describe: (1) what is measured and types and frequencies of measurements; and (2) how out-of-control occurrences are handled, including root cause determination, correction, and verification of corrections.

b. how processes are analyzed and improved to achieve better quality, performance, and cycle time. Describe how the following are considered: (1) process simplification; (2) waste reduction; (3) process research and testing; (4) use of alternative technologies; and (5) benchmark information.

c. how overall product and service performance data are analyzed, root causes determined, and results translated into process improvements

d. how the company integrates process improvement with day-to-day process management: (1) resetting process characteristics to reflect the improvements; (2) verification of improvements; and (3) ensuring effective use by all appropriate company units

Notes:

(1) For manufacturing and service companies which have specialized measurement requirements, a description of the method for measurement quality assurance should be given. For physical, chemical, and engineering measurements, describe briefly how measurements are made traceable to national standards.

(2) The distinction between 5.2b and 5.2c is as follows: 5.2b addresses ongoing improvement activities of the company; 5.2c addresses performance information related to the use of products and services ("performance in the field"), including customer problems and complaints. Analysis in 5.2c focuses on the process level—root causes and process improvement.

5.3 Process Management— Business Processes and Support Services

(*35 pts.*) Describe how the company's business processes and support services are managed so that current requirements are met and quality and performance are continuously improved.

AREAS TO ADDRESS

a. how the company maintains the quality of the business processes and support services. Describe: (1) how key processes are defined based upon customer and/or company quality performance requirements; (2) principal indicators used to measure quality and/or performance; (3) how day-to-day quality and performance are determined, including types and frequencies of measurements used; and (4) how out-of-control occurrences are handled, including root cause determination, correction, and verification of corrections.

b. how processes are improved to achieve better quality, performance, and cycle time. Describe how the following are used or considered: (1) process performance data; (2) process and organizational simplification and/or redefinition; (3) use of alternative technologies; (4) benchmark information; (5) information from customers of the business processes and support services—inside and outside the company; and (6) challenge goals.

Notes:

(1) Business processes and support services might include activities and operations involving finance and accounting, software services, sales, marketing, information services, purchasing, personnel, legal services, plant and facilities management, basic research and development, and secretarial and other administrative services.

(2) The purpose of this Item is to permit applicants to highlight separately the quality activities for functions that support the product and service production and delivery processes the applicant addressed in Item 5.2. The support services and business processes included in Item 5.3 depend on the applicant's type of business and quality system. Thus, this selection should be made by the applicant. Together, Items 5.1, 5.2, 5.3, 5.4, and 5.5 should cover all operations, processes, and activities of all work units.

5.4 Supplier Quality
(***20 pts.***) Describe how the quality of materials, components, and services furnished by other businesses is assured and continuously improved.

AREAS TO ADDRESS

a. approaches used to define and communicate the company's quality requirements to suppliers. Include: (1) the principal quality requirements for key suppliers; and (2) the principal indicators the company uses to communicate and monitor supplier quality.

b. methods used to assure that the company's quality requirements are met by suppliers. Describe how the company's overall performance data are analyzed and relevant information fed back to suppliers.

c. current strategies and actions to improve the quality and responsiveness (delivery time) of suppliers. These may include: partnerships, training, incentives and recognition, and supplier selection.

Notes:

(1) The term "supplier" as used here refers to other-company providers of goods and services. The use of these goods and services may occur at any stage in the production, delivery, and use of the company's products and services. Thus, suppliers include businesses such as distributors, dealers, and franchises as well as those that provide materials and components.

(2) Methods may include audits, process reviews, receiving inspection, certification, testing, and rating systems.

5.5 Quality Assessment
(***15 pts.***) Describe how the company assesses the quality and performance of its systems, processes, and practices and the quality of its products and services.

AREAS TO ADDRESS

a. approaches the company uses to assess: (1) systems, processes, and practices; and (2) products and services. For (1) and (2), describe: (a) what is assessed; (b) how often assessments are made and by whom; and (c) how measurement quality and adequacy of documentation of processes and practices are assured.

b. how assessment findings are used to improve: products and services, systems, processes, practices, and supplier requirements. Describe how the company verifies that assessment findings are acted on and that actions are effective.

Notes:

(1) The systems, processes, practices, products, and services addressed in this Item pertain to all company unit

activities covered in Items 5.1, 5.2, 5.3, and 5.4. If the approaches and frequency of assessments differ appreciably for different company activities, this should be described in this Item.

(2) Adequacy of documentation should take into account legal, regulatory, and contractual requirements as well as knowledge preservation and transfer to help support all quality-related efforts.

6.0 QUALITY AND OPERATIONAL RESULTS (*180 pts.*)

The ***Quality and Operational Results*** Category examines the company's quality levels and improvement trends in quality, company operational performance, and supplier quality. Also examined are current quality and performance levels relative to those of competitors.

6.1 Product and Service Quality Results (*75 pts.*) Summarize trends in quality and current quality levels for key product and service features; compare the company's current quality levels with those of competitors.

AREAS TO ADDRESS

a. trends and current levels for all key measures of product and service quality

b. current quality level comparisons with principal competitors in the company's key markets, industry averages, industry leaders, and others as appropriate. Briefly explain bases for comparison such as: (1) independent surveys, studies, or laboratory testing; (2) benchmarks; and (3) company evaluations and testing. Describe how objectivity and validity of comparisons are assured.

Notes:

(1) Key product and service measures are measures relative to the set of all important features of the company's products and services. These measures, taken together, best represent the *most important factors that predict customer satisfaction and quality in customer use*. Examples include measures of accuracy, reliability, timeliness, performance, behavior, delivery, after-sales services, documentation, and appearance.

(2) Results reported in Item 6.1 should reflect the key product and service features described in the Overview.

(3) Data reported in Item 6.1 are intended to be results of company ("internal") measurements—not customer satisfaction or other customer data, reported in Items 7.4 and 7.5. If the quality of some key product or service features cannot be determined effectively

through internal measures, external data may be used. Examples include data collected by the company, as in Item 7.1c, data collected by third-party organizations on behalf of the company, and data collected by independent organizations. Such data should provide information on the company's performance relative to *specific product and service features*, not on levels of overall satisfaction. These data, collected regularly, are then part of a system for measuring quality, monitoring trends, and improving processes.

6.2 Company Operational Results (*45 pts.*) Summarize trends and levels in overall company operational performance and provide a comparison of this performance with competitors and appropriate benchmarks.

AREAS TO ADDRESS

a. trends and current levels for key measures of company operational performance.

b. comparison of performance with that of competitors, industry averages, industry leaders, and key benchmarks. Give and briefly explain basis for comparison.

Note: Key measures of company operational performance include those that address productivity, efficiency, and effectiveness. Examples should include generic indicators such as use of manpower, materials, energy, capital, and assets. Trends and levels could address productivity indices, waste reduction, energy efficiency, cycle time reduction, environmental improvement, and other measures of improved *overall company performance*. Also include company-specific indicators the company uses to monitor its progress in improving operational performance. Such company-specific indicators should be defined in tables or charts where trends are presented. Trends in financial indicators, properly labeled, may be included in this Item. If such financial indicators are used, there should be a clear connection to the quality and performance improvement activities of the company.

6.3 Business Process and Support Service Results (*25 pts.*) Summarize trends and current levels in quality and performance improvement for business processes and support services.

AREAS TO ADDRESS

a. trends and current levels for key measures of quality and performance of business processes and support services

b. comparison of performance with appropriately selected companies and benchmarks. Give and briefly explain basis for comparison.

Note: Business processes and support services are those as covered in Item 5.3. Key measures of performance should reflect the principal quality, productivity, cycle time, cost, and other effectiveness requirements for business processes and support services. Responses should reflect relevance to the company's principal quality and company performance objectives addressed in company plans, contributing to the results reported in Items 6.1 and 6.2. They should also demonstrate broad coverage of company business processes, support services, and work units and reflect the most important objectives of each process, service, or work unit.

6.4 Supplier Quality Results (*35 pts.*) Summarize trends in quality and current quality levels of suppliers; compare the company's supplier quality with that of competitors and with key benchmarks.

AREAS TO ADDRESS

a. trends and current levels for the most important indicators of supplier quality

b. comparison of the company's supplier quality levels with those of competitors and/or with benchmarks. Such comparisons could be industry averages, industry leaders, principal competitors in the company's key markets, and appropriate benchmarks. Describe the basis for comparisons.

Note: The results reported in Item 6.4 derive from quality improvement activities described in Item 5.4. Results should be broken down by major groupings of suppliers and reported using the principal quality indicators described in Item 5.4.

7.0 CUSTOMER FOCUS AND SATISFACTION (*300 pts.*)

The ***Customer Focus and Satisfaction*** Category examines the company's relationships with customers and its knowledge of customer requirements and of the key quality factors that determine marketplace competitiveness. Also examined are the company's methods to determine customer satisfaction, current trends and levels of satisfaction, and these results relative to competitors.

7.1 Customer Relationship Management (*65 pts.*) Describe how the company provides effective management of its relationships with its customers and uses information gained from customers to improve customer relationship management strategies and practices.

AREAS TO ADDRESS

a. how the company determines the most important factors in maintaining and building relationships with customers and develops strategies and plans to address them. Describe these factors and how the strategies take into account: fulfillment of basic customer needs in the relationship; opportunities to enhance the relationships; provision of information to customers to ensure the proper setting of expectations regarding products, services, and relationships; and roles of all customer-contact employees, their technology needs, and their logistics support.

b. how the company provides information and easy access to enable customers to seek assistance, to comment, and to complain. Describe types of contact and how easy access is maintained for each type.

c. follow-up with customers on products, services, and recent transactions to help build relationships and to seek feedback for improvement.

d. how service standards that define reliability, responsiveness, and effectiveness of customer-contact employees' interactions with customers are set. Describe how standards requirements are deployed to other company units that support customer-contact employees, how the overall performance of the service standards system is monitored, and how it is improved using customer information.

e. how the company ensures that formal and informal complaints and feedback received by all company units are aggregated for overall evaluation and use throughout the company. Describe how the company ensures that complaints and problems are resolved promptly and effectively.

f. how the following are addressed for customer-contact employees: (1) selection factors; (2) career path; (3) special training to include: knowledge of products and services; listening to customers; soliciting comments from customers; how to anticipate and handle problems or failures ("recovery"); skills in customer retention; and how to manage expectations; (4) empowerment and decision-making; (5) attitude and moral determination; (6) recognition and reward; and (7) attrition

g. how the company evaluates and improves its customer relationship management practices. Describe key indicators used in evaluations and how evaluations lead to improvements, such as in strategy, training, technology, and service standards.

Notes:

(1) Information on trends and levels in indicators of complaint response time and trends in percent of complaints resolved on first contact should be reported in Item 6.3.

(2) In addressing empowerment and decision making in 7.1f, indicate how the company ensures that there is a common vision or basis guiding customer-contact employee action.

7.2 Commitment to Customers (*15 pts.*)
Describe the company's explicit and implicit commitments to customers regarding its products and services.

AREAS TO ADDRESS

a. types of commitments the company makes to promote trust and confidence in its products, services, and relationships. Describe how these commitments: (1) address the principal concerns of customers; and (2) are free from conditions that might weaken customer confidence.

b. how improvements in the quality of the company's products and services over the past three years have been translated into stronger commitments. Compare commitments with those of competitors.

c. how the company evaluates and improves its commitments, and the customers' understanding of them, to avoid gaps between expectations and delivery.

Note: Commitments may include product and service guarantees, product warranties, and other understandings with the customer, expressed or implied.

7.3 Customer Satisfaction Determination (*35 pts.*) Describe the company's methods for determining customer satisfaction and customer satisfaction relative to competitors; describe how these methods are evaluated and improved.

AREAS TO ADDRESS

a. how the company determines customer satisfaction. Include: (1) a brief description of market segments and customer groups and the key customer satisfaction requirements for each segment or group; (2) how customer satisfaction measurements capture key information that reflects customers' likely market behavior; and (3) a brief description of the methods, processes, and scales used; frequency of determination; and how objectivity and validity are assured.

b. how customer satisfaction relative to that for competitors is determined. Describe: (1) company-based comparative studies; and (2) comparative studies or evaluations made by independent organizations, including customers. For (1) and (2) describe how objectivity and validity are assured.

c. how the company evaluates and improves its overall processes and measurement scales for determining customer satisfaction and customer satisfaction relative to that for competitors. Describe how other indicators (such as gains and losses of customers) and customer dissatisfaction indicators (such as complaints) are used in this improvement process.

Notes:

(1) Customer satisfaction measurement may include both a numerical rating scale and descriptors assigned to each unit in the scale. An effective customer satisfaction measurement system is one that provides the company with reliable information about customer views of specific product and service features and the relationship between these views or ratings and the customer's likely market behaviors.

(2) Indicators of customer dissatisfaction include complaints, claims, refunds, recalls, returns, repeated services, litigation, replacements, downgrades, repairs, warranty work, and warranty costs. If the company has received any sanctions under regulation or contract during the past three years, include such information in the Item. Briefly summarize how sanctions were resolved or give current status.

(3) Company-based or independent organization comparative studies in 7.3b may take into account one or more indicators of customer dissatisfaction.

7.4 Customer Satisfaction Results (*75 pts.*) Summarize trends in the company's customer satisfaction and trends in key indicators of dissatisfaction.

AREAS TO ADDRESS

a. trends and current levels in indicators of customer satisfaction, segmented as appropriate

b. trends and current levels in indicators of customer dissatisfaction. Address all indicators relevant to the company's products and services.

Notes:

(1) Results reported in this Item derive from methods described in Item 7.3 and 7.1c and e.

(2) Indicators of customer dissatisfaction are listed in Item 7.3, Note 2.

7.5 Customer Satisfaction Comparison (*75 pts.*) Compare the company's customer satisfaction results with those of competitors.

AREAS TO ADDRESS

a. trends and current levels in indicators of customer satisfaction relative to that for competitors, based upon methods described in Item 7.3. Segment by customer group, as appropriate.

b. trends in gaining or losing customers, or customer accounts, to competitors

c. trends in gaining or losing market share to competitors

Note: Competitors include domestic and international ones in the company's markets, both domestic and international.

7.6 Future Requirements and Expectations of Customers (*35 pts.*) Describe how the company determines future requirements and expectations of customers.

AREAS TO ADDRESS

a. how the company addresses future requirements and expectations of customers. Describe: (1) the time horizon for the determination; (2) how data from current customers are projected; (3) how customers of competitors and other potential customers are considered; and (4) how important technological, competitive, societal, economic, and demographic factors and trends that may bear upon customer requirements and expectations are considered.

b. how the company projects key product and service features and the relative importance of these features to customers and potential customers. Describe how potential market segments and customer groups are considered. Include considerations that address new product/service lines as well as current products and services.

c. how the company evaluates and improves its processes for determining future requirements and expectations of customers. Describe how the improvement process considers new market opportunities and extension of the time horizon for the determination of customer requirements and expectations.

INDEX

D

E

F

Q

T

U

V

W

X–Z

Other Business One Irwin Books of Interest to You:

THE CORPORATE GUIDE TO THE MALCOLM BALDRIGE NATIONAL QUALITY AWARD
Proven Strategies for Building Quality Into Your Organization
Marion Mills Steeples

The Baldrige guidelines are an objective yardstick any organization can use to measure a comprehensive quality process. Steeples explains the application process with tables, diagrams, and step-by-step instructions. The guidelines also benefit nonapplicants with competitive strategies to cut development and production costs, streamline processes and eliminate waste, and enhance worker morale and customer relations. (383 pages)
ISBN: 1-55623-653-0

QUALITY IN AMERICA
How to Implement a Competitive Quality Program
V. Daniel Hunt

Dramatically improve your firm's market share, performance, and profitability! Hunt, the author of several award-winning productivity books, analyzes the present state of the practice of quality in America and helps you understand the theories, basic tools, and techniques that can improve quality in your organization. (308 pages)
ISBN: 1-55623-536-4

TOTAL QUALITY
An Executive's Guide for the 1990s
The Ernst & Young Quality Improvement Consulting Group

This ground-breaking book explains not only what you must know, but also what you must do to implement a world-class quality model for success. You will refer to this succinct reference time and time again for all of your quality questions. *Total Quality* shows how to implement an optimal balance of technology and automation, measure and reward performance, reinforce a quality commitment throughout the workforce, and much more! (248 pages)
ISBN: 1-55623-188-1

CONTINUOUS IMPROVEMENT AND MEASUREMENT FOR TOTAL QUALITY
A Team-Based Approach
Dennis C. Kinlaw
Copublished by University Associates/Business One Irwin

Discover how to make continuous improvement and measurement a constant part of your organization's environment! By making TQM a team-centered and team-driven activity, you can improve your competitive position and profitability. This useful resource provides the quality tools you'll need—including methods to assess and measure continuous improvement, strategies to design improvement projects, and more. (225 pages)
ISBN: 1-55623-778-2